# From Calculus to Computers

### Using the last 200 years of mathematics history in the classroom

© 2005 by

The Mathematical Association of America (Incorporated)

Library of Congress Control Number 2005908422

ISBN 0-88385-178-4

Printed in the United States of America

Current Printing (last digit):

10 9 8 7 6 5 4 3 2 1

# From Calculus to Computers

## Using the last 200 years of mathematics history in the classroom

Edited by

Amy Shell-Gellasch
*Formerly of the United States Military Academy*

and

Dick Jardine
*Keene State College*

*Published and Distributed by*
The Mathematical Association of America

The MAA Notes Series, started in 1982, addresses a broad range of topics and themes of interest to all who are involved with undergraduate mathematics. The volumes in this series are readable, informative, and useful, and help the mathematical community keep up with developments of importance to mathematics.

## Council on Publications
Roger Nelsen, *Chair*

### Notes Editorial Board
Sr. Barbara E. Reynolds, *Editor*

Paul E. Fishback, *Associate Editor*

Jack Bookman    Annalisa Crannell    Rosalie Dance
William E. Fenton    Mark Parker    Sharon C. Ross
David Sprows

### MAA Notes

11. Keys to Improved Instruction by Teaching Assistants and Part-Time Instructors, Committee on Teaching Assistants and Part-Time Instructors, *Bettye Anne Case,* Editor.

13. Reshaping College Mathematics, Committee on the Undergraduate Program in Mathematics, Lynn A. Steen,Editor.

14. Mathematical Writing, by *Donald E. Knuth, Tracy Larrabee, and Paul M. Roberts.*

16. Using Writing to Teach Mathematics, *Andrew Sterrett,* Editor.

17. Priming the Calculus Pump: Innovations and Resources, Committee on Calculus Reform and the First Two Years, a sub-committee of the Committee on the Undergraduate Program in Mathematics, *Thomas W. Tucker,* Editor.

18. Models for Undergraduate Research in Mathematics, *Lester Senechal,* Editor.

19. Visualization in Teaching and Learning Mathematics, Committee on Computers in Mathematics Education, *Steve Cunningham and Walter S. Zimmermann,* Editors.

20. The Laboratory Approach to Teaching Calculus, *L. Carl Leinbach et al.,* Editors.

21. Perspectives on Contemporary Statistics, *David C. Hoaglin and David S. Moore,* Editors.

22. Heeding the Call for Change: Suggestions for Curricular Action, *Lynn A. Steen,* Editor.

24. Symbolic Computation in Undergraduate Mathematics Education, *Zaven A. Karian,* Editor.

25. The Concept of Function: Aspects of Epistemology and Pedagogy, *Guershon Harel and Ed Dubinsky,* Editors.

26. Statistics for the Twenty-First Century, *Florence and Sheldon Gordon,* Editors.

27. Resources for Calculus Collection, Volume 1: Learning by Discovery: A Lab Manual for Calculus, *Anita E. Solow,* Editor.

28. Resources for Calculus Collection, Volume 2: Calculus Problems for a New Century, *Robert Fraga,* Editor.

29. Resources for Calculus Collection, Volume 3: Applications of Calculus, *Philip Straffin,* Editor.

30. Resources for Calculus Collection, Volume 4: Problems for Student Investigation, *Michael B. Jackson and John R. Ramsay,* Editors.

31. Resources for Calculus Collection, Volume 5: Readings for Calculus, *Underwood Dudley,* Editor.

32. Essays in Humanistic Mathematics, *Alvin White,* Editor.

33. Research Issues in Undergraduate Mathematics Learning: Preliminary Analyses and Results, *James J. Kaput and Ed Dubinsky,* Editors.

34. In Eves' Circles, *Joby Milo Anthony,* Editor.

35. You're the Professor, What Next? Ideas and Resources for Preparing College Teachers, The Committee on Preparation for College Teaching, *Bettye Anne Case,* Editor.

36. Preparing for a New Calculus: Conference Proceedings, *Anita E. Solow,* Editor.

37. A Practical Guide to Cooperative Learning in Collegiate Mathematics, *Nancy L. Hagelgans, Barbara E. Reynolds, SDS, Keith Schwingendorf, Draga Vidakovic, Ed Dubinsky, Mazen Shahin, G. Joseph Wimbish, Jr.*

38. Models That Work: Case Studies in Effective Undergraduate Mathematics Programs, *Alan C. Tucker,* Editor.

39. Calculus: The Dynamics of Change, CUPM Subcommittee on Calculus Reform and the First Two Years, *A. Wayne Roberts,* Editor.

40. Vita Mathematica: Historical Research and Integration with Teaching, *Ronald Calinger,* Editor.

41. Geometry Turned On: Dynamic Software in Learning, Teaching, and Research, *James R. King and Doris Schattschneider,* Editors.

MAA Service Center
P.O. Box 91112
Washington, DC 20090-1112
1-800-331-1MAA    FAX: 1-301-206-9789

# Preface

In the summer of 2001, the editors of this volume organized a contributed papers session at the MAA MathFest in Madison, Wisconsin. The topic was ways to use the history of mathematics in teaching. We received many wonderful abstracts. Due to time limitations, we had to make some cuts. Two or three of the abstracts dealt with more recent topics. The majority of the topics however, were more "traditional" in that they addressed ways to use the history of Greek mathematics or calculus for example. We decided to use talks which dealt with topics up to the development of the calculus. The abstracts that addressed more recent mathematics were very intriguing, and made us realize that even though using the history of mathematics in the classroom is becoming very popular, the topic of recent history may be overlooked.

We decided to organize a follow-up session the next year in Burlington. This session was entitled The Use of History in the Teaching of Mathematics, with a focus on roughly the last two hundred years. Again, we had many excellent abstracts. The talks covered a range of topics from Galois theory to using the history of women and minorities in teaching.

The interest at MathFest in the topic of recent mathematical history, and a noticeable lack of information for educators on how to incorporate that history into the classroom, was the motivation for this volume. Several of the papers are from that 2002 session, while others are new. The papers cover a range of topics, from logic and computer science to Galois theory to the evolution of statistical thought in education. There are a variety of levels of detail, from general course outlines to ideas for projects and units in specific courses. Some papers focus on the history, while others focus on teaching.

This volume is by no means an exhaustive survey, simply a first step toward building a body of knowledge and ideas for incorporating recent history into the classroom. Whether you teach lower or upper division courses, mathematics, computer science or statistics, there should be something of interest to you.

We thank Andrew Sterrett for his support of this project from the beginning. Additionally, Barbara Reynolds and Donald Albers were instrumental in guiding our effort. Of course we appreciate the determination of all the contributing authors, many of whom were speakers at the contributed papers sessions we chaired at the MathFests. Both of us are very thankful for the mentorship provided by V. Frederick Rickey and Victor Katz. Finally, special thanks goes to our spouses, Chris Gellasch and Deb Jardine, who provided support on the home front, which is immeasurably important.

# Contents

# Introduction

Using the history of mathematics to motivate interest and understanding in the mathematics classroom has become an acknowledged pedagogical option for mathematics instructors. The knowledge base on how to incorporate history into the classroom has grown by leaps and bounds in the last few decades. A look at the number of recent books published by many academic publishers, to include the MAA, is indicative of this effort to improve mathematics teaching and learning.

However, due to the nature of mathematical advances in the last few centuries, the history of topics such as logic and Galois theory are much less accessible. The level of mathematical sophistication needed to grasp these topics, even at an introductory level, is much higher than that needed for most older subjects. Given that the development of a subject is usually harder to follow than the final theory, the history of more recent topics may be even harder to comprehend. For exactly these reasons, it is important to present the historical evolution of and motivation for these topics. Many higher-level concepts seem remote and arbitrary to students. Presenting the historical development of these more recent and advanced topics gives the students the connections needed to gain a deeper understanding of higher mathematics.

This volume is intended to be a resource for undergraduate mathematics teachers, providing ideas and materials for immediate adoption and proven examples to motivate innovation by the reader. The book is divided into sections by subject area to make it easier for teachers to find a relevant article for a course they have in mind.

Section One focuses on courses commonly found in the undergraduate mathematics core and elective courses: calculus, abstract algebra, numerical methods and number theory. David Pengelley uses a seminal paper by Cayley to motivate student learning in an abstract algebra course. Bob Rogers offers activities which utilize the differential to solve calculus problems, an alternative approach to motivate students' geometric and historical insights. Matt Lunsford's contribution answers the question: How can the ideas of Evariste Galois enhance the teaching of undergraduate abstract algebra? The study of elliptic curves relates algebra and geometry, and Lawrence D'Antonio ties together those concepts in describing some deep and current mathematics in a way accessible to upper-level undergraduate students. Through a presentation of the historical development of predator-prey models, Holly Hirst provides activities to ensure that students understand that mathematical modeling is an evolving process.

Section Two contains articles on the subject of geometry. Jeff Johannes relates how advances in that subject area dispelled fears of prominent mathematicians of the late eighteenth century that the future of mathematics was bleak since all the work was done and no new great minds were apparent to carry on the mathematical tradition. Daina Taimina and David W. Henderson integrate the history of mathematics in undergraduate geometry instruction and include historical presentation of non-Euclidean geometries. Eisso Atzema and Homer White survey a few of Euler's late geometrical papers and offer ideas for exercises in a geometry class or problems for investigation by students.

Section Three includes additional topics found in other courses of the undergraduate mathematics curriculum, to include statistics and the mathematics of computation. Dick Jardine offers the use of historically based projects to engage students and deepen their learning of numerical analysis topics. For instructors of a course in logic for undergraduate mathematics or computer science majors, Francine Abeles provides a paper for those who wish to include an historical approach to the significant developments of modern logic. In a similar vein, William Calhoun connects logic and computing to motivate both mathematics and computer science majors in discrete mathematics and theory of computation courses. Linda McGuire describes important issues concerning semester-long research projects given in an upper division combinatorics and graph theory class. The mathematics of cryptography has growing interest among our students, and Shai Simonson writes about classroom activities that work for him, his colleagues, and his students in discrete mathematics and number theory courses. Patti Hunter uses a historical perspective to address the controversy of just how much rigor to include in undergraduate statistics courses. Jerry Lodder offers ideas for implementing the history of logic and programming into computer science courses.

Section Four addresses some issues relevant to the history of mathematics and methods for conducting a history of mathematics course, to include those courses targeted at pre-service school teachers. Sarah Greenwald notes that students identify more readily with current mathematicians, and she offers ideas to include projects involving the history of current mathematicians, with examples of women and minority practitioners. Dave Roberts offers mathematical topics from a history of science course that would apply to portions of a history of mathematics course, relating the interconnection between mathematics and other disciplines. Amy Shell-Gellasch "personalizes" the history of mathematics course by suggesting ways to include the history of one's own institution as a significant component. The protractor is an oft-used tool in middle and secondary education; Amy Ackerberg-Hastings outlines the historical development of that instructional tool. A history of the slow acceptance of the metric system in American mathematics education is given by Peggy Kidwell. Peter Ross offers some ideas to spark interest and discussion and to motivate student interest in the history of mathematics. John Prather concludes the volume with a description of a problem-solving approach to the history of mathematics in a course intended for pre-service middle and secondary teachers.

This collection, then, is an eclectic mix of ideas expressed by colleagues with interests similar to ours: the improvement of mathematics education through the use of the history of mathematics. There are other references more definitive on the specific mathematical topics addressed, and we invite the reader to take advantage of those available from the MAA collection and their local college and university library. We hope the reader will be inspired to apply the history of mathematics for the betterment of teaching and learning, and then share the experience with the mathematics community.

# I

# Algebra, Number Theory, Calculus, and Dynamical Systems

*For the things of this world cannot be made known without a knowledge of mathematics*

—Roger Bacon

# 1

# Arthur Cayley and the
# First Paper on Group Theory

**David J. Pengelley**
*New Mexico State University*

## Introduction

Arthur Cayley's 1854 paper *On the theory of groups, as depending on the symbolic equation* $\theta^n = 1$ inaugurated the abstract idea of a group [2]. I have used this very understandable paper several times for reading, discussion, and homework in teaching introductory abstract algebra, where students are first introduced to group theory. Many facets of this short paper provide wonderful pedagogical benefits, not least of which is tremendous motivation for the modern theory. More information on teaching with original historical sources is available at our web resource [8], which includes information on the efficacy of teaching with original sources [3, 4, 5, 9, 10], about courses based entirely on original sources [6, 7], and which provides access to books and other resources for teaching with original sources.

Cayley introduces groups as a unifying framework for several critical examples current in the early nineteenth century. These include "substitutions" (permutations, with reference to Galois on solutions of equations), "quaternion imaginaries," compositions arising in the theory of elliptic functions, and trans-

Figure 1. Arthur Cayley (1821–1895) [11]

3

position and inversion of matrices. These phenomena are either familiar to or easily understandable for our students. Cayley observes that they all have common features, which he abstracts in a first attempt at defining a group axiomatically. He proceeds to develop initial steps of a theory, including stating and using what Lagrange's theorem tells us on orders of elements, which classifies groups of prime order. Then he provides a detailed classification of all groups of order up to six. Cayley gives a brief description of how to form what we call a group algebra, and comments that when applied to the six-element symmetric group on three letters, the group algebra does not appear to have anything analogous to the modulus for the quaternions. He ends with a description of two nonabelian groups of higher order.

The richness of Cayley's paper provides a highly-motivated foundation for the beginnings of the theory of groups in an abstract algebra course. Numerous questions and contrasts arise with students as one studies the paper. First I will display some excerpts and will footnote features that I have focused on with students. Then I will discuss specific pedagogical aspects of incorporating Cayley's paper in the classroom, and the accruing benefits.

## Cayley's paper

I provide here selected extracts from Cayley's paper, with my own pedagogical comments as footnotes in order not to interrupt unduly the beautiful flow of Cayley's exposition. I will let his paper speak for itself as an inspiring and powerful tool for learning.

### On the theory of groups, as depending on the symbolic equation $\theta^n = 1$

#### Arthur Cayley

Let $\theta$ be a symbol of operation, which may, if we please, have for its operand, not a single quantity $x$, but a system $(x, y, \ldots)$, so that

$$\theta(x, y, \ldots) = (x', y', \ldots),$$

where $x', y', \ldots$ are any functions whatever of $x, y, \ldots$, it is not even necessary that $x', y', \ldots$ should be the same in number with $x, y, \ldots$. In particular, $x', y'$, &c. may represent a permutation of $x, y$, &c., $\theta$ is in this case what is termed a substitution; and if, instead of a set $x, y, \ldots$, the operand is a single quantity $x$, so that $\theta x = x' = f x$, $\theta$ is an ordinary functional symbol.[1] It is not necessary (even if this could be done) to attach any meaning to a symbol such as $\theta \pm \phi$, or to the symbol $0, \ldots$ but the symbol 1 will naturally denote an operation which ... leaves the operand unaltered .... A symbol $\theta\phi$ denotes the compound operation, the performance of which is equivalent to the performance, first of the operation $\phi$, and then of the operation $\theta$; $\theta\phi$ is of course in general different from $\phi\theta$. But the symbols $\theta, \phi, \ldots$ are in general such that $\theta.\phi\chi = \theta\phi.\chi$, &c., so that $\theta\phi\chi, \theta\phi\chi\omega$, &c. have a definite signification independent of the particular mode of compounding the symbols;[2] this will be the case even if the functional operations involved in the symbols $\theta, \phi$, &c. contain parameters such as the quaternion imaginaries[3] $i, j, k$; but not if these functional operations contain parameters such as the imaginaries which enter into the theory of octaves[4], and for which, e.g., $\alpha.\beta\gamma$ is something different from $\alpha\beta.\gamma$, a supposition which is altogether excluded from the present paper[5]. The order of the factors of a product $\theta\phi\chi \ldots$ must of course be attended to, since even in the case of a product of two factors the order is material;

---

[1]It is delightfully unclear just how Cayley's initial general notion of operation really differs from that of function, and this makes good classroom discussion.

[2]One can discuss here why we know this, e.g., for function composition, and whether this is exactly Cayley's actual context.

[3]The eight-element quaternion group is a wonderful example to motivate students.

[4]Also called the octonions, or Cayley numbers, these are an eight-dimensional non-associative real division algebra [1].

[5]This contrast offers opportunity for discussing the sense in which quaternions or octaves are or are not acting as functions, and how their multiplication is connected to function composition.

it is very convenient to speak of the symbols $\theta$, $\phi$ ... as the first or furthest, second, third, &c., and last or nearest factor. What precedes may be almost entirely summed up in the remark, that the distributive law has no application to the symbols $\theta\phi\ldots$; and that these symbols are not in general convertible,[6] but are associative. It is easy to see that $\theta^0 = 1$, and that the index law $\theta^m.\theta^n = \theta^{m+n}$, holds for all positive or negative integer values, not excluding zero.[7]

A set of symbols,

$$1, \alpha, \beta, \ldots$$

all of them different, and such that the product of any two of them (no matter what order), or the product of any one of them into itself, belongs to the set, is said to be a group[8] [Cayley's own footnote]. It follows that if the entire group is multiplied by any one of the symbols, either as further or nearer factor, the effect is simply to reproduce the group[9]; or what is the same thing, that if the symbols of the group are multiplied together so as to form a table,[10] thus:

Further factors

|  | 1 | $\alpha$ | $\beta$ | .. |
|---|---|---|---|---|
| 1 | 1 | $\alpha$ | $\beta$ | .. |
| $\alpha$ | $\alpha$ | $\alpha^2$ | $\beta\alpha$ | |
| $\beta$ | $\beta$ | $\alpha\beta$ | $\beta^2$ | |
| : | | | | |

Nearer factors

that as well each line as each column of the square will contain all the symbols $1, \alpha, \beta, \ldots\ldots$ Suppose that the group

$$1, \alpha, \beta, \ldots$$

contains $n$ symbols, it may be shown[11] that each of these symbols satisfies the equation

$$\theta^n = 1;$$

so that a group may be considered as representing a system of roots of this symbolic equation.[12] It is, moreover, easy to show that if any symbol $\alpha$ of the group satisfies the equation $\theta^r = 1$, where $r$ is less than $n$, then that

---

[6]I.e., what we today call commutative.

[7]By indicating that negative exponents are always allowed, Cayley here implicitly seems to assume that his "operations," now metamorphosing seamlessly into "symbols," always have inverses. ...

[8][Cayley's footnote]: The idea of a group as applied to permutations or substitutions is due to Galois, and the introduction of it may be considered as marking an epoch in the progress of the theory of algebraical equations.

[9]This raises the question of what Cayley is really requiring for something to be a group. Although not mentioned explicitly, it seems clear from above and below that he has in mind associativity and existence of inverses as implicit. Is this a good example of axiomatization? Will students agree, and think this necessary? All Cayley specifies literally is closure and the existence of an identity. Cayley does not justify his claim that translation always reproduces the group, but it clearly follows from existence of inverses and associativity.

[10]Here we see why we call such a "Cayley table" today.

[11]This fascinating claim of Cayley's is usually proven from Lagrange's theorem on orders of subgroups, which comes from an analysis of cosets. How did Cayley know this? Did he have a proof? Students can look ahead for this result in a text.

[12]Note Cayley's interesting point of view here, that this result on orders of elements can be interpreted as an "equation" for which the group is a solution set.

$r$ must be a submultiple of $n$; it follows that when $n$ is a prime number, the group is of necessity of the form[13]

$$1, \alpha, \alpha^2, \ldots \alpha^{n-1}, \ (\alpha^n = 1);$$

and the same may be (but is not necessarily) the case, when $n$ is a composite number. But whether $n$ be prime or composite, the group, assumed to be of the form in question, is in every respect analogous to the system of the roots of the ordinary binomial equation[14] $x^n - 1 = 0 \ldots$.

The distinction between the theory of the symbolic equation $\theta^\mu = 1$, and that of the ordinary equation $x^n - 1 = 0$, presents itself in the very simplest case, $n = 4$. $\ldots$[15]

$\ldots$ and we have thus a group[16] essentially distinct from that of the system of roots of the ordinary equation $x^4 - 1 = 0$.

Systems of this form are of frequent occurrence in analysis, and it is only on account of their extreme simplicity that they have not been expressly remarked. For instance, in the theory of elliptic functions, if $n$ be the parameter, and

$$\alpha(n) = \frac{c^2}{n} \ \beta(n) = -\frac{c^2 + n}{1 + n} \ \gamma(n) = -\frac{c^2(1 + n)}{c^2 + n},$$

then $\alpha, \beta, \gamma$ form a group of the species in question[17]. $\ldots$[18]

Again, in the theory of matrices, if $I$ denote the operation of inversion, and tr that of transposition, $\ldots$ we may write[19]

$$\alpha = I, \ \beta = \text{tr}, \ \gamma = I.\text{tr} = \text{tr}.I.$$

I proceed to the case of a group of six symbols, $\ldots$[20]

An instance of a group of this kind is given by the permutation of three letters[21]; $\ldots$

It is, I think, worth noticing, that if, instead of considering $\alpha, \beta$, &c. as symbols of operation, we consider them as quantities (or, to use a more abstract term, 'cogitables') such as the quaternion imaginaries; the equations expressing the existence of the group are, in fact, the equations defining the meaning of the product of two complex quantities of the form

$$w + a\alpha + b\beta + \ldots;$$

thus, in the system just considered, $\ldots$[22]

It does not appear that there is in this system anything analogous to the modulus $w^2 + x^2 + y^2 + z^2$, so important in the theory of quaternions.[23]

---

[13]Students may relate this form to the group of numbers on a clock, i.e., cyclic groups.

[14]By this Cayley means the group of $n$th roots of unity in the complex plane. One can examine with students why these complex roots form a cyclic group.

[15]At this point Cayley makes a detailed logical analysis of all the possibilities for the nature of a group of order four, shows that there are only two essential types (i.e., up to isomorphism), and displays their multiplication tables. This is excellent material for a class to work through.

[16]The non-cyclic "Klein Viergruppe."

[17]Students may check that these functions compose to create the Viergruppe.

[18]Cayley also mentions how the theory of quadratic forms provides another example.

[19]And this, too, may be verified to create the Viergruppe.

[20]Cayley's exhaustive analysis now shows that there are exactly two types of groups of order six, the cyclic one and another kind.

[21]This is a nice segue to permutation groups for students.

[22]Cayley now shows how to multiply in the six-dimensional system based on his noncommutative group of order six. Here he is presenting the general idea of a group algebra, and he makes analogy and contrast with quaternions. In fact, while the quaternions are not actually a group algebra, the real group algebra of the eight-element quaternion group is a direct sum of two copies of the four-dimensional algebra of quaternions over the real numbers. This can be shown from the modern theories of semi-simple rings or of Clifford algebras. I thank Pat Morandi and Ray Mines for showing me this.

[23]Cayley's interest in a modulus is presumably related to the question of whether such a system supports division, since division in the complex numbers, quaternions, and octonions (Cayley numbers) goes hand in hand with a modulus.

...I conclude for the present with the following two examples of groups of higher orders. The first of these is a group of eighteen, viz...; and the other a group of twenty-seven, ....

## Pedagogy

What are the challenges and rewards Cayley's paper offers in an abstract algebra course? The challenges include that, while Cayley's writing is very expansive and accessible, it is hard to understand in places, partly due to older terminology, but more due to its pioneering nature, in which his assumptions and contexts are not always crystal clear to us today. For instance, reading Cayley with students raises the following questions:

- Are all his group operations tacitly formed from composing functions or not, and what are the implications for associativity (e.g., quaternions, octonions)?

- Are inverses always assumed to exist? Does one need to assume this?

- Is Cayley tacitly assuming that his groups are finite? Why?

- How did Cayley know that every element in a (finite) group satisfies the "symbolic equation" for the order of the group? This is a nontrivial result, usually proved today by studying the theory of cosets of a subgroup.

But these very challenges are also precisely the pedagogical strengths of Cayley's paper, leading directly to rewards. These questions lead instructor and student to grapple with many of the key motivations and issues for group theory: What is the appropriate role of function composition, associativity, inverses, orders of elements? Through studying this paper and these questions, via the diverse and rich examples Cayley raises, students will quickly become intimately engaged by the most fundamental issues about groups, showcasing their unifying role in algebra and other mathematical contexts. Relevant assignments can include:

- Find Cayley's claims in their context in a modern textbook, and contrast their place today with their role for Cayley. For instance, Cayley's claim about elements satisfying the "symbolic equation" for the group will lead students ahead to cosets and Lagrange's theorem on orders of subgroups.

- Contrast Cayley's "symbolic equation" $\theta^n = 1$ with his "ordinary equation" $x^n - 1 = 0$, with its complex $n^{\text{th}}$ roots of unity as solutions, and consider which group it represents.

- Study and describe the general "classification" question in mathematics, algebra, and group theory, e.g., the classification results for finite simple groups, and finitely generated abelian groups, with Cayley's classification of groups of certain orders as inspiration.

- Develop the concepts of normal subgroups and quotient groups by examining Cayley's tables for groups of order six, including how his layout and heavily drawn lines begin to display the quotient group formed by certain cosets.

- Explore a new concept raised by Cayley, e.g., the notion of a group algebra, with examples.

## Conclusion

Arthur Cayley's paper produces group theory in a fully motivated context, emerging from important mathematical currents of the mid-nineteenth century. Working with the paper inserts students, as personal witnesses and even as vicarious participants, into the dynamic process of mathematical research, at a formative moment in the birth of one of the most essential concepts of modern mathematics. Students' own struggles with Cayley's ideas in comparison with their modern textbook will leave them with a more profound technical comprehension, while simultaneously initiating them into thinking and acting like mathematicians.

## References

1. J. C. Baez, "The octonions," *Bull. Amer. Math. Soc.,* 39 (2002) 145–205, and at `math.ucr.edu/home/baez`.

2. A. Cayley, "On the theory of groups, as depending on the symbolic equation $\theta^n = 1$," *Philosophical Magazine,* 7 (1854) 40–47, and in *The Collected Mathematical Papers of Arthur Cayley,* Cambridge University Press, Cambridge, 1889, vol. 2, pp. 123–130.

3. M. de Guzmán, "Origin and Evolution of Mathematical Theories: Implications for Mathematical Education," *Newsletter of the International Study Group on the History and Pedagogy of Mathematics,* 8 (March, 1993) 2–3 (and excerpted in `math.nmsu.edu/~history/guzman.html`).

4. J. Fauvel, J. van Maanen, editors, *History in Mathematics Education,* Kluwer, Dordrecht & Boston, 2000.

5. R. Laubenbacher, D. Pengelley, M. Siddoway, "Recovering motivation in mathematics: Teaching with original sources," *Undergraduate Mathematics Education Trends,* 6, No. 4 (September, 1994) 1,7,13 (and at `math.nmsu.edu/~history`).

6. R. Laubenbacher, D. Pengelley, "Great problems of mathematics: A course based on original sources," *American Mathematical Monthly,* 9 (1992) 313–317 (and at `math.nmsu.edu/~history`).

7. R. Laubenbacher, D. Pengelley, "Mathematical masterpieces: teaching with original sources," in *Vita Mathematica: Historical Research and Integration with Teaching,* R. Calinger, ed., pp. 257–260, MAA, Washington, D.C., 1996 (and at `math.nmsu.edu/~history`).

8. R. Laubenbacher, D. Pengelley, *Teaching with Original Historical Sources in Mathematics,* a resource web site, `math.nmsu.edu/~history`, New Mexico State University, Las Cruces, 1999–.

9. D. Pengelley, "A graduate course on the role of history in teaching mathematics," in *Study the Masters: The Abel-Fauvel Conference, 2002* Otto Bekken et al, eds., pp. 53–61, National Center for Mathematics Education, University of Gothenburg, Sweden, 2003 (and at `math.nmsu.edu/~davidp`).

10. M. Siu, "The ABCD of using history of mathematics in the (undergraduate) classroom," *Bulletin of the Hong Kong Mathematical Society, 1* (1997) 143–154; reprinted in *Using History To Teach Mathematics: An International Perspective,* V. Katz, ed., pp. 3–9, MAA, Washington, D.C., 2000 (and at `math.nmsu.edu/~history`).

11. D.E. Smith, *Portraits of Eminent Mathematicians: With Brief Biographical Sketches, part 2,* Scripta Mathematica, New York, 1938–1946.

# 2

# Putting the Differential Back Into Differential Calculus

**Robert Rogers**
*SUNY Fredonia*

## 2.1  Introduction

The topic of the differential of a function of a single variable is typically relegated to a single section in calculus textbooks and is usually only applied in linear approximation applications. Though the notion of linearization is certainly important, this brief treatment seems to belie the fact that the subject itself is called the differential calculus. While the present treatment of calculus focuses on the derivative as the main topic of study, this was not always the case, as illustrated by the following crude chronology of the ideas encountered in calculus. The reader must bear in mind that it is often incorrect to attribute an idea to just one person.

Calculus Chronology

- (ca. 250 B.C.): Rudiments of integration — Archimedes
- (1666): Instantaneous velocity (fluxions) — Newton
- (1676): $\int y\, dx$ — Leibniz
- (1684): First paper on differential calculus — Leibniz
- (1691): L'Hôpital's rule — Bernoulli
- (1693): Fundamental theorem of calculus — Leibniz
- (1715): Taylor series — Taylor
- (1754): Replacing infinitesimals with limits — D'Alembert
- (1772): Derivatives — Lagrange
- (1797): Mean-value theorem — Lagrange
- (1817): Continuity and the intermediate-value theorem — Bolzano
- (1823): Riemann integral — Cauchy
- (1861): Extreme-value theorem — Weierstrass
- (1874): Limit of a function — Weierstrass
- (1960): Rigorization of infinitesimals — Robinson

As the chronology indicates, differentials in fact predate formal limits and derivatives. The author contends that it is not only possible to recapture the power of differentials in our current calculus sequence, but that it is fruitful for our students to exploit this tool in applications of the calculus. Furthermore, this can be done intuitively without delving into the rigor of non-standard analysis. This is consistent with our current teaching philosophy of exploiting but not studying foundational issues such as the structure of the real number line in the calculus sequence.

Before we approach the issue of implementing these ideas, it will be useful to provide a brief history of differentials and infinitesimals.

## 2.2   Brief History of Differentials

One of the first to exploit the infinitesimally small was Archimedes. See [4, p. 29–76] for a more detailed account. In his *Method*, Archimedes describes his procedure for computing areas and volumes by mentally slicing objects into infinitesimally thin slices which could be balanced on a lever. By comparing the corresponding objects, theorems about areas and volumes could be deduced. As Archimedes pointed out, these theorems would need to be proved by geometrical means (typically the method of exhaustion) since the infinitesimally small had not been rigorously defined. However, as Archimedes states, "It is of course easier, when we have previously acquired, by the method, some knowledge of the questions, to supply the proof than it is to find it without any previous knowledge ... I am persuaded that it will be of no little service to mathematics; for I apprehend that some, either of my contemporaries or of my successors, will, by means of the method when once established be able to discover other theorems in addition, which have not yet occurred to me." [4, p. 68]

The prophecy of Archimedes was realized in the seventeenth century by mathematicians such as Kepler, Galileo, Cavalieri, Torricelli, Roberval, Fermat, Pascal, Barrow, and Huygens who exploited infinitesimals to solve a number of specific problems of areas, volumes, and tangents [20, Chapter IV]. The work of these pioneers paved the way for the invention of calculus by Newton and Leibniz in the late seventeenth century. While Newton's fluxions were based on velocities and infinitesimal instances of time, Leibniz's differential and integral calculi worked directly with differences and sums and provided the tools for applications by eighteenth century continental mathematicians.

The first paper on the calculus, Leibniz's *A New Method for Maxima and Minima as well as Tangents, Which is Impeded Neither by Fractional Nor by Irrational Quantities, and a Remarkable Type of Calculus for This* (1684), introduces differentials in a manner which is similar to our modern definition [20, p. 272]. Leibniz' idea (in more modern terms) is to draw a tangent line to the curve $v = v(x)$ at an arbitrary point as in Figure 1.

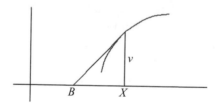

Figure 1.

Given an arbitrary length $dx$, Leibniz defines the differential of $v$, $dv$, to satisfy $\frac{dv}{dx} = \frac{v}{BX}$. In present day calculus textbooks, this is written as $dv = v'(x)\,dx$. As with our modern definition, Leibniz does not state that $dx$ is infinitesimally small (perhaps for the controversy it would create). Leibniz does state later that in dealing with tangents these differentials can represent infinitesimally small distances. In his applications, these differentials were always taken as infinitesimally small.

This reluctance to mention the infinitesimally small does not appear in the first calculus textbook, L'Hôpital's *Analysis of the Infinitely Small, for the Understanding of Curve Lines* (1696) [20, p. 312]. In this work L'Hôpital provides the following:

*Definition II:* The infinitely small part whereby a variable quantity is continually increased or decreased, is called the *differential* of that quantity.

*Scholium:* The differential of a variable quantity is expressed by the note or characteristic $d$, and to avoid confusion this note $d$ will have no other use in the sequence of this calculus.

*Postulate I:* Grant that two quantities, whose difference is an infinitely small quantity, may be taken (or used) indifferently for each other: or (which is the same thing) that a quantity, which is increased or decreased by only an infinitely small quantity, may be considered as remaining the same.

*Postulate II:* Grant that a curve line may be considered as the assemblage of an infinite number of infinitely small right lines: or (which is the same thing) as a polygon of an infinite number of sides, each of an infinitely small length, which determine the curvature of the line by the angles they make with each other.

Despite the foundational questions raised by critics such as George Berkeley [5, p. 555–58 and 9, p. 86–93], the exploitation of the calculus led to astounding advances in mathematics and physics in the eighteenth century. Infinitesimal quantities were routinely used with great success. The question of their existence was overshadowed by the correctness of the results obtained from using them. However these foundational issues surfaced more and more throughout the century. One of the first people to seriously address this issue was d'Alembert in 1754 [20, p. 341 and 9, p. 91]. D'Alembert's idea was to replace an infinitesimally small quantity with a limiting process.

Another attempt was made by Lagrange in 1772 and later years to settle the foundational issue of calculus by considering the coefficient of the linear term in the Taylor series expansion of a function. By considering this *fonction dérivée*, Lagrange contended that he had placed the foundational questions of calculus on a firm algebraic footing [8, p. 38–46].

Both of these approaches were utilized by Cauchy. While much of our modern approach to calculus as analysis can be attributed to Cauchy, it is incorrect to assume that Cauchy dropped all references to infinitesimals. After all, Cauchy's principal audience was students of engineering and science and most French textbooks took a practical point of view in teaching calculus [15, p.274]. In the forward to his *Résumé of Lessons Given at the École Royale Polytechnique on the Infinitesimal Calculus*, Cauchy wrote:

> "… My principal aim has been to reconcile rigor, which I have made a law to myself in my *Cours d'analyse*, with the simplicity which the direct consideration of infinitely small quantities produces. …" [4, p. 309].

In contrast to L'Hôpital, Cauchy defines an infinitesimal to be a variable quantity converging to zero rather than a fixed quantity. This definition not only appears in calculus books into the twentieth century [7 and 14], but more importantly made limits the fundamental concept in the study of calculus. However, Cauchy was not ignoring the accomplishments made in the eighteenth century and it is my belief that neither should we in our calculus sequence. With this in mind, I have reworked the differentials section of my own calculus course.

## 2.3    Implementation of Differentials in a Calculus Class

My approach to calculus starts traditionally, with limits and derivatives. What follows is my treatment of differentials.

Given a quantity $x$, an **increment** of $x$, denoted by $\Delta x$, is a measurable (finite) change in $x$. The idea is that no matter how small $\Delta x$ may be, it is possible to magnify the picture sufficiently to perceive the change. By contrast, a **differential** of $x$, denoted by $dx$, is an instantaneous change in $x$. Here the idea is that no matter how large the picture is magnified, $dx$ cannot be perceived. It is often convenient to think that a differential can be perceived if the graph is "infinitely magnified". In my experience, students have no difficulty with the lack of rigor in this definition nor its intuitive meaning.

The author's rules for dealing with differentials are, to a large part, based on L'Hôpital's work and the definition of infinitesimal given by Cauchy.

**Rule I:** One should not try to measure differentials, only compare them with other differentials.

**Rule II:** An increment cannot equal a differential, but a differential can be thought of as a sequence of increments converging to zero. One should not write $\Delta x = dx$, but instead write $\Delta x \approx dx$.

**Rule III:** When a graph is "infinitely magnified", a differential change along the curve is the same as the differential change along the tangent line (local linearity).

As a consequence, we have the following familiar definition.

**Rule IV:** If $y = y(x)$, then $dy\Big|_{x=a} = y'(a)\,dx$.

For example, if $y = x^2$, then $dy\Big|_{x=a} = 2a\,dx$.

**Rule V:** Differential triangles can be treated as regular triangles and in fact may be similar to regular triangles.

As I tell my students, these heuristic definitions and rules are really secondary to the application of differentials. Here is where students can see the power and utility of this tool and hopefully will become comfortable enough to use differentials in subsequent courses such as differential equations and physics. In my opinion, these applications provide a simpler, more natural approach than the traditional approaches do.

## Differentiation of sine and cosine

This is an adaptation of an argument given by Leibniz concerning simple harmonic motion [12, p. 529].

Consider the unit circle in Figure 2. Since the tangent to the circle is perpendicular to the radius, then the differential triangle is similar to $\triangle OPQ$.

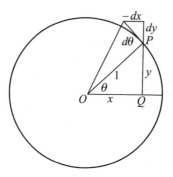

Figure 2.

Thus we have the proportion

$$\frac{dy}{d\theta} = \frac{x}{1}.$$

Notice that we are using the local linearity to identify the hypotenuse of the differential triangle as the change along the circle. Since $y = \sin\theta$ and $x = \cos\theta$, we obtain

$$\frac{d(\sin\theta)}{d\theta} = \cos\theta.$$

Similarly, the proportion $\frac{-dx}{d\theta} = \frac{y}{1}$ yields $\frac{d(\cos\theta)}{d\theta} = -\sin\theta$. The negative sign is due to the fact that $dx$ itself is negative in the first quadrant; the other quadrants will work out similarly.

I personally find this derivation more appealing than the standard one found in many calculus textbooks. It eliminates the need for demonstrating $\lim_{\theta\to 0} \frac{\sin\theta}{\theta} = 1$, though this can be presented as a corollary of the derivative formula if one wishes. It also emphasizes the importance of radian measure.

## Curvature

One can define the curvature, $\kappa$, of a plane curve by $\kappa = \frac{d\alpha}{ds}$ where $s$ denotes arc length and $\alpha$ is the angle formed by the tangent line to the curve and a fixed line (typically the $x$-axis). Many present-day calculus books show that the curvature of a circle is constant. Few present textbooks mention the converse: if a plane curve has a constant curvature, then it must be either a line ($\kappa = 0$) or a circle of radius $\frac{1}{\kappa}$ ($\kappa \neq 0$). This result can be found as an exercise in [21, p. 958]. The solution manual solves this problem by parameterizing the curve with respect to arc length, and the problem appears fairly late in the book. The problem also appears as an exercise in [14, p. 367] and [11, p. 299], though I do not know how it was intended to be solved. The following solution is based on the presentation in [16, p. 472] where it is shown that a curve, in general, is determined by its arc length and curvature. It should also be noted that this solution is accessible to first semester calculus students.

We will solve the case where curvature is nonzero and leave the other case as an exercise for the reader. Consider Figure 3 which shows a piece of the curve in question and the corresponding tangent differential triangle.

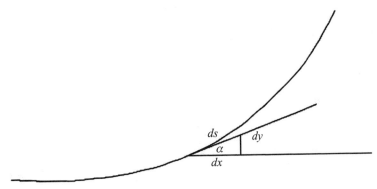

Figure 3.

We have the equations $\sin\alpha = \frac{dy}{ds}$, $\cos\alpha = \frac{dx}{ds}$, $ds = \frac{1}{\kappa}d\alpha$ which in turn yield

$$dy = \frac{1}{\kappa}\sin\alpha\, d\alpha, \qquad dx = \frac{1}{\kappa}\cos\alpha\, d\alpha.$$

Antidifferentiation gives us

$$y + c_2 = -\frac{1}{\kappa}\cos\alpha, \qquad x + c_1 = \frac{1}{\kappa}\sin\alpha$$

which results in the circle $(x + c_1)^2 + (y + c_2)^2 = \frac{1}{\kappa^2}$.

## The fundamental theorem of calculus

There is the oft-told story of Leibniz's first encounter with Christian Huygens who was then the most renowned scientist and mathematician on the continent [4, p. 236-38]. To "test his mettle," Huygens essentially gave Leibniz the following problem: compute the sum of the series $\frac{1}{1\cdot2} + \frac{1}{2\cdot3} + \frac{1}{3\cdot4} + \cdots$. By rewriting the series as $\left(\frac{1}{1} - \frac{1}{2}\right) + \left(\frac{1}{2} - \frac{1}{3}\right) + \left(\frac{1}{3} - \frac{1}{4}\right) + \cdots$ and canceling like terms, Leibniz arrived at the correct answer of 1. Geometrically, this can be seen in Figure 4.

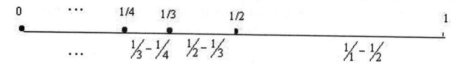

Figure 4.

More important than the answer is Leibniz's observation that the *sum of consecutive differences equals the difference of the extreme terms*. Applying this to differentials we obtain (essentially) Leibniz's proof of the fundamental theorem of calculus [20, p. 282]: if $\frac{dF}{dx} = f$, then $f\,dx = dF$ and $\int_{x=a}^{b} f\,dx = \int_{x=a}^{b} dF = F(b) - F(a)$. Figure 5 illustrates this geometrically.

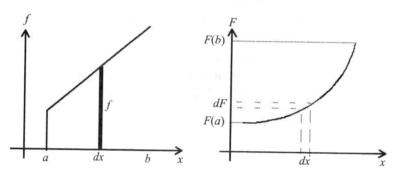

Figure 5.

Comparing the curves $f$ and $F$, we have that the area of the rectangle $(f\,dx)$ is equal to the length $dF$. As the sum of the rectangles fills out the area under the curve $f$, the sum of the differences $dF$ yields the difference of the extremes.

From my experience, this argument is appealing to students and is readily understood. This argument can be "rigorized" by using the mean-value theorem [6, p. 248–49], but the value of this rigor for students in the calculus sequence is not apparent.

This notion of an integral as the sum of infinitely many infinitesimally small parts is often ignored in modern treatments of the calculus in favor of Riemann sums. However, for applications, this infinitesimal approach provides a simpler, more streamlined, and holistic approach. For example, consider arc length and surface area of a surface of revolution. Rather than having students memorize a formula involving an integral, it is advantageous to compute arc length $s$ by summing infinitesimally small pieces of arc length and applying the Pythagorean Theorem to a differential triangle as follows:

$$s = \int_{x=a}^{b} ds = \int_{x=a}^{b} \sqrt{(dx)^2 + (dy)^2}.$$

This integral has the advantage of being reformulated into either

$$\int_{x=a}^{b} \sqrt{1 + \left(\frac{dy}{dx}\right)^2}\,dx \quad \text{or} \quad \int_{x=a}^{b} \sqrt{\left(\frac{dx}{dy}\right)^2 + 1}\,dy.$$

Furthermore replacing infinitely small segments of the curve with the corresponding tangent lines is consistent with the later treatment of the surface area of general surfaces where using inscribed polyhedra does not work and approximating tangent planes must be used. (See [3, p. 341–42] for an example of a sequence of approximating polyhedra inscribed in a cylinder with their surface areas diverging.) Moreover, this holistic approach works well in treating the surface area of a surface of revolution as a sum of infinitely thin frusta. Although this seems natural to a working mathematician, this approach seems foreign to many calculus students and their textbooks. As a test of this assertion, ask yourself how many calculus students could solve the following problem and then ask yourself if you have ever seen such a problem in any calculus text. The problem is to find the surface area of the surface of revolution generated by revolving the curve $y = x^2$, $0 \leq x \leq 1$ about the line $y = x$. The solution using differentials [See Figure 6.] makes the problem as natural as one where the axis of revolution is a coordinate axis and requires students to think about the process instead of applying a formula.

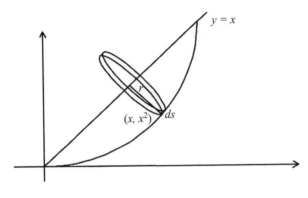

Figure 6.

Noting that the total surface area is the sum of the areas of the infinitely thin frusta, it is straightforward to see that the surface area is given by

$$\int_{x=1}^{1} 2\pi r\, ds = 2\pi \int_{x=0}^{1} \frac{1}{\sqrt{2}} (x - x^2)\sqrt{1 + 4x^2}\, dx.$$

The idea that an integral is a sum of infinitesimally small parts can be infused throughout the integral calculus course. Instead of having separate chapters for various applications, these could be presented as variations on the same theme. However, integration is not the only place where differentials provide a valuable tool for physical applications. My classroom presentation on the utility of differentials typically starts with the following example.

### The cable on a suspension bridge and the catenary

What is the shape of a cable of negligible weight supporting a uniform horizontal load of weight density $w$?

To solve this, consider one half of the cable and assume that the lowest point of the cable is at the origin as in Figure 7. Let $h$ denote the magnitude of the (constant) horizontal tension throughout the cable.

Since the cable is at rest, the vertical and horizontal components of the tension $T$ at the point $P$ must equal the weight of the horizontal load and $h$ respectively. Drawing a differential triangle [Figure 8] and using similar triangles, one has $\frac{dy}{dx} = \frac{wx}{h}$. This is readily seen to be the differential equation of a parabola.

The differential equation for the *catenary* (a chain of weight density $w$ hanging under its own weight) can be derived in the same manner, except that the vertical force is now $ws$ where $s$ denotes the arc length of the chain between the origin and the point $P$. Differentiating both sides of the equation $\frac{dy}{dx} = \frac{ws}{h}$ results

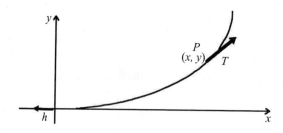

Figure 7.

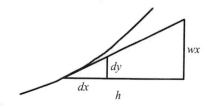

Figure 8.

in the differential equation

$$\frac{d^2 y}{dx^2} = \frac{w}{h}\frac{ds}{dx} = \frac{w}{h}\sqrt{1 + \left(\frac{dy}{dx}\right)^2}.$$

The derivation of the final equation can be given as an exercise to first semester calculus students as well as showing that $y = \left(\frac{h}{2w}\right)\left(e^{wx/h} + e^{-wx/h}\right)$ provides a solution to the equation. Students can also graph this for various values of $h$ and $w$ to see that this answer makes sense physically.

A similar problem is to find the shape of a stone arch bridge if the force due to the weight of the stone is to equal the vertical component of the tangential force at any point on the arch (so that the weight of the bridge is directed toward the base of the bridge).

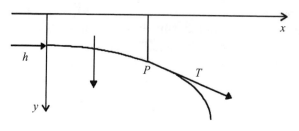

Figure 9.

Using Figure 9 and an analysis similar to before, we obtain that the differential equation of the arch is $\frac{dy}{dx} = \frac{w}{h}\int_0^x y(x)\,dx$ where $w$ is the weight density of the stone (per cross sectional area) and $h$ is the horizontal force from the matching half of the bridge. Differentiation and the fundamental theorem of calculus yield $d^2 y/dx^2 = wy/h$. As in the catenary this leads to the hyperbolic cosine as the solution. This also provides a nice application of the differentiation of an integral.

The stone arch bridge problem can be found in [13, p. 171] while the previous two problems can be found in [19, p. 58] along with an analysis that does not involve differentials. This more modern solution (of the catenary for example) involves labeling the angle of elevation, say $\alpha$, and noting that $\sin \alpha = \frac{ws}{|T|}$ and $\cos \alpha = \frac{h}{|T|}$. Division yields $\frac{dy}{dx} = \tan \alpha = \frac{ws}{h}$.

I gave the catenary problem to a group of students a number of years ago, thoroughly intending them to provide this (standard) solution. After we discussed the forces involved I told them that the first step was to obtain the equation $\frac{dy}{dx} = \frac{ws}{h}$. Immediately, one student said, "But that's obvious!" He proceeded to draw the differential triangle and provided the first solution I mentioned. I must admit I was a bit stunned. The student's solution not only rendered the trigonometry and slope superfluous, but his solution turned out to be (essentially) the solution given by Johann Bernoulli (1692). [10, p. 249] This episode and the student's insight led me to notice that over the years we have sought to replace the simplicity of differentials with "more rigorous" approaches rather than to reconcile it with rigor as in the words of Cauchy. For our students who struggle with calculus, including superfluous items such as angles of elevation, etc. when not necessary, makes many applications unduly complicated. A good example of this can be seen in the following application.

## Brachistochrone

James McKenna of SUNY Fredonia devised a very nice exercise for his students where they are to examine the time it takes for a ball to roll down various curves from rest only under the influence of gravity. Differentials provide a nice analysis of this situation. In Figure 10, $y = y(x)$ will denote a curve, $g$ will denote the acceleration due to gravity, $a$ will denote the tangential component of this acceleration, $t$ will denote time, $v$ will denote velocity, and $s$ will denote arc length.

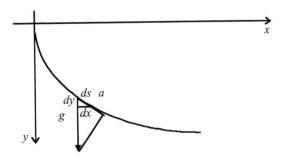

Figure 10.

Using similar triangles we have $\frac{a}{g} = \frac{dy}{ds}$ which leads to the following calculations:

$$g\,dy = a\,ds$$
$$g\,dy = \frac{dv}{dt}\,ds$$
$$g\,dy = dv\,\frac{ds}{dt} = v\,dv.$$

Antidifferentiating and applying the initial condition $v = 0$ when $y = 0$, we obtain

$$g \cdot y = \frac{v^2}{2}$$
$$\frac{ds}{dt} = v = \sqrt{2gy}$$
$$dt = \frac{ds}{\sqrt{2gy}}.$$

Therefore the time it takes for the ball to roll from the origin to a point with abscissa $b$ is given by

$$\int_{x=0}^{b} dt = \int_{x=0}^{b} \frac{ds}{\sqrt{2gy}}.$$

Students can proceed to examine this integral for various curves. Note that this derivation makes no mention of kinetic energy as one might find in a typical modern calculus text. This problem also lends itself to Johann Bernoulli's solution of the brachistochrone problem [17, p. 315–17] or [20, p. 391–96]. To find this path of fastest descent, Bernoulli starts with the fact that this path must satisfy Snell's law of refraction, namely $\frac{\sin\alpha}{v} = c$ where $\alpha$ is the angle the tangent line makes with the vertical and $c$ is a constant. We already know that the velocity $v$ is given by $\sqrt{2gy}$. Using a differential triangle, it is readily seen that $\sin\alpha = \frac{dx}{ds}$. Substituting these into Snell's law yields the differential equation $1/2gc^2 = y\left(1 + (\frac{dy}{dx})^2\right)$.

Although solving this equation is beyond the scope of a first semester calculus course, its derivation is not. Furthermore, students can show that the cycloid

$$x = \frac{t - \sin t}{4gc^2}$$
$$y = \frac{1 - \cos t}{4gc^2}$$

provides a solution. [Note that to Leibniz and his school, $\frac{dy}{dx} = \frac{dy/dt}{dx/dt}$ would be obvious.]

Not only can these examples be used to develop meaningful exercises, but the instructor and students will find that quite often in physical and geometric applications (for example, the tractrix or pursuit curves), the use of differentials and similar triangles will render the use of slopes superfluous and rather unnatural. This is particularly exemplified in the solution to this problem found in [18, p. 235]: If a cylindrical container partially filled with a liquid is rotated about its axis with a constant angular velocity of $w$, then the liquid will rise along the walls and sink in the center. Show that the surface formed is that of a parabola revolved about the axis. [Hint: The force $R$ perpendicular to the surface at point $P$ has as its horizontal component the centripetal force $mxw^2$ and as its vertical component the force required to counteract gravity $mg$, where $m$ is the point mass at $P$ and $g$ is the acceleration due to gravity.]

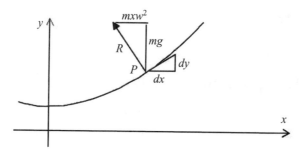

Figure 11.

It is readily seen in Figure 11 (without mentioning slopes of perpendicular lines) that $\frac{dy}{dx} = \frac{w^2 x}{g}$ which is the differential equation for a parabola. The Steward Observatory Mirror Laboratory at the University of Arizona employs a rotating furnace to cast parabolic mirrors in this fashion. [See 23.] Students becoming proficient in using differentials in this manner will appreciate the power that made differential and integral calculus a major tool for mathematics in the eighteenth century.

## 2.4   Conclusion

Of course the issue of adding more to an already crowded curriculum must be addressed. It has been suggested by others that the mean-value theorem be dropped from the calculus sequence (sometimes in favor of more intuitive theorems) [22]. The mean-value theorem itself acts as a bridge between the finite and the infinitesimally small (or more appropriately the average and the instantaneous). This notion provides an answer to the previous question of why inscribed polyhedra do not work for computing surface area; there is no such bridge for surfaces [3, p. 268]. The direct use of instantaneous changes renders this bridge unnecessary in any case. As was seen earlier, presentations of the fundamental theorem of calculus and arc lengths and surface areas of surfaces of revolution can be presented more quickly and cleanly with infinitesimals than by applying the mean-value theorem.

The theorem ($f' > 0 \Rightarrow f$ increasing) is typically given as a corollary of the mean-value theorem. Interestingly, Cauchy uses this result to prove his generalized mean-value theorem [4, p. 314]. While Cauchy's proof of the first result leaves much to be desired, a perfectly rigorous indirect proof can be given by repeatedly bisecting the (finite) interval in question and applying the nested interval property of the real number system (equivalent to the least upper bound property). See [22] for details. I present this proof in my intermediate real analysis class; in calculus I prefer appealing to a diagram to convince students.

As for L'Hôpital's rule, the following special case can be proved without use of the generalized mean-value theorem. See [1, p. 500] for such a presentation.

If $f$ and $g$ are differentiable on an open interval containing $a$ with $g' \neq 0$ on that interval and $f(a) = g(a) = 0$, then $\lim_{x \to a} \frac{f(x)}{g(x)} = \frac{f'(a)}{g'(a)}$.

Not only is the proof of this result readily seen by students, but it is essentially Bernoulli's original version as given in L'Hôpital's calculus book [20, p. 315–16].

For non-majors, foundations such as the mean-value theorem could be argued as unnecessary. This is consistent with our chronology showing that the calculus was utilized with much success long before the foundations of the subject became central. Mathematics majors will see these foundations in a subsequent real analysis course. Indeed a curriculum that presents this more intuitive geometric approach in calculus provides a springboard for a more rigorous approach in introductory real analysis. See [2] for a presentation of this subject put into its historical context.

Throughout the calculus sequence, many places where we stop to supply nineteenth and twentieth century rigor can be eliminated in favor of differentials with the understanding that this rigor will surface later in the curriculum. This model more closely matches the evolution of the subject. At the very least, it is an injustice to ignore the very tool that revolutionized mathematics and science throughout the eighteenth century.

## References

1. H. Anton, I. Bivens, and S. Davis, *Calculus, 7th ed.*, John Wiley & Sons, New York, 2002.

2. D. Bressoud, *A Radical Approach to Real Analysis*, MAA, Washington, DC, 1994.

3. R. Courant, *Differential and Integral Calculus, Vol. II*, Interscience Publishers, Inc., New York, 1936.

4. C. Edwards, *The Historical Development of the Calculus*, Springer-Verlag, New York, 1979.

5. J. Fauvel and J. Gray, *The History of Mathematics, a Reader*, The Open University, U.K., 1987.

6. A. Goodman, *Analytic Geometry and the Calculus, 3rd ed.*, MacMillan Publishing Co., Inc., New York, 1974.

7. E. Goursat, *A Course in Mathematical Analysis, Vol. I — Derivatives and Differentials, Definite Integrals, Expansion in Series, Applications to Geometry*, Dover Publications, New York, 1904.

8. J. Grabiner, *The Origins of Cauchy's Rigorous Calculus*, MIT Press, Cambridge, MA., 1981.

9. I. Gratton-Guiness, editor, *From the Calculus to Set Theory, 1630–1910, an Introductory History*, Princeton University Press, Princeton, NJ, 1980.

10. ——, *The Norton History of the Mathematical Sciences*, W.W. Norton & Co., New York, 1997.

11. G. Hardy, *A Course of Pure Mathematics, 10th ed.*, Cambridge University Press, London, 1952.

12. V. Katz, *A History of Mathematics, an Introduction, 2nd ed.* Addison Wesley, Reading, MA, 1998.

13. L. Kells, *Elementary Differential Equations, 3rd ed.*, McGraw-Hill, New York, 1947.

14. H. Lamb, *An Elementary Course of Infinitesimal Calculus, Revised Edition*, Cambridge University Press, London, 1961.

15. D. Laugwitz, "Comments on the Paper 'Two Letters by N. N. Luzin to M. Ya. Vygodskii'," *American Mathematical Monthly* 107 (2000), 267–76.

16. A. Schwartz, *Calculus and Analytic Geometry, 2nd ed.*, Holt, Reinhart, & Winston, New York, 1967.

17. G. Simmons, *Calculus Gems*, McGraw-Hill, New York, 1992.

18. ——, *Calculus with Analytic Geometry, 2nd ed.*, McGraw-Hill, New York, 1996.

19. ——, *Differential Equations with Applications and Historical Notes*, McGraw-Hill, New York, 1972.

20. D. Struik, *A Source Book in Mathematics, 1200-1800*, Harvard University Press, Cambridge, MA, 1969.

21. E. Swokowski, M. Olinick, and D. Pence, *Calculus, 6th ed.*, PWS Publishing Co., Boston, MA, 1994.

22. T. Tucker, "Rethinking Rigor in Calculus: The Role of the Mean Value Theorem," *American Mathematical Monthly* 104 (1997), 231–40.

23. http://medusa.as.arizona.edu/mlab/mlab.html

# 3

# Using Galois' Ideas in the Teaching of Abstract Algebra

**Matt D. Lunsford**
*Union University*

## Introduction

Joseph Gallian [3, p. 47] in his popular text *Contemporary Abstract Algebra* states, "The goal of abstract algebra is to discover truths about algebraic systems (that is, sets with one or more binary operations) that are independent of the specific nature of the operations." While this modern approach, because of its generality, provides a unifying element to the study of mathematics, it conceals from the student many of the great ideas generated by significant problems in the history of the discipline. B. Melvin Kiernan [4, p. 40] asserts that "without a clear historical perspective it is difficult to see or even imagine the connection between the [abstract] algebra of the present day and the computational problems from which it arose." Similarly, John Stillwell [6, p. 143] in his book *Elements of Algebra* says that, "it seems to be one of the laws of mathematical history that if a concept can be detached from its origins, it will be."

One approach to teaching a first course in abstract algebra that attempts to address the concerns raised by Kiernan and Stillwell without compromising sound pedagogy will be discussed here. In particular, this approach introduces the basic algebraic structures of group, ring, and field in a natural way using the theory of polynomial equations and the ideas of several mathematicians including Lagrange, Abel, and, especially, Evariste Galois. The historical problem of the solvability of polynomial equations provides a context for motivating the abstract structures as well as a link between this advanced course and elementary courses in algebra.

## Historical Background

Polynomial equations have been an object of study since antiquity [7, Chapter 1]. Babylonian mathematicians around 1600 B.C. gave a verbal recipe for the solution of various forms of quadratic equations. Greek mathematicians added a geometric flair to solutions of polynomial equations by using standard straight-edge and compass constructions along with conic sections. The Arab mathematician al-Khwarizmi provided a systematic account of the solution of all known quadratic equations in his work *The Condensed Book on the Calculation of al-Jabr and al-Muqabala*. It is from this work that we have derived the modern term *algebra*. The Persian mathematician (and poet) Omar Khayyam presented conic section solutions to all known cubic equations in his text *Treatise on Demonstrations of Problems of al-Jabr and al-Muqabala*.

Polynomial equations that can be solved using only the algebraic operations of arithmetic and the extraction of roots are said to be *solvable by radicals*. Many mathematicians of antiquity and the Middle Ages believed that, unlike quadratic equations, an algebraic formula for the solution of all types of cubic equations was not possible. However, sixteenth century Italian mathematicians proved this wrong. Cardano, in his monumental work *Ars Magna* (*The Great Art*), demonstrated that both cubic and quartic equations are solvable by radicals. An interesting account of the major figures who were involved in these discoveries can be found in Dunham [1, Chapter 6]. With Cardano's publication, the quest for a general formula for solving quintic equations began. By the end of the eighteenth century, no such formula had been found. As in the case of the cubic, some began to doubt the existence of such a formula.

In 1770–71, Lagrange published *Reflections on the Resolution of Algebraic Equations*. In this work, Lagrange recounted the various known methods for solving polynomial equations of degree 2, 3, and 4 and attempted to synthesize these methods into a coherent theory to address the solvability of higher degree equations. He focused on relationships that exist between the roots of a polynomial equation and its coefficients. In particular, Lagrange used the fundamental theorem of symmetric polynomials and the fact that the coefficients of a polynomial equation are given by the elementary symmetric polynomials in the equation's roots to give a necessary condition for the roots to be rationally expressible in terms of the *resolvent* of the equation and other known quantities (i.e. the coefficients of the given equation, rational numbers, and roots of unity) [2, pp. 32–35].

In 1826, Abel built upon the work of Lagrange and was able to prove that the general quintic equation is not solvable by radicals [7, Chapter 13]. His work not only answered the question for the quintic but also demonstrated that the general polynomial equation of degree $n$, $n > 4$, is not solvable by radicals. Thus the conjecture that no algebraic formula existed was proven correct. Of course, specific polynomial equations might be solvable by radicals. In fact, Gauss had shown previously that cyclotomic polynomial equations of arbitrary degree are solvable by radicals [2, pp. 23–30].

Evariste Galois addressed the question of solvability by radicals in his *Memoir on the Conditions for Solvability of Equations by Radicals*. The memoir, which was published posthumously in 1846, not only provided a theoretical basis for answering the solvability question for any polynomial equation but also developed a framework for a mathematical theory with far-reaching applications. Galois translated the solvability problem into the language of field theory using a primitive form of the idea of an extension field. He then associated a set of permutations, which he called a *group* and which is now known as a Galois group, with a polynomial equation over its base field of coefficients. He coupled these new ideas by showing that the Galois group of a polynomial over an extension field either remains the same or is a normal subgroup of the original Galois group. Conversely, he showed that if the Galois group of a polynomial has a normal subgroup, then this subgroup is in fact the Galois group of the polynomial over a particular extension field [2, pp. 59–65]. (Galois assumed the presence of roots of unity in his field extensions.)

Recall that a splitting field for a polynomial is a field containing all of the polynomial's roots. Using the Galois group, Galois was able to analyze a particular splitting field for a given polynomial. His results completely characterized the situation: a polynomial is solvable by radicals precisely when its Galois group admits a sequence of normal subgroups, beginning with the whole group and ending with the trivial subgroup, for which the consecutive quotient groups are cyclic. A finite group with this property is now called *solvable*. Complete details on Galois work can be found in the works by Edwards [2] and Tignol [7, Chapter 14].

## Applications To Pedagogy

A historically focused first course in abstract algebra can be designed using the solvability of polynomial equations as its cornerstone. Students come to a first course with some understanding of polynomials and their solutions. Building upon this prior knowledge, the instructor can formulate the question of solvability by radicals and use this concept to introduce the basic abstract structures of group, ring, and field in a concrete way. Four pertinent questions bring about the structure of the course.

### 1.   How are the set of integers and the set of polynomials with rational coefficients similar?

As with any mathematics course, a certain amount of the material is preliminary. In this design, the first connection to discover is the existence of similarities between the set $Z$ of integers and the set $Q[X]$ of polynomials with rational coefficients. A thorough discussion of the basic notions of divisibility, unique factorization, prime (or irreducible) elements, and zero divisors is essential and leads smoothly to the axiomatic definitions of *ring* and *integral domain*. Moreover, Galois used the ideas of divisibility and irreducibility early in his memoir. In particular, Galois associated to a given polynomial $f(x) \in Q[X]$ of degree $n$ an auxiliary polynomial $F(X) \in Q[X]$ of degree $n!$ which he factored into a product of irreducible polynomials over the rational field $Q$ [2, p. 38]. The factorization of $F(X)$ played a crucial role in Galois' determination of whether or not the polynomial $f(x)$ is solvable by radicals.

### 2.   Where can we find non-rational solutions to a polynomial equation with rational coefficients?

Galois assumed that all of the roots of a polynomial $f(x) \in Q[x]$ exist in some set containing $Q$ and that this set is closed under the four arithmetical operations of addition, subtraction, multiplication, and division [2, p. 37].

Galois referred to the elements of this set as *rational* quantities since they could be expressed as a rational function of the coefficients of $f(x)$ and of a certain number of *adjoined* quantities. Clearly, Galois was articulating, in a rudimentary form, the idea of an extension field of $Q$ and, more specifically, a splitting field of the polynomial $f(x)$. These observations inspire the current definitions of *field*, *extension field*, and *splitting field*.

To illustrate Galois' thoughts, consider the simple example of a monic quadratic polynomial $f(x) = x^2 + px + q$, $p, q \in Q$. We assume that the equation $f(x) = 0$ has *distinct* roots $a \neq b$ (an assumption that Galois made which causes no loss of generality [2, pp. 35–36]). Then, the two expressions $t_1 = a - b$ and $t_2 = b - a$ in the roots are also distinct and are known as the *resolvents* of the equation. If $a, b \notin Q$, then $t_1, t_2 \notin Q$ and the associated auxiliary polynomial

$$F(X) = (X - t_1)(X - t_2) = X^2 - (a - b)^2 \in Q[X],$$

is irreducible over $Q$. The roots

$$a = \frac{t_1 - p}{2} \quad \text{and} \quad b = \frac{-t_1 - p}{2}$$

are elements of $K = Q(t_1)$ and therefore $K$ is a splitting field of $f(x)$.

### 3. What relationships exist between a polynomial equation and its roots?

To exploit the connection between solutions of polynomial equations and their coefficients, we need the fundamental theorem of symmetric polynomials. This theorem, which states that any symmetric rational function can be expressed as a rational function of elementary symmetric polynomials, implies that any rational symmetric function of the roots of $f(x) = 0$ can be expressed in terms of the coefficients of $f(x)$. This crucial result not only introduces the notion of symmetry into the theory of polynomial equations but also forces the definition of the general symmetric group $S_n$.

Galois brilliantly generalized the idea that rational symmetric functions of the roots of a polynomial equation can be expressed in terms of its coefficients. He associated to a given polynomial $f(x) \in Q[X]$ of degree $n$ a set of permutations of the $n$ distinct roots of the equation $f(x) = 0$. This set of permutations, denoted by $G(f, Q)$, is a subgroup of $S_n$ and is known as the Galois group of the polynomial $f(x)$ over $Q$. He then showed that *any* rational function (not necessarily symmetric) of the roots which is invariant under all of the permutations in $G(f, Q)$ is, in fact, expressible in terms of the elements of the coefficient field $Q$ of $f(x)$ [2, pp. 52–53].

Consider again the monic quadratic polynomial $f(x) = x^2 + px + q$, $p, q \in Q$. The coefficients of $f(x)$ are given by the elementary symmetric functions in the roots $a, b$: $p = -(a + b)$ and $q = ab$. Applying the fundamental theorem of symmetric polynomials, we conclude that *any* symmetric polynomial in the roots $a$ and $b$ can be expressed in terms of the coefficients $p$ and $q$. So, in particular, the discriminant $(a - b)^2$, being symmetric in the roots, can be expressed simply as $(a - b)^2 = p^2 - 4q \in Q$. Hence the splitting field $K$ is obtained from $Q$ by the adjunction of a square root of the discriminant $(a - b)^2 \in Q$. Moreover, since the roots $a$ and $b$ are not in $Q$, the Galois group $G(f, Q)$ is not trivial; hence it is the symmetric group $S_2$.

This introduction to finite permutation groups allows for an easy transition to the definition of an abstract *group*. The course would then focus on the discovery and exploration of elementary properties of abstract groups.

### 4. When can we express solutions of a polynomial equation using only arithmetical operations and the extraction of roots?

To answer this question, we must discuss the concept of a *normal* subgroup. Galois was the first mathematician to recognize the significance of those subgroups of a finite permutation group that are invariant under conjugation. In particular, he showed that if an extension field $K'$ is obtained from the field $K$ by the adjunction of a $p$th root of an element from $K$, where $p$ is prime, then the associated Galois group $G(f, K')$ for the polynomial $f(x)$ over $K'$ either remains equal to the Galois group $G(f, K)$ or reduces to a normal subgroup of $G(f, K)$ of index $p$ [2, p. 59].

Next, he showed that for a polynomial equation with rational coefficients to be solvable by radicals, the splitting field must be obtainable from the rational field $Q$ by a finite sequence of field extensions in which each intermediate field extension $K'$ is obtained from the previous field $K$ by the adjunction of a $p$th root of an element from $K$. Combining these ideas, he proved that a polynomial with rational coefficients is solvable by radicals precisely when its Galois group $G(f, Q)$ is a *solvable* group [2, pp. 64–65].

Let's return to our example of the monic quadratic polynomial $f(x) = x^2 + px + q$, $p, q \in Q$. As stated earlier, the Galois group $G(f, Q) = S_2$. The adjunction of the square root of the discriminant $(a - b)^2 \in Q$ yields the splitting field $K$. Moreover, $G(f, K)$, which contains only the identity permutation, is the only proper normal subgroup of $G(f, Q)$. It follows easily that $G(f, Q) = S_2$ is a solvable group. Finally, the roots $a$ and $b$ are given succinctly by the radical expression

$$a, b = \frac{(a+b) \pm \sqrt{(a-b)^2}}{2} = \frac{-p \pm \sqrt{p^2 - 4q}}{2}.$$

Therefore, exposing the answer to the question of when the roots of a polynomial equation are expressible by purely algebraic methods naturally leads to a discussion of *normal subgroups*. This conversation permits a straightforward introduction to the related topics of *homomorphism*, *isomorphism*, *ideal*, *quotient group*, and *factor ring*.

## Conclusion

The solvability of polynomial equations, with its interplay of polynomial rings, field extensions, and permutation groups, provides a wonderful context for introducing the basic topics of abstract algebra. In addition, the use of concrete problems from the history of the discipline motivates the subject matter and interjects humanity into the enterprise. Highly motivated students may be interested in reading Galois' *Memoir on the Conditions for Solvability of Equations by Radicals*. An English translation of the memoir can be found in Edwards [2, Appendix 1].

The approach presented here, while far from unique, presents major topics in a context that respects both the origins of the discipline and the pedagogical needs of the student, and motivates algebraic structures through the examination of concrete mathematical problems. A different approach to teaching abstract algebra from a historical perspective can be found in Kleiner [5].

## Acknowledgements

The author gratefully acknowledges the editors of this volume for their valuable comments and suggestions. The author also wishes to thank Union University for its support of this research through the Pew Summer Research and Teagle Course Development grants. Finally, the author wishes to express his gratitude to Craig Fraser and the Institute for the History and Philosophy of Science and Technology, University of Toronto and to Delinda Buie, Curator, Department of Rare Books and Special Collections, University of Louisville.

## References

1. William Dunham, *Journey Through Genius*, John Wiley and Sons, New York, 1990.

2. H. M. Edwards, *Galois Theory*, Springer-Verlag, New York, 1984.

3. Joseph Gallian, *Contemporary Abstract Algebra*, Houghton Mifflin, Boston, 1998.

4. B. M. Kiernan, "The Development of Galois Theory from Lagrange to Artin," *Archives for the History of the Exact Sciences*, 8 (1971) 40–154.

5. Israel Kleiner, "A Historically Focused Course in Abstract Algebra," *Mathematics Magazine*, 71 (1998) 105–111.

6. John Stillwell, *Elements of Algebra*, Springer-Verlag, New York, 1994.

7. J. P. Tignol, *Galois' Theory of Algebraic Equations*, World Scientific, Singapore, 2001.

# 4

# Teaching Elliptic Curves Using Original Sources

**Lawrence D'Antonio**
*Ramapo College of New Jersey*

## Introduction

In this paper we give an overview of the subject of elliptic curves and examine some of the original sources that can be used to teach this very important topic in a history of mathematics course. Portions of this material are also appropriate for courses in abstract algebra, number theory, or geometry.

There are two purposes for introducing students to elliptic curves. First, the topic of elliptic curves is a beautiful one combining significant ideas from algebra, geometry, number theory, and analysis; the elements of this topic can be understood by the undergraduate mathematics major. Second, this topic gives students an exposure to mathematics with a rich history, that is still an active area of research. As L.J. Mordell writes, the subject of elliptic curves is fascinating "not only because of the simplicity of enunciation, but also because it appeals to the natural curiosity of persons who have anything at all to do with numbers" [31, p. 41]

The subject of elliptic curves is both ancient and very modern. The roots of the subject lie in the work of Diophantus, and Diophantine problems remain to this day a primary source for research in elliptic curves. Fermat's Last Theorem is a well-known example of a Diophantine problem and elliptic curves played a central role in its proof. Andrew Wiles' proof is beyond the scope of the present paper, but the text by Hellegouarch [17] has an excellent discussion of the mathematical background needed for the proof.

The use of original sources is, in general, an excellent method to allow students to see actual examples of how mathematicians think and to observe the process of discovery in mathematics. The historical development of the subject of elliptic curves is mirrored in the order of the sections in the paper. We consider a progression of topics from rational points on conics to rational points on cubics, from the group law on elliptic curves to Mordell's finite basis theorem, ending with two applications of elliptic curves, the Beha Eddn problem and Euler's conjecture.

In parallel with this historical exposition, a series of original sources are considered. These sources were chosen for their intrinsic interest and for their accessibility to undergraduate majors. Most of the sources discussed are in English; a few are only available in French. We begin with Newton's very interesting but rather overlooked work on the classification of third order curves. Next we consider an 1879 paper by Édouard Lucas that shows how the chord and tangent process can be used to generate additional solutions for a cubic from given solutions. The first intimations of an algebraic structure to the set of rational points on a cubic curve can be found in the series of four articles that J.J. Sylvester published in the *American Journal of Mathematics*. The structure of the group of these rational points is investigated in a paper of Mordell. The Beha Eddn problem is solved in a paper from 1983 by Horst Zimmer (based on earlier work of Lucas, [26]). We end with a discussion of a counterexample to Euler's conjecture from a 1988 paper by Noam Elkies.

## Preliminaries

In this section we give a brief overview of the basic concepts in this subject, including the concepts of algebraic curve, singular point, genus, inflection point, cubic curve, homogeneous coordinates, birational equivalence and elliptic curve.

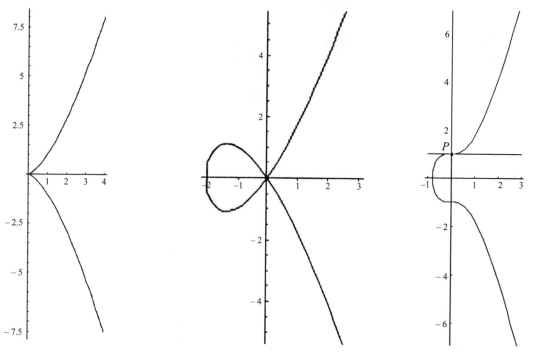

**Figure 1.** Cubic curve with a cusp           **Figure 2.** Double point           **Figure 3.** Flex point

We consider an *algebraic curve* to be defined by an equation of the form $f(x, y) = 0$, where the left side is a polynomial, with coefficients usually taken to be rational. The independent variables $x$, $y$ may be taken to be real or complex. The resulting curve is then called real algebraic or complex algebraic. A *cubic curve* is simply an algebraic curve of degree 3 (i.e., the largest sum of $x$ and $y$ powers in any term is 3). We say that the point $P$ is on the curve if $f(P) = 0$.

A *singular point* $P$ is one for which

$$\frac{\partial f}{\partial x}(P) = \frac{\partial f}{\partial y}(P) = 0.$$

A *singular curve* is one with at least one singular point (otherwise we call the curve nonsingular).

An alternative definition for a singular point considers the multiplicity of intersections of lines with the curve. By Bezout's Theorem [44, pp. 242-251], two algebraic curves of degree $m$ and $n$ respectively, will have $m \cdot n$ points of intersection (counting multiplicity). Any line intersects an algebraic curve of degree 2 or higher in at least two points (counting multiplicity). We can then define a point to be singular if more than one line intersects the curve at that point with multiplicity at least 2. So for a nonsingular point, there will be a unique tangent line.

For example the curve $y^2 = x^3$ has a cusp at the origin, since lines such as $y = x$, $y = 2x$ intersect the curve at the origin with multiplicity two. See Figure 1.

Another example of a singularity is a double point. Consider $y^2 = x^2(x + 2)$. The curve has a double point at the origin. See Figure 2.

For the curves of the form $y^2 = f(x)$, there is a simple criterion for singularity. The curve is nonsingular if and only if $f(x)$ has no repeated roots.

The *genus g* of a curve is defined by the equation

$$g = \frac{d(d - 1)}{2} - \sum \frac{m_p(m_p - 1)}{2}$$

where $d$ is the degree of the curve and the sum is over all singular points $P$, where $m_p$ is the multiplicity of the singularity. So a line or a conic has genus 0, and a nonsingular cubic has genus 1.

An *inflection point* (or *flex point*) is a point on a cubic whose tangent meets the curve with multiplicity 3. Flex points may be defined analytically as points where the determinant of the Hessian of the curve is zero. A simple example of a flex point is shown in Figure 3. This is the curve $y^2 = 1 + 2x^3$ which has a flex point at P(0,1). Also, the point (-1,0) is a flex point.

Many problems in this subject are more easily handled by changing to *homogeneous coordinates*. This is equivalent to looking at the curve as being defined over the projective plane (either the real or complex plane, depending on the context). In general, given a curve $f(x, y) = 0$ we change to the homogeneous form by letting

$$F(x, y, z) = f\left(\frac{x}{z}, \frac{y}{z}\right) = 0.$$

For example, the cubic curve $y^2 - 2y = x^3 + 2x - 4$ becomes the *projective cubic curve* $y^2 z - 2yz^2 = x^3 + 2xz^2 - 4z^3$.

The concept of *birational equivalence* is very important when working with elliptic curves. Two curves $C_1, C_2$ are birationally equivalent if there is a rational function, with rational coefficients, $\psi : C_1 \to C_2$ which is invertible (except possibly for a finite set of points). In such a case there is a 1-1 correspondence between the rational points on the curves with, again, a few possible exceptions.

So now we may ask, what is an elliptic curve? It's not an ellipse! The origins of the term "elliptic curve" are fairly obscure. Leonhard Euler and Count Giulio di Fagnano in the 18th century defined a class of integrals, the so-called elliptic integrals, which includes the integral giving the arclength of an ellipse. Translations into English of selections from the original texts of Euler and Fagnano can be found in Struik [45, pp. 374–383]. These integrals are defined by $u(x) = \int_0^x \frac{R(t)}{f(t)} dt$ where $f$ is a polynomial of third or fourth degree, without multiple roots and $R(t)$ is a rational function of t.

The inverses to these integrals are called elliptic functions. That is, a function $v$ such that $x = v(u)$. This terminology is due to C.G. Jacobi [21], although Legendre had earlier used the phrase "elliptic functions" for what we call elliptic integrals.

Then in 1865, Alfred Clebsch published an article showing that curves of genus 1 can be parametrized using elliptic functions [8]. After this, the phrase "elliptic curve" gradually came into use for such curves, although it is not clear who actually first used it. Clebsch made several important observations. One being that three points on a cubic are collinear if and only if their associated parameters sum to zero.

To agree with modern usage, we will restrict the phrase "elliptic curve" to mean a nonsingular curve of genus 1 having a rational point.

There are several equivalent ways to define an elliptic curve.

1. A nonsingular cubic $f(x, y) = 0$ having a rational point.

2. A nonsingular projective cubic $F(x, y, z) = 0$ having a rational point.

3. A nonsingular curve of genus 1 having a rational point.

4. A curve $y^2 = f(x)$, where $f(x)$ is a cubic with rational coefficients and distinct roots. This implies that at least one of the roots must be rational.

5. A curve $y^2 = x^3 + ax + b$ with $4a^3 + 27b^2 \neq 0$ (which guarantees that the curve is nonsingular). Such curves are said to be in canonical or Weierstrass normal form.

We are in fact asserting that all elliptic curves are birationally equivalent to a curve in canonical form. Two typical examples are found in Figures 4 and 5. Note that an elliptic curve in the canonical form may have one or two branches.

An important point to make is that we are focusing on the set of rational points on the cubic, denoted by $E(\mathbb{Q})$, but one can also consider the cubic defined over other fields, such as the set of real points $E(\mathbb{R})$ or the set of

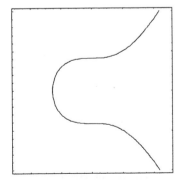

Figure 4. Elliptic curve — one branch

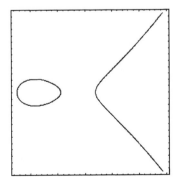

Figure 5. Elliptic curve — two branches

complex points $E(\mathbb{C})$. Our primary interest is in examining the group of rational points that is defined below. But also both $E(\mathbb{R})$ and $E(\mathbb{C})$ can be made into Lie groups. In particular, $E(\mathbb{C})$ is isomorphic to the direct sum of two circles, i.e. the complex torus. This leads to another way of looking at elliptic curves, since the complex torus is also isomorphic to $\mathbb{C}/L$, where L is a lattice in the complex plane. This viewpoint has been fundamental in the development of the theory of elliptic functions; see the text by Akhiezer [1, Chapter 1].

## Cubic Curves

*Original source: Isaac Newton, Enumeration of Lines of the Third Order, 1667*

The general form of a cubic curve in two variables is,

$$ay^3 + bxy^2 + cx^2y + dx^3 + ey^2 + fxy + gx^2 + hy + kx + l = 0$$

where we assume that the coefficients are rational (although we sometimes we need to assume they are integral).

Isaac Newton (1642–1727) was the first to give a detailed analysis of cubic curves, using the well-established theory of conics as a starting point. See his paper [33] from circa 1667 or the excellent study [47] by C.R.M. Talbot of Newton's later revision [35] of this text. The paper that we are discussing is contemporary with Newton's investigations in the calculus.

Newton had studied the analytical geometry of René Descartes in 1664 (see [33, p. 1-9] for Whiteside's introduction to this paper). Newton encountered the problem that, at this time, few examples of curves were known, beyond the category of conics. Descartes, John Wallis, and Thomas Neil had given a handful of examples of cubic curves, but without any satisfactory analysis of their properties. Newton set himself the task to investigate cubics as thoroughly as the ancients had studied conics.

> On this basis a line of the first order will be the straight line alone, those of the second or quadratic order will be conics and the circle, and those of third or cubic order the cubic parabola, Neilian parabola, cissoid of the ancients and the rest which we here undertake to enumerate. [35, p. 589]

His analysis is based on the number and type of branches the curve possesses and the nature of the asymptotes to the curve. Newton distinguishes between infinite hyperbolic or parabolic branches, depending on whether or not the branches have an asymptote. Hyperbolic cubics may have up to three asymptotes. Figures 6 and 7 show examples of some of the more unusual curves studied by Newton.

Newton begins his analysis by showing that every cubic can be reduced to the form

$$bxy^2 = dx^3 + gx^2 + hy + kx + l.$$

The further analysis that Newton makes is based on Figure 8. In this figure we take $AB$ to be the x-axis, and $Ad$ to be the y-axis. He then states that:

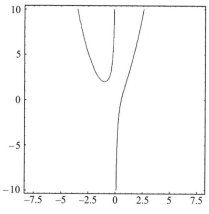

Figure 6. Cleft parabola
$$xy = x^3 + x^2 + x - 1$$

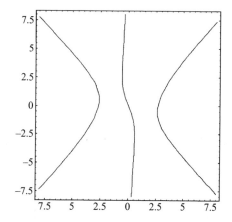

Figure 7. Three-limbed adiametral hyperbola
$$xy^2 + 3y = x^3 - 7x$$

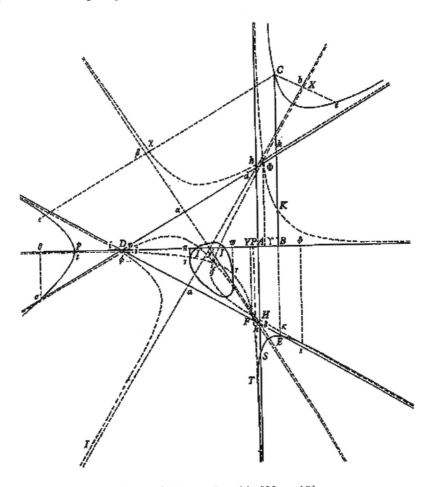

Figure 8. Newton's cubic [33, p. 19]

If the signs of $b$ and $d$ are the same, the curve will be a hyperbola having three asymptotes, none of which are parallel, and round which it crawls in each of its three pairs of infinite branches and on opposite sides. [33, p. 19]

Note that, as shown in Figure 8, the general curve may possess up to three infinite branches along with possibly having an oval component.

Newton goes on to compute the asymptotes. The directrix $Adh$ is one of the asymptotes. Assuming that $g \neq 0$, and letting

$$AD = -\frac{g}{2d}, \quad \frac{Ad}{A\delta} = \pm\frac{g}{2\sqrt{bd}}.$$

The other two asymptotes lie on $Dd$, $Dd$.

There are many interesting observations made by Newton on the geometric properties of this cubic. For example, if we bisect the asymptotes $Dd$, $D\delta$ at $\alpha$, $a$ respectively. Then either of the resulting pairs of lines $\alpha\delta$ and $Dd$, or, $ad$ and $D\delta$, makes up a coordinate system in which the cubic equation has the same form as the original.

Altogether, Newton identified 72 species of cubics. Julius Plücker did a different classification in which he gave 219 species of cubics [39, pp. 220–241]. In any case, it is clear that the geometry of cubics is much more complicated than the more familiar case of conics.

## Rational Points on Conics

We consider a polynomial equation of degree two in two variables, that is,

$$f(x, y) = ax^2 + bxy + cy^2 + dx + ey + k.$$

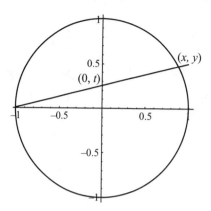

Figure 9. Rational parametrization of the circle

Such an equation defines a conic if the polynomial is not the product of linear factors, which would simply lead to a pair of lines (or a doubled line). We also exclude the case where the solution set is empty or a single point.

Not all rational conics have rational points. For example, the circle $x^2 + y^2 = 3$ has no rational points. Let $(x, y)$ be a rational point on the circle, then by clearing denominators we may obtain the equation $a^2 + b^2 = 3c^2$ where $a, b, c$ are relatively prime integers. Since the only squares in $\mathbb{Z}/3\mathbb{Z}$ are 0 and 1, $a^2 + b^2 \equiv 0 \bmod 3$ implies that $a \equiv b \equiv 0 \bmod 3$. But this implies that $3|c$ which contradicts that $a, b, c$ are relatively prime.

All conics have a rational parametrization. For example, the circle, $x^2 + y^2 = 1$, may be parametrized by

$$x = \frac{1 - t^2}{1 + t^2}, \; y = \frac{2t}{1 + t^2}.$$

This parametrization may be thought as the intersection of the line $y = t(x + 1)$ with the circle. This sets up a one-to-one correspondence between rational points on the circle and rational points on the y-axis, as in Figure 9. Note that the rational point $(-1, 0)$ on the circle corresponds to letting the parameter $t \to \infty$.

This defines a birational equivalence between the circle and the line.

More generally, David Hilbert and Adolf Hurwitz proved in an 1890 paper [18] that a curve of genus 0 having a rational point is birationally equivalent to a line. Specifically they proved that such a curve admits a rational parameter, and hence has infinitely many rational points. We have not used an original source for this section, but much of this material is due to Euler and Joseph Louis Lagrange. L.E. Dickson has a general discussion of the problem of finding rational solutions to second degree equations [10, Vol. II, pp. 412-419].

## Rational Points on Cubics

*Original Source: Édouard Lucas, Sur l'analyse indéterminée du troisième degré, 1879*
The basic problem considered in this section is how to generate additional rational points on a cubic from given points. This problem has both an algebraic and geometric component. The algebraic method is to start with a particular solution to the cubic, use that solution to perform a substitution, which will then lead to further rational solutions. The geometric approach is to take a tangent line at a rational point on the cubic or a secant line passing through two rational points, and find additional solutions by intersecting that line with the curve. This uses the fact that a line intersecting a cubic has three intersection points, counting multiplicity. If two of those intersections are rational then so is the third (a fact that seems to have been first mentioned by Newton).

Lucas was the first to explicitly merge these points of view (although Euler and Cauchy had given earlier examples of Lucas' general approach). He recognized that the substitution method employed by Fermat , and tracing back to Diophantus, was equivalent to finding intersections of secant and tangent lines with the cubic. Lucas also gave an historical overview of the importance of cubics in Diophantine problems.

In the work of Diophantus we see the first attempt to find rational solutions to cubics (which is the primary thesis of the work of Bashmakova [3]). For example, Problem 24 of book IV of the *Arithmetic* is to "Divide a given number into two numbers such that their product is a cube minus its side" [3, p. 34]. Diophantus takes the given

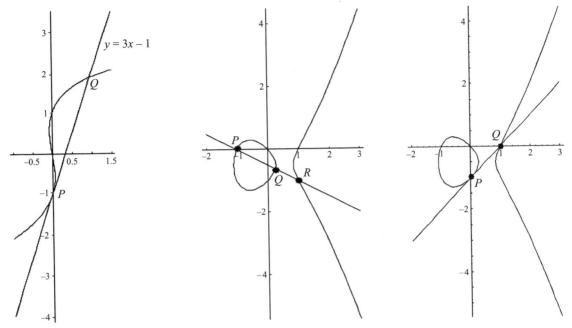

**Figure 10.** Diophantus Problem 25     **Figure 11.** Chord method     **Figure 12.** Tangent method

number to be 6. Despite this, the method of Diophantus is entirely general. The problem to be solved is then,

$$y^3 - y = x(6 - x).$$

Using the linear substitution $y = kx - 1$, representing an arbitrary line passing through the solution $(0, -1)$, Diophantus obtains the equation

$$x(6 - x) = k^3 x^3 - 3k^2 x^2 + 2kx. \tag{1}$$

By equating the first power terms, we obtain $2k = 6$. This selects the tangent line through the point $(0, -1)$ as shown in Figure 10. Solving for x in equation (1), we get the solution $x = 26/27$ that corresponds to point $Q$ in the figure.

It must be noted that the description of the above solution as using a method of tangents is of course anachronistic. It is not a valid historical interpretation of the methods of Diophantus; rather it is a way of describing his method that connects his work to the later tradition of Fermat, Euler, Cauchy, and Lucas.

Viète and Fermat extended the algebraic methods of Diophantus (see the surveys by Bashmakova [3], Norio [36], and Weil [48, Chapter 2]). It should be noted that all of these approaches implicitly used the tangent line at a point on a cubic.

Newton discovered the basic geometric laws for generating rational points on a cubic [34]. In particular, Newton discovered the method known as the chord or secant method. The chord method states that given two rational points on a cubic, the line connecting them intersects the cubic at a third rational point.

Here is an example of the chord method, illustrated in Figure 11:

$$y^2 + y = x^3 - x.$$

$P = (-1, 0)$ and $R = (1, -1)$ are rational points. $\overline{PR}$ intersects the curve at $Q = (\frac{1}{4}, -\frac{5}{8})$.

Given one rational point on an elliptic curve, the tangent line at that point will intersect the curve at another rational point. Here is an example of the tangent method, shown in Figure 12.

$$y^2 + y = x^3 - x.$$

$P = (0, -1)$ is a rational point. The tangent line at $P$ intersects the curve at another rational point $Q = (1, 0)$.

Now let us turn to the paper of Lucas [27]. The French mathematician François Édouard Anatole Lucas (1842–1891) was one of the major figures in 19th century number theory (see [16] for a study of the mathematical research of Lucas). We are interested in his work on Diophantine equations, but he is perhaps best known for his work on primality testing. He originated the procedure now known as the Lucas–Lehmer test for primes. Using this method, Lucas showed that the Mersenne number $2^{127} - 1$ is prime. This remains the largest prime number found without use of a computer. Lucas also did considerable work in recreational mathematics. For example, he invented the Tower of Hanoi puzzle (well-known to computer science students).

Lucas begins the paper by considering the work of Pierre de Fermat. Given the equation,

$$x^3 + y^3 = a^3 + b^3,$$

the rational point $(a, b)$ clearly lies on the curve. Using the linear substitution passing through this point,

$$x = a + 2u, \quad y = b + u.$$

If we choose $z = -b^2/a^2$, this will eliminate the first power term in $u$, giving us an equation of the form,

$$Au^3 + Bu^2 = 0.$$

This leads to a first degree equation in $u$, which we can then use to look for additional solutions. Specifically, assuming $a \neq 0, b \neq 0$, we obtain the solution

$$x = \frac{a^4 + ab^3}{a^3 - b^3}, \quad y = -\frac{b^4 + 2a^3b}{a^3 - b^3}.$$

This is all well known. Lucas makes the important contribution of noticing, for the first time, that this procedure is actually finding the intersection of the tangent line at $(a, b)$ with the cubic.

Lucas uses this procedure in studying the homogeneous cubic

$$x^3 + y^3 = Az^3.$$

This example, related to Fermat's Last Theorem, was also investigated by Sylvester. Lucas finds by using the tangent method that if $(x, y, z)$ is a solution, then further solutions can be obtained from the relations,

$$X = x(x^3 + 2y^3), \quad Y = -y(y^3 + 2x^3), \quad Z = z(x^3 - y^3).$$

In the case of $A = 9$, Lucas obtains the sequence of integral solutions,

$$
\begin{array}{lll}
x_1 = 2 & y_1 = 1 & z_1 = 1 \\
x_2 = 20 & y_2 = -17 & z_2 = 7 \\
x_3 = 188479 & y_3 = -36520 & z_3 = 90391.
\end{array}
$$

He notes the rapid growth of these solutions, suggesting the existence of an infinite number of rational points.

Lucas considers a general cubic in homogeneous coordinates, $f(x, y, z) = 0$. Given a solution $(x_1, y_1, z_1)$, the intersection of the tangent at this point with the cubic is found by solving,

$$f(x, y, z) = 0, \quad x\frac{\partial f}{\partial x_1} + y\frac{\partial f}{\partial y_1} + z\frac{\partial f}{\partial z_1} = 0.$$

Given two solutions, $(x_1, y_1, z_1)$, $(x_2, y_2, z_2)$, we find the intersection $(x, y, z)$ of the secant through these points by solving,

$$f(x, y, z) = 0, \quad \begin{bmatrix} x & y & z \\ x_1 & y_1 & z_1 \\ x_2 & y_2 & z_2 \end{bmatrix} = 0.$$

Lucas applies these formulae to the equation,

$$x^3 + 2y^3 + 3z^3 = 6xyz.$$

From a solution $(x, y, z)$ he obtains a new solution $(X, Y, Z)$,

$$
\begin{aligned}
X &= x(2y^3 - 3z^3) \\
Y &= y(3z^3 - x^3) \\
Z &= z(x^3 - 2y^3)
\end{aligned}
$$

[27, pp. 179–180].

From the obvious solution $(x_0, y_0, z_0) = (1, 1, 1)$, he obtains additional solutions $(x_1, y_1, z_1) = (1, -2, 1)$, $(x_2, y_2, z_2) = (19, 4, -17)$. Then the secant through $(x_0, y_0, z_0)$, $(x_2, y_2, z_2)$ meets the curve at the point $(143, 113, 71)$.

Lucas goes on to use this method to investigate further properties of the equation $x^3 + y^3 = Az^3$, showing the power of elliptic curves to solve problems in number theory.

One last note for this section; cubics, like conics, need not have any rational points. But unlike conics, there is no known effective method to determine whether or not a cubic has any rational points.

## The Group Law for Cubics

*Original Source: J.J. Sylvester, On Certain Cubic-Form Equations, 1879–80*

Now we examine what is perhaps the key idea associated with elliptic curves. One can define a procedure to "add" rational points on an elliptic curve so that the set of all rational points on the curve forms an Abelian group. Why is this important? Why is it necessary to bring in all the machinery of abstract algebra in order to investigate? Further, why do we single out cubic curves; if we are going to go to all this trouble, why not do it for all algebraic curves?

The significance of defining a group law for elliptic curves is the ability it gives us to answer two fundamental questions. Given a particular rational solution how does one find other solutions? How does one know all possible rational solutions have been found?

The question of how to find other solutions has a simple answer. Use the group law itself to generate further solutions; i.e., if $P, Q$ are rational points on the curve then $P + Q$ will be a new rational point, where we must define how to add points. To answer the second question, we must consider the structure of the group. As we discuss below, Mordell proved that all rational solutions are linear combinations of a finite basis of solutions. Finding a basis is not trivial; in fact, there is no known effective procedure for determining whether or not an elliptic curve has any rational solutions at all. As to why we define a group law only for cubics, the answer is directly related to the degree of the curve. A line intersects a cubic at three points, counting multiplicity. Given a line through two rational points on the cubic we can produce a third rational point by intersection. This geometric process is easily converted into an algebraic one; one simply views the third point as the result of a binary operation performed on the first two points. Although, as we note below, this operation needs to be carefully defined to have a group law.

The chord and tangent method does not directly define a group operation. Given rational points $P, Q$ on a cubic, suppose we try to define a binary operation $\oplus$, where we take the sum $P \oplus Q$ to equal the third point of intersection of the chord $\overline{PQ}$ with the cubic. We would also want to define $P \oplus P$ to equal the intersection of the tangent line with the cubic. This operation does not define a group. For example, if $P \oplus P = Q$ then we must also have $P \oplus Q = P$. But if $P, Q$ were group elements under $\oplus$ then $Q = 0$, which clearly cannot always be the case.

To properly define a group law, we start with a particular rational point $O$ for elliptic curve $E$ that we will use for the identity element. A flex point is a common choice for the identity. Any elliptic curve with rational point $O$ is birationally equivalent to an elliptic curve with $O$ mapped to a flex point. Another useful result is that any elliptic curve with rational point $O$ is birationally equivalent to a curve in Weierstrass normal form where $O$ is mapped to the point at infinity. More generally, for any elliptic curve of the form $y^2 = f(x)$ we may use the point at infinity for the group identity.

We wish to compute the sum of rational points $P, Q$. Let $R$ be the intersection of chord $\overline{PQ}$ with $E$. Define $P + Q$ to be the intersection of chord $\overline{OR}$ with E. Under this operation the set of rational points forms an Abelian group which we denote $E(\mathbb{Q})$. If $Q$ is the intersection of the chord $\overline{OP}$ with $E$, then $Q = -P$. So the only group property that does not immediately follow from this definition is associativity. See Milne [29, pp. 13–14] for a proof of the associative law.

Figure 13 is a visualization of the group law. As we can see from the figure, $(P + Q) + R = O$, thus illustrating the result of Clebsch that the necessary and sufficient condition for three points on the cubic being collinear is that the points sum to zero (Clebsch expressing this in terms of the elliptic parameters associated with the points).

This method also works in the case of the tangent, i.e., when adding $P + P$. One can see this illustrated in Figure 14.

Note that in this case $R = -2P$. It is also worth mentioning that the point $-2P$ is independent of the choice of identity element.

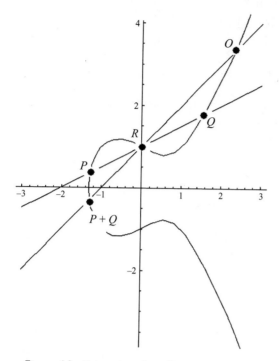

Figure 13. Group law for elliptic curves

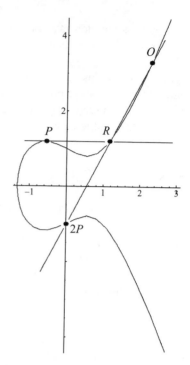

Figure 14. Doubling a point on an elliptic curve

The group $E(\mathbb{Q})$ itself can be either finite or infinite. For example, consider the curve $y^2 + y = x^3 - x$ (the examples in Figures 15 and 16 are adapted from Husemöller, [20, p. 26-28]). The point $P = (0, 0)$ is an obvious rational solution. Some of the elements of the cyclic group generated by $P$ are shown in Figure 15. In this example we are using the point at infinity for the identity element. Note that odd multiples of $P$ are on the left branch and even multiples on the right. So $E(\mathbb{Q})$ must be infinite.

On the other hand consider the example $y^2 + y = x^3 - x^2$. In this case let $P = (0, 1)$. It is easy to then see that the group of rational points is precisely $\{0, P, 2P, 3P, 4P\}$. That is, $E(\mathbb{Q}) = \mathbb{Z}/5\mathbb{Z}$.

Notice that if we start with an arbitrary rational point $P$, then the points that are obtained by the chord and tangent process are of the form $nP$, where $n \equiv 1 \pmod 3$. So the tangent process produces the sequence of points

$$P, -2P, 4P, -8P, \ldots .$$

This fact was first noticed by Hurwitz in 1891 [19, p. 725] and then more generally by Henri Poincaré in 1901 [40, p. 491].

The general structure of the elliptic group is given by,

$$E(\mathbb{Q}) = \underbrace{\mathbb{Z} \oplus \cdots \oplus \mathbb{Z}}_{r \text{ times}} \oplus E(\mathbb{Q})_{\text{tor}}$$

where $r$ is an invariant called the rank of the curve and the torsion subgroup $E(\mathbb{Q})_{\text{tor}}$ is the set of points of finite order. So the group is completely determined by its rank and the torsion group.

The structure of the torsion group was first investigated by Beppo Levi in a series of papers in 1906-08 [24]. Among his many results in this area, he showed that an elliptic curve can have no points of order 16. A survey of Levi's results can be found in the article by Schappacher and Schoof [42].

The study of the torsion group culminated in Barry Mazur's proof in 1976 of the following result [28], previously known as Ogg's conjecture,

**Mazur's Theorem**: $E(\mathbb{Q})_{\text{tor}}$ is isomorphic to either

$$\mathbb{Z}/m\mathbb{Z} \qquad \qquad \text{for } m = 1, 2, 3, \ldots, 10, 12$$
$$\text{or}$$
$$\mathbb{Z}/m\mathbb{Z} \oplus \mathbb{Z}/2\mathbb{Z} \qquad \text{for } m = 2, 4, 6, \text{ or } 8.$$

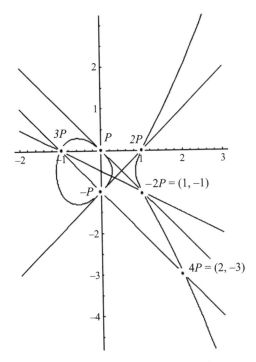

Figure 15. Points of infinite order

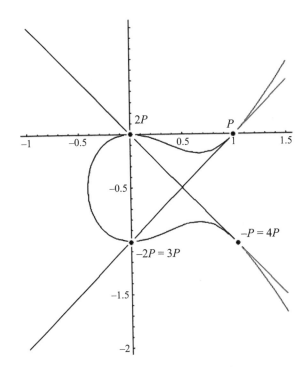

Figure 16. Points of finite order

This theorem gives a complete characterization of the torsion group for an elliptic curve.

The first examination of the rational points of a cubic in algebraic terms is due to the British mathematician J. J. Sylvester [46]. James Joseph Sylvester (1814–1897) stands as a major figure in several areas of mathematics, matrix theory, combinatorics, invariant theory, and number theory. Sylvester is an interesting personality, described in [37, p. 210] as "prolific, gifted, flamboyant, egocentric, and cantankerous." He played a central role in establishment of the mathematical community in the United States. It was during his tenure as the chair of the mathematics department at Johns Hopkins University (1876–1883) that he publishes in the *American Journal of Mathematics* the series of four articles that we are presently considering. Good biographical sources for the life and work of Sylvester are the article in the *American Mathematical Monthly* [37] and the well-received treatise [38, Chapters 2, 3] on the emergence of the American mathematical community.

None of the 19th century investigators such as Clebsch, Hurwitz, Poincaré, or Sylvester actually recognized the group structure for the rational points on elliptic curves, although, in some ways, Sylvester comes the closest. Sylvester, in a series of masterful articles on cubic curves, begins with a study of a classical Diophantine problem going back to Fermat. This is the question of which numbers are resolvable into the sum of two cubes. Sylvester's research into this problem is very interesting and worthy of notice, but we will only give a summary of his results. Building on earlier work by Pepin and Lucas, Sylvester shows that if $p_i, q_i$ are primes respectively of the forms $18n + 5, 18n + 11$, then any number of one of the following forms is not resolvable into the sum of two cubes,

$$p, q, p^2, q^2, pq, p^2, q^2, p_1 p_2^2, q_1 q_2^2$$

together with 9 times each of these classes and further classes $2p, 4q, 4p^2, 2q^2$.

It is in the course of this research that Sylvester develops a law of composition for elliptic curves. Sylvester starts with a rational point and considers the set of points one may obtain using the chord and tangent process, but does not use the group law described above. So the system of points that he obtains do not form a group but satisfy several group properties (closure and analogs for associativity, identity and inverse). In addition to these points Sylvester includes an inflection point, which plays a role that is analogous to a group identity. He calls the points that are obtained "rational derivatives" of the original point and the resulting set a "chain" [46, Part III p. 58]. Sylvester then states that the resulting set of points may be infinite or finite, "winding round and round upon itself so as to include only a finite number of distinct points." [46, Part III p. 59]. The intersection of the tangent line with the curve he calls the tangential.

Sylvester labels the starting point 1, the tangential to the first point 2, the second tangential 4, etc. He then defines the binary operation $(m, n)$, which he calls a "residue," to be the point on the cubic collinear with $m, n$. So $(1, 4) = 5$, $(2, 5) = 7$, $(1, 7) = 8$ and so on. Sylvester defines $m \ddagger n$ to be whichever of $m + n, m - n$ is not divisible by 3. He then proves that $(m, n)$ equals $m \ddagger n$. This differs from Poincaré's result since Sylvester is only allowing positive labels for his points.

The chord and tangent process produces a "numbered chain" of points, $1, 2, 4, 5, 7, \ldots$. Sylvester calculates the coordinates of the points produced in this chain, each of which will of course be rational functions of the coordinates of the original point. The terms of this chain "constitute what may properly be called a self-contained *group*, infinite or finite" [46, Part III p. 61]. His use of the word "group" is not in the formal algebraic sense.

He considers two cases in which the chain is finite. First, the closed polygon case occurs when some tangential goes back to the original point of the chain. The second case, the open chain, occurs when a tangential falls on a point of inflection; succeeding tangentials remain at this point.

Sylvester never considers the associative law for his rational derivatives, but does prove what he calls the law of double decomposition. This states that $((a, b), (c, d)) = ((a, c), (b, d))$. The extensive use that he makes of this law shows the typical combinatorial technique that Sylvester was famous for, as opposed to the more purely geometric approach of Poincaré. Sylvester uses this law to simplify calculations, as for example, $(1, 7) = (1, 2)$, $(2, 5) = (2, 2)$, $(1, 5) = (4, 4)$.

If $I$ is a given point of inflection, then Sylvester defines points collinear with $I$ to be "opposites." He points out that $(I, I) = I$, so that $I$ is its own opposite. He denotes the opposite of point $p$ to be $p'$, So $(p, p') = I$, $(p')' = p$, and he shows the identity $(p', q)' = (I, (p', q)) = (I, I), (p'q) = (I, p'), (I, q) = (p, q')$. Note that $(I, p') = p$, which of course is the "wrong" formula for a group identity.

Sylvester now defines the residues $(I, 2) = 3$, $(I, 5) = 6, \ldots, (I, 3i - 1) = 3i$. This means that the system of residues produced by the chord and tangent process includes the chain of points $1, 2, 3, 4, 5, 6 \ldots$ together with their opposites and the inflection point. He then proves that this system is closed under the chord and tangent operations.

This gives only a glimpse into the many interesting aspects of Sylvester's paper on cubics. There are other sections of this paper worth studying; for example, the long article on divisors of cyclotomic functions. Sylvester states, with great enthusiasm, "that Cyclotomy is to be regarded not as an incidental application, but as the natural and inherent centre and core of the arithmetic of the future." [46, Part II, p. 380]

## Mordell's finite basis theorem

*Original source: L.J. Mordell, Indeterminate Equations of the Third Degree, 1923*

Once the group law had been introduced, the next important question is the structure of the group. The first result was the finite basis theorem of Mordell, published in 1922 [30]. This theorem states that for any elliptic curve $E$, the group $E(\mathbb{Q})$ is finitely generated. This means that there is a finite set of rational solutions, the generators, such that **all** rational solutions are linear combinations of the generators. Mordell's proof only states the existence of this generating set, but gives no method for finding it.

Louis Joel Mordell (1888–1972) was born and raised in Philadelphia, but attended Cambridge University and spent his subsequent professional career first at Manchester University and then returned to Cambridge. It was during his stay in Manchester that the research that we are now considering was done. Two sources of further information about Mordell's life and work are the article by Cassels [5] and Mordell's own reminiscences [32].

A useful summary of Mordell's proof was presented in Mordell's paper [31] and examined in more detail by Cassels [6]. Weil generalized Mordell's proof to number fields [49]. That is, he proved that for any elliptic curve $E$ over a number field $K$, $E(K)$ is finitely generated.

Mordell begins the paper, which was based on a lecture delivered to the Manchester Mathematical Society in 1922, with a historical overview of elliptic curves. He examines the work of Fermat and Euler. For example, he looks at Fermat's method of infinite descent to show that $x^4 + y^4 = z^4$ has no positive integral solutions (which implies there are no positive rational solutions). The method of infinite descent generates a strictly decreasing sequence of positive solutions, which is impossible.

He then looks at the methods of Euler and Lagrange used to deduce new solutions from given ones. For example, Euler analyzed the equation

$$ax^2 + by^2 = z^3$$

for integer solutions. First he writes,

$$z = pa^2 + qb^2$$

and takes

$$x\sqrt{a} + y\sqrt{-b} = (p\sqrt{a} + q\sqrt{-b})^3$$

which then implies

$$x = ap^3 - 3bpq^2, \ y = 3ap^2q - bq^3.$$

Turning to the 19th century, Mordell looks at the work of Lucas and Sylvester that we examined previously. Also, he considers results of Hurwitz and Lebesgue on cubics that have no integer solutions [31, pp. 49–50]. One such example is

$$y^2 - 7 = x^3.$$

Mordell first notes that $x$ cannot be even, for a number of the form $8n + 7$ is not a square. Also $x$ cannot be of the form $4n + 1$, for then

$$y^2 = (4n + 3)^2 + 7.$$

The left side is then divisible by 4, but the right side is not. So $x$ is of the form $4n + 1$. In this case,

$$y^2 + 1 = (x + 2)(x^2 - 2x + 4).$$

But the factor $x + 2$, which is of the form $4n + 3$, must divide $y^2 + 1$, which is impossible.

Mordell next examines the work of Axel Thue who had proven in 1908–17 a series of results about integer points on cubics. For example, $ax^3 + by^3 = c$ has finitely many integer solutions (see [43, Chapter V] for a thorough discussion of this result). This result was later generalized by Siegel, who showed that any nonsingular cubic has finitely many integer solutions. Mordell examines the equation $y^2 - k = x^3$, showing by three different methods that the equation has only finitely many solutions. Let us look briefly at one of the methods. Mordell factors the equation as

$$(y + \sqrt{k})(y - \sqrt{k}) = x^3.$$

He then deduces, using the theory of ideals, that there are a finite number of solutions to

$$c(y + \sqrt{k}) = (a + b\sqrt{k})(p + q\sqrt{k})^3,$$

where $a, b, c$ are known integers and $p, q, y$ are unknown. Since $c$ can be expressed as a cubic in $p, q$, Mordell is able to use Thue's theorem to conclude the argument.

Mordell uses these ideas to first show that the cubic

$$y^2 = ax^3 + bx^2 + cx + d$$

has finitely many integer solutions. When Mordell tries to apply the same argument to the quartic

$$z^2 = ax^4 + bx^3y + cx^2y^2 + dxy^3 + ey^4 \tag{2}$$

his application of Thue's theorem breaks down. "But I could not prove that there were only a finite number of solutions, as I thought I had at one time proved," [31, p. 54].

To briefly summarize the proof of Mordell, he shows that an infinite descent argument applies. Given any specific solution to (2), a descending sequence of solutions are produced. Each solution in this sequence is a linear combination of the next, smaller, solution. We can also guarantee that in every case the last solution in our descent is smaller in magnitude than some fixed constant. So every integer solution of (2) is a linear combination of a finite set of solutions. A similar argument applies to rational solutions of a homogeneous cubic, thus proving Mordell's theorem.

It is interesting to see the historical basis for Mordell's proof. Mordell himself makes that basis very clear in his exposition.

## The Beha Eddn Problem

*Original Sources:*

*Baha' al-Din 'Amili, Khulasat al-hisab, tr. 1846*
*Édouard Lucas, Recherches sur plusieurs ouvrages de Léonard de Pise, 1877*
*Horst Zimmer, On the problem of Beha Eddn 'Amuli and the computation of height functions, 1983*

The Islamic mathematician Baha' al-Din Muhammad ibn Husain 'Amili (1547–1621) wrote an important algebra text entitled "Khulasat al-hisab" or "The Essence of Computing" (see [2] for a French translation and for a biographical introduction). Note that we are using the older spelling "Beha Eddn" for the name of the problem, primarily because that is the spelling used in the secondary literature. Baha' al-Din was the author of works in many different areas; law, religion, grammar, and astronomy. He wrote a treatise on the astrolabe and left an unfinished work, "Bahr al-hisab" or "The Ocean of Computing."

At the end of his algebra text, Baha' al-Din poses seven problems which he claims to be unsolvable, including Fermat's Last Theorem in the cubic case. Another of these problems, now known as the Beha Eddn problem, seeks to find all rational solutions to the pair of Diophantine equations,

$$x^2 + x + 2 = y^2, \; x^2 - x - 2 = z^2.$$

The problem clearly does have solutions, for example, $(x, y, z) = (-2, 2, 2)$. Additional solutions $\left(-\frac{17}{16}, \frac{23}{16}, \frac{7}{16}\right)$, $\left(\frac{34}{15}, \frac{46}{15}, \frac{14}{15}\right)$ were found in the nineteenth century [50, p. 180]. Lucas was the first to transform the problem into one involving elliptic curves. Using the following substitution [50, pp. 180–181]

$$u = x - z, v = y - x$$

and eliminating x, we obtain the cubic

$$-(2u - 1)v^2 - 2(u^2 + 2)v + (u^2 + 4u) = 0.$$

This can then be put into normal form by the substitution,

$$u = \frac{Y - 3X + 207}{6X + 342}, \; v = \frac{-3Y + X^2 + 69X - 450}{-6Y + 36X - 216}.$$

We end up with the Beha Eddn curve (see Figure 17),

$$Y^2 = X^3 - 5211X + 31050.$$

The right hand side of this cubic has roots $-75$, $6$, and $69$. So the following points are torsion points of order 2, $P_1 = (-75, 0)$, $P_2 = (6, 0)$, $P_3 = (69, 0)$ (we are using the point at infinity as the identity). In fact one can show that these are the only non-zero torsion points. Zimmer proves that $E(\mathbb{Q})_{tor} = \{0, P_1, P_2, P_3\} = \mathbb{Z}/2\mathbb{Z} \times \mathbb{Z}/2\mathbb{Z}$. These

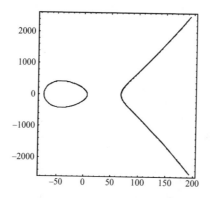

Figure 17. Beha Eddn cubic

points correspond respectively to solutions $\{\infty, (-2, -2, 2), \infty, (-2, 2, -2)\}$ of the original Beha Eddn problem (where $\infty$ is the point at infinity).

So any other rational point on the curve must have infinite order. The point $Q = (-3, 216)$ is such a point. It corresponds to the solution $\left(\frac{34}{15}, \frac{46}{15}, \frac{14}{15}\right)$ to the original problem. Zimmer shows that $Q$ generates all points of infinite order (i.e., all points of infinite order are multiples of $Q$). So this establishes that the Beha Eddn curve has rank 1. The description of the torsion group and the rank of the curve completely determines the solutions to this problem. It is interesting to see how much of the machinery of elliptic curves was needed to solve the Beha Eddn problem. This is a good illustration for students of how ideas in mathematics get connected together.

## Euler's Conjecture

*Original Source: Noam Elkies, On $A^4 + B^4 + C^4 = D^4$, 1988*

Euler's conjecture concerns the sums of like powers. As a generalization of Fermat's Last Theorem in the cubic case, Euler conjectured in 1769 that the sum of three biquadrates is never a biquadrate (for the original Euler reference see [13] and a general discussion of the background of the problem see Dickson [10, Vol. II, pp. 648–649]). That is, the equation

$$A^4 + B^4 + C^4 = D^4$$

has no solutions in positive integers. More generally he conjectured that no fifth power is the sum of four fifth powers and similarly for higher powers.

The first counterexample to Euler's conjecture was found by a computer search of fifth powers [23],

$$27^5 + 84^5 + 110^5 + 133^5 = 144^5.$$

A significant breakthrough in this problem was made by Noam Elkies [11]. Elkies, born in 1966, received his PhD in mathematics from Harvard at the age of twenty. His paper on Euler's conjecture came only a year after receiving his PhD. In addition to his work in mathematics, Elkies is also an accomplished classical composer (see the Web page [12] for more biographical data).

Elkies transformed the problem into one involving elliptic curves. In this way he found the fourth power counterexample,

$$2682440^4 + 15365639^4 + 18796760^4 = 20615673^4.$$

It's not surprising that it took more than 200 years to find this counterexample!

We now consider a sketch of Elkies' proof. It is remarkable how much of this recent paper is at a level that can be understood by an undergraduate major.

First note that integer solutions of $A^4 + B^4 + C^4 = D^4$ correspond to rational solutions to,

$$r^4 + s^4 + t^4 = 1 \tag{3}$$

where $(r, s, t) = \left(\pm\frac{A}{D}, \pm\frac{B}{D}, \pm\frac{C}{D}\right)$. Elkies not only shows that (3) has rational solutions, but in fact the rational solutions are dense in the set of real solutions.

Elkies introduces the following parametrization:

$$r = x + y, \ s = x - y \tag{4}$$
$$(2m^2 + n^2)y^2 = -(6m^2 - 8mn + 3n^2)x^2 - 2(2m^2 - n^2)x - 2mn \tag{5}$$
$$\pm(2m^2 + n^2)t^2 = 4(2m^2 - n^2)x^2 + 8mnx + (n^2 - 2m^2). \tag{6}$$

This defines a pencil of conics parametrized by $m, n$. We may take $m, n$ to be relatively prime, with $n$ odd. Elkies shows that $m, n$ must satisfy certain conditions (e.g., $m$ must be divisible by 4).

The general strategy is to start with a pair of values for $m, n$, find a particular rational solution of the resulting conic (5), use that point to parametrize the conic, and substitute the parametrization into (6), thus obtaining an elliptic curve whose group of rational points may be studied (to find, for example, if the group is infinite).

Consider the case $(m, n) = (0, 1)$. The resulting conic $y^2 = -3x^2 + 2x$ has a rational point $(x, y) = (0, 0)$. This gives rise to the parametrization

$$x = \frac{2}{k^2 + 3}, \; y = kx$$

which, after further substitutions, leads to the elliptic curves

$$Y^2 = X^3 + X \mp 2.$$

These curves have finite groups of rational points (specifically $\mathbb{Z}/2\mathbb{Z}$, $\mathbb{Z}/4\mathbb{Z}$ respectively). These points give rise only to trivial solutions $(\pm 1, 0, 0)$ of (3).

A "better" choice of parameters is $(m, n) = (3/14, 1/42)$. After some further algebraic manipulation, which is straightforward, Elkies obtains the elliptic curve

$$Y^2 = -31790X^4 + 36941X^3 - 56158X^2 + 28849X + 22030.$$

A computer search turns up the rational solution

$$(X, Y) = \left(-\frac{31}{467}, \frac{30731278}{467^2}\right)$$

which in turn leads to the earlier mentioned counterexample

$$2682440^4 + 15365639^4 + 18796760^4 = 20615673^4.$$

Furthermore, Elkies shows that the difference of the solutions

$$\left(-\frac{31}{467}, \pm\frac{30731278}{467^2}\right)$$

is a point $Q$ of infinite order in the group of rational points. Elkies uses Mazur's theorem, discussed above, to simplify the computation. Namely, it suffices to show that $nQ \neq 0$ for $n = 2, 3, \ldots, 12$. This then gives rise to an infinite number of counterexamples to Euler's conjecture.

The above counterexample is not minimal. Taking $(m, n) = (20, -9)$ leads to the counterexample,

$$95800^4 + 217519^4 + 414560^4 = 422481^4$$

which was proven to be minimal by exhaustive computer search. With the computing power available to students today, finding this minimum solution can be assigned as a student exercise.

Further work on Euler's conjecture can be found in the paper by Yu Li, [25]. Using elementary methods, Li finds a necessary condition for integer solutions to the equation $x^4 + y^4 + z^4 = 2c^4$. If $(x, y, z)$ is such a solution and $c$ is a product of primes of the form $4k + 1$ and $d = \gcd(x, y, z)$, then one of $x/d$, $y/d$, $z/d$ is divisible by 4.

## Further Reading

There are several textbooks on elliptic curves that give a good introduction to the subject. Particularly worth noting are the texts by Cassels [7], Husemöller [20], Knapp [22], Silverman [43], and Silverman and Tate [44]. A very useful survey for someone who wants to see an overview of the subject is the online text by Milne, [29]. The work by Bix [4] has a very good discussion of the geometric properties of elliptic curves and only assumes first year calculus.

Texts that have a historical bent are those by Hellegouarch [17], Prasalov and Solovyev [41], and Weil [48]. Weil's book has an in-depth study of the work of Fermat and Euler. The other texts have a good discussion of more modern material.

As far as other original sources that one may wish to use, the classic papers in the field are those of Clebsch [8] and Poincaré [40], although these are less accessible than the papers discussed above. Another source paper worth reading is that by Desboves [9], which gives a thorough discussion of the chord and tangent methods from a coordinate perspective.

## Conclusion

Now that we have seen a selection of topics from elliptic curves, how are they to be used in an actual course? Here are five scenarios that an instructor may use to fit their own courses.

| Course | Sections Covered | Lectures |
|---|---|---|
| History of Mathematics (Only Calculus prerequisite) | *Cubics,* *Rational Points on Conics,* *Rational Points on Cubics,* *Euler's Conjecture* | 4 |
| History of Mathematics (Senior level course) | *Cubics,* *Rational Points on Conics,* *Rational Points on Cubics,* *Group Law for Cubics,* *Beha Eddn Problem* | 6 |
| Abstract Algebra | *Rational Points on Cubics,* *Group Law for Cubics,* *Mordell's Finite Basis Theorem* | 4 |
| Geometry | *Cubics,* *Rational Points on Conics,* *Rational Points on Cubics* | 4 |
| Number Theory | *Cubics,* *Rational Points on Cubics,* *Euler's Conjecture* | 3 |

With regard to the use of original sources, there are several options for the instructor. One can use some or all of the cited sources, use full-text, excerpts, or summaries. The author teaches a senior level history of mathematics course for which all the students have had a course in linear algebra, and most have had abstract algebra. For such an audience, I discuss excerpts of these sources and then assign the students to read and report on a particular source of their choice.

In conclusion, I hope that I have conveyed some sense of the beauty of the ideas in elliptic curves and given instructors some ideas for how they may be included in a course.

## References

1. N.I. Akhiezer, *Elements of the Theory of Elliptic Functions*, American Mathematical Society, Providence, 1990.

2. Beha Eddn al-'Amuli, "Khelasat al Hisáb ou Essence du calcul de Behâ-eddin Mohammed ben al-Hosain al-Aamoulli," Marre, A. tr. *Nouvelles annales de mathématiques*, Ser. 1, 5(1846), 263–323.

3. I.G. Bashmakova, *Diophantus and Diophantine Equations*, Mathematical Association of America, Washington D.C., 1997.

4. Robert Bix, *Conics and Cubics*, Springer-Verlag, New York,1998.

5. J.W.S. Cassels, "J. L. Mordell," *Bull. London Math. Soc.*, 6(1974), 69–96.

6. ——, "Mordell's finite basis theorem revisited," *Math. Proc. Camb. Phil. Soc.*, 100(1986), 31–41.

7. ——, *Lectures on Elliptic Curves*, Cambridge University Press, Cambridge,1991.

8. Alfred Clebsch, "Ueber diejenigen Curven, deren Coordinaten sich als elliptische Functionen eines Parameters darstellen lassen," *J. reine angew. Math.*, 64(1865), 210–270.

9. Adolphe Desboves, "Résolution, en nombres entiers et sous la forme la plus générale, de l'équation cubique, homogéne, a trois inconnues," *Nouv. ann. math.*, (3) 5(1886), 545–579.

10. L.E. Dickson, *History of the Theory of Numbers*, Chelsea Press, New York, 1971.

11. Noam Elkies, "On $A^4 + B^4 + C^4 = D^4$," *Math. Comp.*, 51(1988), 825–835.

12. ——, personal Web page, http://www.math.harvard.edu/~elkies.

13. Leonhard Euler, "Observationes circa bina biquadrata quorum summam in duo alia biquadrata resolvere liceat," *Novi Comm. Acad. Sci. Petropolitanae* 17 (1772), 1773, 64–69; Opera Omnia, ser. 1, v. 3, 211–217, B.G. Teubner, Leipzig, 1917, E. 428.

14. ——, *Elements of Algebra*, tr. J. Hewlett (translation of *Vollständige Anleitung zur Algebra*, 1770), Springer-Verlag, New York, 1984.

15. Angelo Genocchi, Brano d'una lettera dal Sig. Angelo Genocchi a D. Baldassare Boncompagni in data dei 18 di Aprile 1855, *Annali di scienze matematiche e fisiche*, 6(1855), 129–134.

16. Duncan Harkin, "On the Mathematical Work of François-Édouard-Anatole Lucas," *Enseign. math.*, ser. 2, 3(1957), 276–288.

17. Yves Hellegouarch, *Invitation to Fermat-Wiles,* Academic Press, London, 2002.

18. David Hilbert and Adolf Hurwitz, "Über die diophantischen Gleichungen vom Geschlecht Null," *Acta Math.*, 14(1890–91), 217–224.

19. Adolf Hurwitz, "Über die Schröter'sche Konstruktion der ebenen Kurven dritter Ordnung," *J. Reine Angew. Math.*, 107(1891), 141–147.

20. Dale Husemöller, *Elliptic Curves*, Springer-Verlag, New York, 1987.

21. C.G.J. Jacobi, "De usu theoriae integralium ellipticorum et integralium abelianarum in analysi diophantea," *J. Reine Angew. Math.*, 13(1834) (=*Gesammelte Werke* II, 51–55).

22. A.W. Knapp, *Elliptic Curves*, Princeton University Press, Princeton , 1992.

23. L.J. Lander and T.R. Parkin, "Counterexamples to Euler's conjecture on sums of like powers," *Bull. Amer. Math. Soc.*, 72(1966), 1079.

24. Beppo Levi, Saggio per una teoria aritmetica delle forme cubiche ternarie, *Atti Reale Acc. Sci. Torino*, 42(1906), 739–764, 43(1908), 99–120, 413–434, 672–681.

25. Yu Li, "On the Diophantine Equation $x^4 + y^4 \pm z^4 = 2c^4$," *Sichuan Daxue Xuebao*, 33(1996), 366–368 MR 97f:11023.

26. Édouard Lucas, "Recherches sur plusieurs ouvrages de Léonard de Pise et sur diverses questions d'arithmétiques supérieure," *Bullettino die Bibliografia e di Storia delle Scienze Matematiche e Fisiche*, 10(1877), 129-193.

27. ——, "Sur l'analyse indéterminée du troisième degré - Demonstration de plusieurs théorèmes de M. Sylvester," *Amer. Jour. Math.*, 2(1879), 178–185.

28. Barry Mazur, "Modular curves and the Eisenstein ideal," *IHES Publ. Math.* 47(1977), 33–186.

29. J.S. Milne, *Elliptic Curves*, http://www.jmilne.org/math, 1996.

30. L.J. Mordell, "On the rational solutions of the indeterminate equations of the third and fourth degrees," *Proc. Cambridge Philos. Soc.*, (2) 21(1923), 415–419.

31. ——, "Indeterminate Equations of the Third Degree," *Science Progress*, 18(1923), 39–55.

32. ——, "Reminiscences of an octogenarian mathematician," *American Mathematical Monthly*, 78(1971), 952–961.

33. Isaac Newton, "Analysis of the Properties of Cubic Curves and their Classification by Species," in *The Mathematical Papers of Isaac Newton*, Vol. II, 10–89, ed. D. Whiteside, Cambridge University Press, 1968.

34. ——, "The Generation of Rational Solutions from Given Instances," in *The Mathematical Papers of Isaac Newton*, Vol. IV, 110–115, ed. D. Whiteside, Cambridge University Press, Cambridge, 1971.

35. ——, "Enumeratio Linearum Tertii Ordinis," in *The Mathematical Papers of Isaac Newton*, Vol. VII, 588, ed. D. Whiteside, Cambridge University Press, Cambridge, 1974.

36. Adachi Norio, "Elliptic Curves: From Fermat to Weil," *Historia Scientarium*, 9(1999), 27–35.

37. Karen Hunger Parshall and Eugene Seneta, "Building an International Reputation: The Case of J. J. Sylvester," *American Mathematical Monthly*, 104(1997), 210–222.

38. Karen Hunger Parshall and David E. Rowe, *The Emergence of the American Mathematical Research Community: J. J. Sylvester, Felix Klein, and E. H. Moore*, American Mathematical Society, Providence, 1994.

39. Julius Plücker, *System der analytischen Geometrie*, Duncker, Berlin, 1835.

40. Henri Poincaré, "Sur les propriétés arithmétiques des courbes algébriques," *Jour. de Math.*, ser. 5, 7(1901), 161–233 (=*Oeuvres* V, 483–550).

41. Viktor Prasalov and Yuri Solovyev, *Elliptic Functions and Elliptic Integrals*, American Mathematical Society, Providence, 1997.

42. Norbert Schappacher and René Schoof, "Beppo Levi and the Arithmetic of Elliptic Curves," *Mathematical Intelligencer*, 18(1996), 57–69.

43. Joseph H. Silverman, *The Arithmetic of Elliptic Curves*, Springer-Verlag, New York, 1986.

44. Joseph H. Silverman and John Tate, *Rational Points on Elliptic Curves*, Springer-Verlag, New York, 1992.

45. D.J. Struik, *A Source Book in Mathematics*, Princeton University Press, Princeton, N.J., 1986.

46. J.J. Sylvester, "On Certain Cubic-Form Equations," Part I, *Amer. Jour. Math.*, 2(1879), 280–285, Part II, *Amer. Jour. Math.*, 2(1879), 357–393, Part III, *Amer. Jour. Math.*, 3(1880), 58–88, Part IV, *Amer. Jour. Math.*, 3(1880), 179–189. (= *Mathematical Papers* Vol. III, 312–391).

47. C.R.M. Talbot, *Sir Isaac Newton's Enumeration of Lines of the Third Order*, Bohn, London, 1861.

48. André Weil, *Number Theory: An approach through history from Hammurapi to Legendre*, Birkhäuser, Boston, 1983.

49. ——, Sur un théorème de Mordell, *Bull. Sci. Math.*, (2) 54(1929), 182–191 (=*Coll. Papers* I, 47–56).

50. Horst G. Zimmer, "On the problem of Beha Eddin 'Amuli and the computation of height functions," in *Lecture Notes in Computer Science 162, Computer Algebra Eurocal '83*, 180–193, Springer-Verlag, Berlin, 1983.

# 5

# Using the Historical Development of Predator-Prey Models to Teach Mathematical Modeling

**Holly P. Hirst**
*Appalachian State University*

## Introduction

Many differential equations texts introduce the classic Lotka-Volterra predator-prey model as an application of coupled systems of differential equations. The model is based on a set of very simple premises:

- In the absence of predators, the prey population grows exponentially.
- Some fraction of predator-prey interactions end in death for prey.
- In the absence of prey, the predator population decreases exponentially.
- The predator birthrate is dependent on predators interacting with the prey (as the food source).

Using these assumptions as starting points for proportionality arguments, the four terms in the model (prey birth and death, predator birth and death) can be built in a variety of ways. The classic Lotka-Volterra form of the model is written as

$$\frac{dx}{dt} = ax - bxy$$

$$\frac{dy}{dt} = cxy - dy,$$

where $x$ is the current prey population, $y$ is the current predator population, and $a$, $b$, $c$, and $d$ are constants of proportionality that can be interpreted as birth and death rates.

This model, first proposed in the 1920s, is the most famous of many models developed during the twentieth century. In attempts to correct the model's unnatural behavior (such as predators never becoming sated or prey growing without bound), scientists proposed alternatives to one or more of the four terms. Studying the historical development of these models, including variations and extensions proposed in the 1930s, 1950s, and 1970s, is an excellent way to introduce students to the issues associated with building mathematical models involving rates of change.

Access to technology such as computer algebra systems or spreadsheets makes these models accessible to students at a much lower level in the undergraduate curriculum, motivating the study of calculus and differential equations, rather than serving as an "afterthought" example.

In this paper, two teaching sequences are presented: The first outline has been used with students who have not studied calculus, such as liberal arts majors in a freshman mathematics course, to motivate mathematical modeling of complex systems. The second outline has been used with students who have had basic calculus, such as sophomore mathematics majors, in order to motivate further study of modeling with differential equations and dynamical systems. These outlines differ mainly in the tools used to explore the models once built. In the first case, discrete models in the form of difference equations are explored with a spreadsheet; in the second, continuous models in the form of differential equations are explored with a computer algebra system.

## The Historical Background

Before presenting the teaching sequences, a review of the historical development of the predator-prey equations is in order. This survey is by no means complete; the models chosen for discussion here have been selected because they are easily described as differential equations and because the rationales are understood by students with little experience in model development. For a more complete discussion of the mathematical evolution of predator-prey models, complete with suggestions for using the models with advanced students, see [9]. For a different point of view, including more historical information on the scientists involved in the lively development of the discipline of population ecology, see [3].

For all models, we will assume the following notation:

- $x(t)$: The size of the population of prey at time $t$, i.e., the number of individuals or some other measure of their biomass.

- $y(t)$: The size of the population of predators at time $t$, again either the number of individuals or some other measure of their biomass.

Using this notation, a general compartmental model of the changes in these populations is

$$\frac{dx}{dt} = f_1(t, x, y) - f_2(t, x, y)$$

$$\frac{dy}{dt} = f_3(t, x, y) - f_4(t, x, y),$$

where $f_1(t, x, y)$ is the function governing the increase in prey population, $f_2(t, x, y)$ the decrease in prey population, $f_3(t, x, y)$ the increase in predator population, and $f_4(t, x, y)$ the decrease.

The most famous and the earliest model of interest to us is the Lotka-Volterra system, named for the two scientists who developed it independently, Alfred Lotka (1880–1949) publishing the equations in 1925 [6] and Vito Volterra (1860–1940) in 1926 [10]. Using modern notation, the equations can be written in the form

$$\frac{dx}{dt} = ax - bxy$$

$$\frac{dy}{dt} = cxy - dy,$$

for non-negative rate constants $a$, $b$, $c$, and $d$. There are several ways to justify the choices for the four functions, and it is interesting to look at how Lotka and Volterra each justified their choices.

First consider Lotka's argument concerning the prey increase term: In the absence of predators, the prey should increase without bound at a rate proportional to the current population level, hence the $ax$ term. Likewise, Lotka assumed that in the absence of prey, the predators should decrease at a rate proportional to the current population level, hence the $dy$ term. When the two species co-exist, Lotka used a chemical kinetics analogy to explain the change in the populations resulting from interactions: When a reaction occurs by mixing chemicals, the law of mass action states that rate of the reaction is proportional to the product of the quantities of the reactants. If two quantities are denoted $A$ and $B$, then the reaction rate would be proportional to $AB$. Using this as a rough analogy, Lotka argued that prey should decrease and predators should increase at rates proportional to the product of the number of prey and predators present, hence the terms $bxy$ for the prey decrease and $cxy$ for the predator increase.

This analogy is not surprising, given Lotka's background. An American educated in Germany and France, Lotka had degrees in physics and chemistry and was extemely interested in the emerging field of physical chemistry, the application of physical principles, particularly thermodynamics, to chemistry. He believed that one could apply physical principles to biological systems as well, and his work on predator-prey interactions is just a small part of extensive work in this area, work he published in 1925 in the text, *Elements of Physical Biology* [6]. For more information on Lotka's life and his contributions to population ecology, see chapter two in [3].

Volterra appears to have arrived at the interaction terms using somewhat different reasoning, namely a competition argument to suggest that in the presence of predators the prey's effective growth rate should be less than $a$, and how much less should depend on the predator population, giving

$$\frac{dx}{dt} = (a - by)x.$$

A similar argument for the predator increase yields

$$\frac{dy}{dt} = (cx - d)y.$$

As with Lotka's ideas, this approach is not surprising given Volterra's background and connections. Born and educated in Italy, Volterra was a physicist whose daughter and son-in-law were biologists. While looking for a mathematical explanation for a problem his son-in-law was working on, Volterra became very interested in interactions of species and spent the rest of his professional life looking for a mathematical theory of evolution. For more information on Volterra's life and work in this field, including an account of the lively interplay between Lotka and Volterra, see chapters five and six in [3].

The Lotka and Volterra rationales yield models that are algebraically equivalent. For most values of the rate constants, the solution of these equations leads to periodic behavior for both species. It is interesting to note that these continuous, quantitative attempts to explain species interaction were met with great skepticism by biologists of the day, but led to a huge increase in research related to the field of population dynamics by both physical and natural scientists as well as mathematicians.

An obvious limitation of these equations is the unrealistic behavior of the prey population in the absence of predators, so one of the first changes to the system proposed above was to limit the growth of the prey by incorporating a maximum sustainable population, $M$. One of the strongest proponents of the notion that intra-species competition is important was Alexander Nicholson (1895–1969), an Australian entomologist who studied predator-prey models applied to parasite-host relationships. Developed in the 1930s, his models were strictly discrete. See chapter 5 of [3].

Adding intra-species competition to the prey equation can be accomplished using Belgian mathematician Pierre Verhulst's (1804–1849) notion that population growth should slow at a rate proportional to the ratio of the excess population to the total population. We can incorporate a birth slowing term into the prey equation, yielding with a little algebra the modified system

$$\frac{dx}{dt} = a\left(1 - \frac{x}{M}\right)x - bxy$$

$$\frac{dy}{dt} = cxy - dy.$$

This generalization appears to have arisen almost immediately after Lotka's and Volterra's work.

Much work was done in the 1930s, 1940s, and 1950s on population ecology and population dynamics. The Bureau of Animal Population was established at Oxford University in 1932, and many articles were published in the *Journal of Animal Ecology*, established the same year. In 1935, physiologist and biomathematican Patrick Leslie (1900–1974) joined the Bureau and helped steer the research in a mathematical direction, including development of the matrix approach to modeling age distributions in animal populations that now bears Leslie's name. For more information on the Bureau and Leslie's role, see chapter six of [3].

In 1958, Leslie [4] discussed several objections to the modified Lotka-Volterra equations, the most interesting of which is that there is "no upper limit to the relative rate of increase of the predator." The predator ($y$) equation above can be interpreted as indicating that predator births increase as prey become more plentiful. Leslie felt that more realism would be reflected if a reduction in the number of prey caused more predator deaths rather than fewer predator births. He suggested that the predator should do worse as the ratio of predators to prey increases. Altering the predator equation accordingly gives the new system

$$\frac{dx}{dt} = a\left(1 - \frac{x}{M}\right)x - bxy$$

$$\frac{dy}{dt} = \left(c - d\frac{y}{x}\right)y.$$

This model yields damped oscillation of the predator and prey populations over time for suitable values of the rate constants.

Since the 1950s there have been many more attempts to improve the predator-prey equations. We will examine two more here. For more examples of the mathematical development of the dynamics of interacting species, see [9].

In his work on using mathematical modeling to investigate the complexity of interacting species, physicist Lord Robert May (1936– ) suggested several modifications to predator-prey models, one of which addressed the following problem with Leslie's model: The prey death term implies that for a fixed number of predators $y$, the number of prey eaten is proportional to the number of prey present. This implies that predators are always hungry, eating a set proportion of the prey present no matter how large the number of prey gets [7]. May addressed this flaw by replacing the prey death term $bxy$ with a term that approximates $bxy$ for small $x$ and approaches $by$ as the number of prey $x$ becomes large, giving

$$\frac{dx}{dt} = a\left(1 - \frac{x}{M}\right)x - b\frac{xy}{x+1}$$

$$\frac{dy}{dt} = \left(c - d\frac{y}{x}\right)y.$$

May's model has behavior similar to Leslie's model, namely decaying oscillations over time for both populations. Many biologists liked these models, since in all but the most contrived laboratory experiments, the predator kills off the prey over time or the two populations seem to reach stable levels. It is interesting to note, however, that these equations can exhibit a wide range of behavior, from stable points to chaotic fluctuations. As a result of studying these behaviors, May is credited with creating the new field of "chaotic dynamics" in biology.

The last model of interest to us is one developed by Jagannathan Gomatam in 1974 [2]. A physicist and applied mathematician, Gomatam used Benjamin Gompertz' (1799–1865) model for limiting population growth to the interaction terms in the predator-prey model. A self-educated mathematician, Gompertz was employed as an actuary by a British assurance company where he developed his mortality model. Gompertz' model implies that deaths from interactions between members of a population of size $N$ should be proportional to $N \log(N)$.

Gomatam applied this mortality model to the interactions between predators and prey, proposing

$$\frac{dx}{dt} = ax - bx \log(y)$$

$$\frac{dy}{dt} = cy \log(x) - dy.$$

The main advantage of this model is that it is explicitly solvable for suitable values of the rate constants.

In all of the above models, a discrete version of the model can be explored, i.e., replacing $\frac{dx}{dt}$ with $x(t+1) - x(t)$, and similarly for $y$. In fact, Nicholson, Leslie and May were primarily interested in these difference equations while Lotka, Volterra, and Gomatam studied the continuous versions presumably assuming that the differential equations were suitable approximations to the more realistic discrete versions. For other variations of predator-prey models, see [1], [5], and [9].

## Teaching Dynamical Systems in a Liberal Arts Mathematics Course

Studying population growth with liberal arts majors is an excellent way to review high school mathematics in a context that the students have not encountered before. If students have access to a spreadsheet, they can build and analyze fairly sophisticated models. These problems lend themselves well to extended writing assignments in which students can practice incorporating data, mathematics, and graphs into reports.

Before examining predator-prey interactions, we begin with a study of one-species population growth, examining well-known models due to Thomas Malthus (1766–1834) and Pierre Verhulst (1804–1849):

$$\frac{dx}{dt} = kx$$

$$\frac{dx}{dt} = k\left(1 - \frac{x}{M}\right)x.$$

The Malthus model is the classic exponential growth model, justified with the statement "the more individuals in the population, the more potential for reproduction." The Verhulst model adds a linear slowing factor, forcing the population to grow more slowly as a maximum sustainable population $M$ is approached. For an excellent introduction to these models from a historical point of view see [8].

We approach all of the models in discrete format; i.e., these models are viewed as

$$\frac{\Delta x}{\Delta t} = f(t, x).$$

Here the time step, $\Delta t$, is defined to be the time between successive observations, typically one month (thought of as $\frac{1}{12}$ of a year). $\Delta x$ is then rewritten as $x_{new} - x$, where the "new" subscript denotes the population after one time step and the non-subscripted variable the population at the previous time. At this point it is important to address the issue of units, discussing the pros and cons of measuring population by population count versus some sort of density or "biomass," and deciding ahead of time how to handle fractional answers. We also discuss the fact that both the left and right sides of the equations have units "population per unit time." Thus rate constants must be "per unit time" in order for the equations to balance. When modifying models to incorporate new assumptions like harvesting, students need to consider the units involved, so we address them right from the start.

Students build these models using a spreadsheet such as Excel by allocating a column for time and a column for the population $x$. The equations are rearranged to make building the formula for the population easier by solving for $x_{new}$. For example, in the Malthus model, we have

$$x_{new} = x + (kx)\Delta t.$$

Note that this approach is exactly Euler's method for solving differential equations.

After discussing with the students the rationales for these two classic models, we work through the following questions in class.

1. Consider the Malthus model with rate constant 0.2 and initial population 10. What is the population after 1 year? 2 years? How does this compare to the behavior of the Verhulst model with rate constant 0.2, initial population 10, and maximum sustainable population 30?

2. In the Verhulst model, what happens when the rate constant is increased to 0.5? decreased to 0.05? Explain what this indicates about the role of $k$ in the model.

3. In the Verhulst model, experiment with different values for $M$. Does $M$ really give the maximum sustainable population? Explain.

4. In the Verhulst model, what happens when the initial population is greater than the maximum sustainable population? What happens when the initial population is equal to the maximum sustainable population?

5. Add harvesting to the Verhulst model, assuming that the population harvested per unit time is proportional to the amount of the population present at any given time. How would you incorporate a harvesting term like this into the model? What happens when you change the rate constant for your harvesting term?

6. Certain animals cannot survive when the population level falls below a minimum size; if the population gets too small and sparsely distributed, reproduction becomes difficult. Modify the model to incorporate a minimum sustainable population $m$, using an approach similar to that used to add the maximum sustainable population $M$ in the Verhulst model. Experiment with the model to see what happens if the initial population is below $m$, between $m$ and $M$, and above $M$.

Through their work on these problems, students encounter ideas related to equilibrium values, numerical approximation, and discrete versus continuous approaches. They also see first-hand that changing coefficients in mathematical expressions yields predictable results: Increase the birth rate constant, and we increase the speed of population growth.

Turning our attention to two-species models, we start by examining Lotka's and Volterra's ways of building the four birth and death terms in the model, carrying out the algebraic manipulation necessary to show that the models are equivalent. Then we implement a particular version of the model, using

$$\frac{x_{new} - x}{\Delta t} = 0.7x - 0.3xy$$

$$\frac{y_{new} - y}{\Delta t} = 0.08xy - 0.44y,$$

letting $x(0) = 4$ and $y(0) = 2$.

As in the one equation case, to build this model in a spreadsheet such as Excel, students solve these equations for $x_{new}$ and $y_{new}$ arriving at

$$x_{new} = x + (0.7x - 0.3xy)\Delta t$$

$$y_{new} = y + (0.08xy - 0.44y)\Delta t.$$

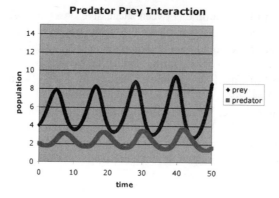

Figure 1. Lotka-Volterra Populations in Excel

Columns are built for time, $x$, and $y$, as in the one species model solution. The populations can be plotted over time yielding a scatter plot similar to Figure 1.

Students are then asked to experiment with the model, answering the following questions.

1. Experiment with changing the rate constants to answer the following questions:

   (a) Describe the effect of increasing the prey birth rate. Is this what you expected to happen? Explain.

   (b) Describe the effect of increasing the prey death rate. Is this what you expected to happen? Explain.

   (c) The predator birth rate is currently much less than the prey birth rate. What effect does having the same birth rate for both species have on the solution?

2. The time step is currently set to one month ($\frac{1}{12}$). Experiment with the time step, increasing it and decreasing it. What happens when the time step is set to 1, i.e., one year? Does the behavior exhibited by the model make sense? This model was proposed as a continuously changing one, i.e., the change of the population over time was assumed to be changing continuously and so a small time step should give the behavior expected by Lotka and Volterra.

3. Experiment with the initial populations for the predators and prey. Are the behaviors of the populations over time sensitive to small changes to the initial populations? Can you force the extinction of one of the species? both?

4. Suppose that we harvest some proportion of both the predators and the prey. Experiment with incorporating harvesting terms into both of the equations. What happens when you change the harvesting rate? Can you harvest and have the populations increase rather than decrease? Is this what you expected? Explain.

After ensuring that the students are comfortable with the Lotka–Volterra model, we then consider adding competition and improving the death terms as in the Leslie and May models described above. In each model, we start with the same values of the rate constants and initial populations, setting maximum sustainable population for prey to 15 in these models. Students are asked to experiment with these models, answering questions similar to those above.

To close the unit on dynamical systems, students are given individual writing projects. A typical example is:

Rabbits-Wolves-Grass: Add grass to the predator-prey models discussed in class. Think about how the rabbits would eat the grass and how the grass would grow back. What equation would you use to model how grass changes over time? What are the units for the grass variable and associated terms? How would you modify the predator and prey equations? List your assumptions, and experiment with modifying the parameters. Think about what the appropriate $\Delta t$ would be: Do you want discrete growth or continuous growth?

This approach has been quite successful with our liberal arts students. The population modeling material is introduced after they have used spreadsheets for financial applications, so they have some facility with building formulas to generate columns of data. The open-ended, experimental nature of the questions and projects requires some coaching on behalf of the instructor at first in order to build the students' confidence, but even the least mathematically inclined students have been successful at interpreting the graphs and drawing reasonable conclusions.

## Teaching Dynamical Systems in a Sophomore Introduction to Modeling Course

The teaching sequence for students who have knowledge of differential calculus is similar to the outline we follow in the freshman class. The equations are approached in a continuous sense as differential equations, and the tool is changed to a computer algebra system such as Maple.

As with the previous outline, we start with an examination of one species models and work through essentially the same questions related to the Malthus and Verhulst models. Next we study the predator-prey models due to Lotka, Volterra, Leslie, May, Ivlev, and Gomatam, answering the same questions as with the liberal arts students and looking at the models using Maple. Maple will produce an image of the solution quite easily as in Figure 2. See the Appendix for example Maple code.

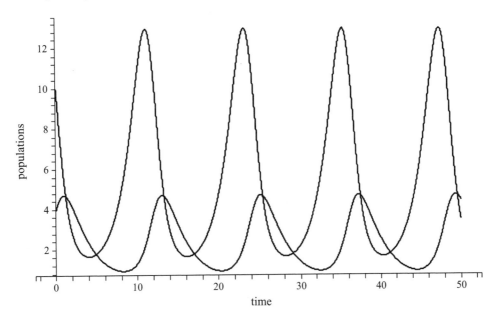

Figure 2. Lotka–Volterra Populations in Maple

Since these students have studied differential calculus, we can introduce slope fields and phase diagrams, and then more formally discuss stable and unstable equilibria. Maple can be used to create nice phase diagrams by changing the viewing scene. Figure 3 shows the result of this Maple command. See the Appendix for example Maple code.

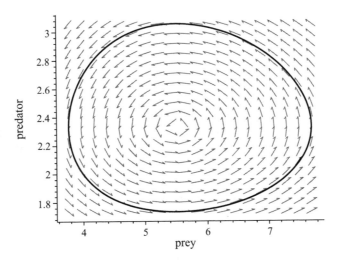

Figure 3. Phase Diagram of Lotka–Volterra Populations

These students can push the models further than the liberal arts students, which raises some issues that need to be addressed. As May realized in his study of these interacting species models, chaotic behavior is very common. Students can become quite frustrated when experimenting with some of these models since certain choices for the rate constants can yield bizarre behavior. It is essential to warn them ahead of time that this is not an error on their part. A good problem for exploring these issues is the competitive hunter model:

Two Species Competition — Competitive Hunters: In the situation where we have two species who compete for the same resources (hunting for the same food rather than one being the food source for the other), we can build a two equation model in a similar fashion. The birth terms do not involve interaction between the species (i.e., only $x$ is present in the birth term for the $x$ equation), but the death terms for each involve both species. Build a model that reflects this situation. What are the parameters? What units are involved in each term? Build this model. What happens to the populations? Experiment with parameter values and report your findings. Beware: This model can exhibit very strange behavior!

Values of the rate constants that give reasonable behavior are given in the Maple code in the Appendix. See Figure 4 for the plot resulting from the solution in Maple. See the Appendix for example Maple code.

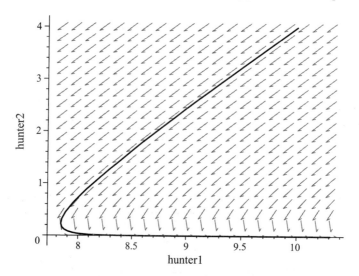

Figure 4. Competitive Hunter Populations in Maple

Students in the sophomore class build their mathematical modeling skills by formulating these population models and then experimenting with different (justifiable) functions for the four terms in the model, then they examine problems from other contexts that also yield differential equations models, such as heating and cooling, pollution flow, basic mechanics, etc. This gives the students a framework from which to study differential equations in more detail in subsequent classes.

## Conclusions

Exploring these classical models with students gives them an appreciation of the mathematical modeling process as it is more commonly applied: Start with a simple model and gradually refine it. The basic models attributed to Lotka, Volterra, Leslie, and May are all sufficiently elementary that students without sophisticated mathematics backgrounds can understand why the algebraic terms in the models were proposed. This progression of models is ideal for demonstrating to students how "behavior" can be modeled algebraically, and how changes to the algebraic formulation in a model often cause predictable changes in the solution.

For advanced students, these models provide a natural segue to the study of dynamical systems through differential equations and/or difference equations. Natural extensions include examining other forms of two-species interactions such as mutualism and host-parasite systems, as well as larger, multi-species ecosystems. Increasing complexity in these models can lead students naturally to the study of stability, bifurcations, and chaos.

For students interested in the interplay between mathematics and science, the development of the mathematics of ecology and population dynamics is well-documented in [3]. The historical information on this topic can be used to motivate many interesting investigations, including a comparison of continuous and discrete models, a study of the origins of chaos in biology, and the role of mathematicians in the development of the subject.

## References

1. M. S. Bartlett, "On Theoretical Models for Competitive and Predatory Biological Systems," *Biometrica*, **44** (1957) 27–42.

2. J. Gomatam, "A New Model for Interacting Populations — I: Two Species Systems," *Bulletin of Mathematical Biology*, 36 (1974) 347–353.

3. S. E. King, *Modeling Nature, Episodes in the History of Population Ecology*, The University of Chicago Press, Chicago, 1988.

4. P. H. Leslie, "A Stochastic Model for Studying the Properties of Certain Biological Systems by Numerical Methods," *Biometrica*, 45 (1958) 16–31.

5. P. H. Leslie and J. C. Gower, "The Properties of a Stochastic Model for the Predator-Prey Type of Interaction Between Two Species," *Biometrica*, 47 (1960) 219–231.

6. A. J. Lotka, *Elements of Physical Biology*, Williams and Wilkins Publishers, Baltimore, 1925.

7. R. M. May, *Stability and Complexity in Model Ecosystems,* Princeton University Press, Princeton NJ, 1973.

8. D. A. Smith, "Human Population Growth: Stability or Explosion?" *Mathematics Magazine*, 80 (1977) 186–197.

9. H. R. van der Vaart, "Some Examples of Mathematical Models for the Dynamics of Several-Species Ecosystems," in *Case Studies in Applied Mathematics,* Committee on the Undergraduate Program in Mathematics, Mathematical Association of America, Washington, DC, 1976.

10. V. Volterra, "Variations and Fluctuations of the Number of Individuals in Animal Species Living Together," in *Animal Ecology*, R. Chapman, ed., McGraw-Hill, 1926.

## Appendix: Maple Code

Using Maple's DEtools and plots packages, we can enter the two differential equations and request numerical solutions using the DEplot command. The DEplot commands give the solutions in the form of plots of time versus prey population and time versus predator population. Finally, the display command shows both of these curves on the same set of axes. See Figure 2 for the plot.

```
> with(DEtools): with(plots):
> eqn1 := diff(prey(t), t) = 0.7*prey(t)
> - 0.3*prey(t) * predator(t):
> eqn2 := diff(predator(t),t) = 0.08*prey(t)*predator(t)-
> 0.44*predator(t):
> a := DEplot([eqn1, eqn2], [prey(t), predator(t)],
> t=0..50, stepsize=.01,
> [[prey(0)=10, predator(0)=4]], scene=[t,prey(t)],
> linecolor=black, thickness=1, labels=[time,populations],
> labelfont=[TIMES, ROMAN, 12]):
> b := DEplot([eqn1, eqn2], [prey(t), predator(t)],
> t=0..50, stepsize=.01,
> [[prey(0)=10, predator(0)=4]], scene=[t,predator(t)],
> linecolor=black, thickness=1):
> display(a,b);
```

Changing the scene to view prey population versus predator population:

```
> DEplot([eqn1, eqn2], [prey(t), predator(t)],
> t=0..50,  stepsize=0.01,
> [[prey(0)=4, predator(0)=2]], scene=[prey(t),predator(t)],
> linecolor=black, labels=[prey,predator], thickness=1,
> labelfont=[TIMES, ROMAN, 12]);
```

The competitive hunter model in Maple:

```
> eqn1 := diff(hunter1(t), t) = 0.07*hunter1(t) -
> 0.3*hunter1(t)*hunter2(t):
> eqn2 := diff(hunter2(t),t) = 0.08*hunter2(t) -
> 0.44*hunter1(t)*hunter2(t):
> DEplot([eqn1, eqn2], [hunter1(t), hunter2(t)],
> t=0..5,  stepsize=0.1,
> [[hunter1(0)=10, hunter2(0)=4]],
> scene=[hunter1(t),hunter2(t)],
> linecolor=black, labels=[hunter1, hunter2], thickness=1,
> labelfont=[TIMES, ROMAN, 12]);
```

# II

# Geometry

*It is easier to square a circle than to get round a mathematician*
—Augustus De Morgan

# 6

# How to Use History to Clarify
# Common Confusions in Geometry

**Daina Taimina and David W. Henderson**
*Cornell University*

Most people judge the size of cities simply from their circumference. So that when one says that Megalopolis is fifty stades in contour and Sparta forty-eight, but that Sparta is twice as large as Megalopolis, what is said seems unbelievable to them. And when in order to puzzle them even more, one tells them that a city or camp with the circumference of forty stades may be twice as large as one of the circumference of which is one hundred stades, what is said seems to them absolutely astounding. The reason of this is that we have forgotten the lessons in geometry we learnt as children.
— Polybius, 2nd century B.C. [56]

## Introduction

We have found that students who take our senior/graduate level geometry course usually have very little background in geometry. We have lead many week-long UFE and PREP workshops (funded by the National Science Foundation) for professors on teaching geometry and we found that even mathematicians are often confused about the history of geometry. In addition, many expository descriptions of geometry (especially non-Euclidean geometry) contain confusing and sometimes-incorrect statements — this is true even in expositions written by well-known research mathematicians. Therefore, we found it very important to give some historical perspective of the development of geometry, clearing up many common misconceptions and increasing people's interest both in geometry and in the history of mathematics.

Questions that often have confusing or misleading (or incorrect) answers:

1. What was the first non-Euclidean geometry?
2. Does Euclid's parallel postulate distinguish the non-Euclidean geometries from Euclidean geometry?
3. Is there a potentially infinite surface in 3-space whose intrinsic geometry is hyperbolic?
4. In what sense are the usual models of hyperbolic geometry "models"?
5. What does "straight" mean in geometry? How can we draw a straight line?

These questions are asked by our students and by mathematicians in our workshops. In the literature we often do not find clear answers.

In order to clarify these questions for ourselves and for our students, we find that it is important to talk about the development of different ideas in mathematics. It also gives us a way to bring the history of mathematics into the geometry course. Talking with our students, we noticed that most confusions related to the above questions come from not recognizing certain strands in the history of geometry. Thus, even though the topic of this volume is the use of more recent history of mathematics in our courses, we will start (as we do with our students) with a quick look back to various historical strands that can help us understand the issues surrounding these questions. More recent history will come later.

It is important to talk with students about deep roots of human experiences of mathematics and how it connects with modern theories. One of our student responses demonstrates how learning is motivated by history:

> I found 451 (Spherical and Euclidian Geometry) to be my favorite math class so far. I feel the way we are learning math is like collecting pieces of a puzzle through our education. We need lots of those pieces to form a picture (or I should say "part of a picture" here), but if we haven't collected "enough" of them, we won't be able to put them together and thus can't see anything. I think this is why my interest in mathematics is growing as I am learning more of it, and want to take [the history of mathematics class]. I want to know how and why people started to think about "mathematics."

## The Four Strands of Geometry

We introduced to our students the notion that the main aspects of geometry today emerged from four strands of early human activity, which seemed to have occurred in most cultures: art/patterns, building structures, motion/machines, and navigation/stargazing. These strands developed more or less independently into varying studies and practices that eventually from the 18th and 19th century on were woven into what we now call *geometry*.

*Art/Patterns Strand*   To produce decorations for their weaving, pottery, and other objects, early artists experimented with symmetries and repeating patterns. Later the study of symmetries of patterns led to tilings, group theory, crystallography, finite geometries, and in modern times to security codes and digital picture compression. Early artists also explored various methods of representing existing objects, and living things. These explorations led to the study of perspective and then projective geometry and descriptive geometry, and (in the 20th century) to computer-aided graphics, the study of computer vision in robotics, and computer-generated movies (for example, Toy Story). For more details, see [1], [3], [15], [18], [21], [23], [24], [26], [27], [40], [41], [54], [57].

*Navigation/Stargazing Strand*   For astrological, religious, agricultural, and other purposes, ancient humans attempted to understand the movement of heavenly bodies (stars, planets, sun, and moon) in the apparently hemispherical sky. Early humans used the stars and planets as they navigated over long distances, thus they were able to solve problems in navigation and to understand the shape of the Earth. Ideas of trigonometry apparently were first developed by Babylonians in their studies of the motions of heavenly bodies. Even Euclid wrote an astronomical work, *Phaenomena*, [4], in which he studied properties of curves on a sphere. Navigation and large-scale surveying developed over the centuries around the world and along with it cartography, trigonometry, spherical geometry, differential geometry, Riemannian manifolds, and thence to many modern spatial theories in physics and cosmology. For more details, see [2], [4], [10], [20], [22], [53], [62], [63].

*Motion/Machines Strand*   Early human societies used the wheel for transportation, making pottery, and in pulleys. In ancient Greece, Archimedes and Hero used linkages and gears to solve geometrical problems

— such as trisecting an angle, duplicating a cube, and squaring a circle. These solutions were not accepted in the Building Structures (Euclid) Strand, which leads to a common misconception that these problems are unsolvable and that Greeks did not allow motion in geometry. See [42] and [49]. We do not know much about the influence of machines on mathematics after Ancient Greece until the advent of machines in the Renaissance. In the 17th century, Descartes used linkages and the ideas of linkages to study curves. This study of curves led to the development of analytic geometry. Computing machines were also developed in the 17$^{th}$ century. As we will discuss below, there was an interaction between mathematics and mechanics that leads to the modern mathematics of rigidity and robotics. For more detail see [14], [16], [19], [52], [58], [65].

**Building Structures Strand**    As humans built shelters, altars, bridges, and other structures, they discovered ways to make circles of various radii, and various polygonal/polyhedral structures. In the process they devised systems of measurement and tools for measuring. The (2000–600 B.C.) *Sulbasutram* [60] is written for altar builders and contains at the beginning a geometry handbook with proofs of some theorems and a clear general statement of the "Pythagorean" Theorem. Building upon geometric knowledge from Babylonian, Egyptian, and early Greek builders and scholars, Euclid (325–265 B.C.) wrote his *Elements,* which became the most used mathematics textbook in the world for the next 2300 years and codified what we now call Euclidean geometry. Using the *Elements* as a basis in the period 300 B.C. to about 1000 A.D., Greek and Islamic mathematicians extended its results, refined its postulates, and developed the study of conic sections and geometric algebra. Within Euclidean geometry, analytic geometry, vector geometry (linear algebra and affine geometry), and algebraic geometry developed later. The *Elements* also started what became known as the axiomatic method in mathematics. For the next 2000 years mathematicians attempted to prove Euclid's Fifth (Parallel) Postulate as a theorem (based on the other postulates); these attempts culminated around 1825 with the discovery of hyperbolic geometry. (See below for discussion of the more recent history of parallel postulates.) Further developments with axiomatic methods in geometry led to the axiomatic theories of the real numbers and analysis, and to elliptic geometries and axiomatic projective geometry. For more detail see [7], [9], [10], [12], [14], [28], [59], [60].

We will now describe how these strands can be used to clarify issues surrounding the five questions identified in the introduction. We include those aspects of the recent history of mathematics that are either related to these issues or are difficult to find in the current literature.

## Spherical Geometry — the first non-Euclidean geometry

Answering the first question: Spherical geometry can be said to be the first non-Euclidean geometry. It developed early in the Navigation/Stargazing Strand. The connections of spherical geometry with other strands will be covered in the following sections. For at least 2000 years humans have known that the earth is (almost) a sphere and that the shortest distances between two points on the earth is along great circles (the intersection of the sphere with a plane through the center of the sphere). For example:

> ...it will readily be seen how much space lies between the two places themselves on the circumference of the large circle which is drawn through them around the earth....[W]e grant that it has been demonstrated by mathematics that the surface of the land and water is in its entirety a sphere,...and that any plane which passes through the center makes at its surface, that is, at the surface of the earth and of the sky, great circles, and that the angles of the planes, which angles are at the center, cut the circumferences of the circles which they intercept proportionately,...
> — Claudius Ptolemy, *Geographia* (ca. 150 A.D.), Book One, Chapter II

Spherical geometry is the geometry of a sphere. The great circles are intrinsically straight on a sphere in the sense that the shortest distances on a sphere are along arcs of great circle and because great circles have the same symmetries on a sphere as straight lines have on the Euclidean plane. The geometry on spheres of different radii is different; however, the difference is only one of scale. In Aristotle we can find evidence that spherical non-Euclidean geometry was studied even before Euclid. (See [30], p. 57 and [64].) Even Euclid in his work on astronomy, *Phaenomena* [4], discusses propositions of spherical geometry. Menelaus, a Greek of the first century A.D., published a book *Sphaerica*, which contains many theorems about spherical triangles and compares them to triangles on the Euclidean plane. (*Sphaerica* survives only in an Arabic version. For a discussion see [45], page 119–120.)

Up into the 19th century, spherical geometry occurred almost entirely in the Navigation/Star-gazing Strand and was used by Brahe and Kepler in studying the motion of stars and planets and by navigators and surveyors. The popular book *Spherical Trigonometry: For the use of colleges and schools* (1886) by Todhunter [62] contains several discussions of the use of spherical geometry in surveying and was used in British schools before hyperbolic geometry was widely known in the British Isles. On the continent, C.F. Gauss (1777–1855) was using spherical geometry in various large-scale surveying projects. This happened before the advent of hyperbolic geometry. Currently, 3-dimensional spherical geometry is being considered as one of the possible shapes for our physical universe. (See [64] and [34].)

## The Parallel Postulate(s)

Let us now discuss Question 2: Does Euclid's parallel postulate distinguish the non-Euclidean geometries from Euclidean geometry? After developing for some 2000 years, spherical geometry was related to the Euclidean postulates of the Building Structure Strand in the 19th and 20th centuries. There are many popular accounts (for example, Katz [42], p. 782, and Davis & Hersh [13], pp. 219–221) that attempt to distinguish between Euclidean and spherical geometries on the basis of the parallel postulate. In Euclid's *Elements*, the "parallel postulate" is the Fifth Postulate (here stated in Heath's translation [29], page 155, but in more modern English as given in [34], Appendix A):

(**EFP**) If a straight line intersecting two straight lines makes the interior angles on the same side less than two right angles, then the two lines (if extended indefinitely) will meet on that side on which the angles are less than two right angles.

The interested reader can check that **EFP** is not needed to assert the *existence* of parallel lines. In fact Euclid in I.31 proves (in modern wording):

Given a line *l* and a point *P* not on *l* there exists a line through *P* parallel to *l*.

In his proof, Euclid constructs first a line *n* through *P* perpendicular to *l* and then another line *m* through *P* and perpendicular to *n*. Then Euclid concludes (I.27) that *m* is parallel to *l* because the alternate interior angles are equal. In none of these constructions or conclusions does Euclid use his Fifth Postulate. Now, in this situation, the reader can easily see that **EFP** immediately implies that any line through *P*, other than *m*, will intersect *l* and thus *m* is the *unique* parallel line through *P* parallel to *l*.

Note that **EFP** is true in spherical geometry (because all lines intersect) and is even provable in the strong sense that the two lines meet closest on that side on which the angles are less than two right angles (see, for example, [34], Chapter 10). However, the *existence* of parallel lines is false in spherical geometry and thus it must be one of the other postulates that fails on the sphere. Thus it is not true that **EFP** distinguishes Euclidean from non-Euclidean geometries.

Over the years, **EFP** has been replaced by many other postulates (see for example, [29], p. 220, [31], p. 203, [13], p. 219). Any of these alternate postulates plus Euclid's first four postulates imply **EFP**;

and **EFP** plus the first four imply each of the alternate postulates. The most common of these alternate postulates used in modern high school textbooks and popular accounts is:

> **(HSP)** Given a line and a point not on the line there is *one and only one* line through the point that is parallel to the given line.

We will call this the High School Postulate (**HSP**). Note that this postulate posits both the *existence* and *uniqueness* of parallel lines. Therefore, unlike **EFP**, the High School Postulate is not true in spherical geometry. Thus, in any context that includes Euclidean and spherical geometry, **EFP** and **HSP** are *not* equivalent.

Often, **HSP** is called by many authors (including us in [32, p. 124]!) "Playfair's Parallel Postulate." However, Playfair actually stated ([55], p. 11, 1839, and in the editions Philadelphia, 1806; Boston, 1814; Edinburgh, 1819; and New York, 1835):

> **(PP)** Two straight lines which intersect one another, cannot be both parallel to the same straight line.

Many other authors (including David Hilbert in his axiomatization of Euclidean geometry [35]) state the Playfair postulate in its logically equivalent form:

> **(PP)** Given a point $P$ and a line $l$, there exists at most one line through $P$ and parallel to $l$.

This postulate only posits *uniqueness* and thus it is true in both spherical and Euclidean geometries.

We see that **EFP** and **PP** are true in spherical geometry, but **HSP** is not true. Nevertheless, many writers are confused, for example: The excellent history text [42] (p. 782) says that one of the "two possible negation of Euclid's parallel postulate" leads to spherical geometry. In the useful paper [31] (p. 203) it is stated (correctly) that **PP** is true on a sphere, but continues "so [**PP**] can only be equivalent to [**EFP**] under some additional condition . . ." The delightful book [13] (pp. 218–221) states **EFP** and then asserts (correctly) that it is "totally equivalent" to **HSP** (but calls it "Playfair's Axiom" and "Playfair's postulate") in the sense that each plus the first four of Euclid's postulates implies the other. However, the text then asserts that elliptic geometry (geometry on a sphere with diametrical opposite points identified) satisfies the first four postulates of Euclid but not **HSP**. It is easy to check that **EFP** holds in elliptic geometry, which contradicts the assertion of total equivalence.

All of these confusions seem to come from replacing **EFP** with **HSP**. We do not know in detail how this happened but there are two hints: Heath's translation of Euclid has been for the past century the de facto standard and in his commentary he states [29] (p. 220) "Playfair's Axiom" as:

> *Through a given point only one parallel can be drawn to a given straight line* or, *Two straight lines which intersect one another cannot both be parallel to one and the same straight line.*

The second statement here clearly posits *uniqueness* while the first statement could be mistaken for the **HSP**. And then Kline, in [45] (p. 865) which was for many years the de facto standard history of mathematics, repeats the first statement of Heath's and then adds:

> This is the axiom used in modern books (which for simplicity usually say there is "one and only one line. . .").

Because of the 2000-year long history of investigations into parallel postulates within the Building Structures Strand, many books misleadingly call hyperbolic geometry "*the* non-Euclidean geometry" or "the first non-Euclidean geometry." Spherical geometry has been studied since ancient times but it did not fit easily into this axiomatic approach and thus was (and still is) left out of many discussions of non-Euclidean geometry.

In the discussions that do include spherical geometry it is called by various names which causes more confusions: Riemannian, projective, elliptic, double elliptic, and spherical. These different labels for spherical geometry usually imply different settings and contexts as summarized in the following:

- Bernhard Riemann (1826–1866) pioneered the intrinsic (and analytic) view for surfaces and space; and, in particular, he introduced an intrinsic analytic view of the sphere that became known as the **Riemann Sphere**. The Riemann Sphere is usually studied in a course on complex analysis. Some writers use the term "Riemannian geometry" to describe spherical geometry (usually in parallel with "Lobatchevskian geometry"), but this practice has led to confusion among students and mathematicians because the term "*Riemannian geometry*" is most often used to describe general manifolds as a part of differential geometry.

- ***Double-elliptic geometry*** usually indicates an axiomatic formalization of spherical geometry in which antipodal points are considered as separate points. This axiom system for spherical geometry is in the spirit of Euclid's Axioms (as embellished by Hilbert [35]) for Euclidean geometry (see for example, [8]), in this context spherical geometry is usually called *double-elliptic geometry*. This axiomatization has seemed not to be very useful.

- ***Elliptic geometry*** usually indicates an axiomatic formalization of spherical geometry in which each pair of antipodal points is identified as one point.

- ***Projective geometry*** originally developed within the Art/Patterns Strand in the study of perspective drawings. If you project space (or a plane) onto a sphere from the center of the sphere (called a *gnomic projection*) then straight lines in space correspond to great circles on the sphere. Thus projective geometry can be thought of as a non-metric version of spherical geometry.

- ***Spherical geometry*** often indicates the geometry of the sphere sitting extrinsically in Euclidean 3-space.

Starting soon after the *Elements* were written and continuing for the next 2000 years mathematicians attempted to either prove Euclid's Fifth Postulate as a theorem (based on the other postulates) or to modify it in various ways. These attempts culminated around 1825, when Nicolai Lobatchevsky (1792–1850) and János Bolyai (1802–1860) independently discovered a geometry that satisfies all of Euclid's Postulates and Common Notions except that the Fifth Postulate does not hold. It is this geometry that is called *hyperbolic geometry*.

## Hyperbolic Surfaces

Historically, one of the first open questions about hyperbolic geometry was whether it is the geometry of any surface in Euclidean space, in the same sense that spherical geometry is the geometry of the sphere in Euclidean 3-space. This is also our Question 3. In the mid-19th century beginnings of differential geometry it was shown that hyperbolic surfaces would be precisely surfaces with constant negative curvature. This aspect of hyperbolic geometry belongs in the Navigation/Star Gazing strand of geometry, in the sense that differential geometry of surfaces (and higher-dimensional manifolds) uses calculus to study the geometric properties that are *intrinsic* — properties of the surface that a bug crawling on the surface could detect. Intrinsic properties include geodesics (intrinsically straight lines), length of paths, shortest paths (which are almost always segments of geodesics), angles, surface area, and so forth.

Students (and mathematicians) desire to touch and feel a hyperbolic surface in order to experience its intrinsic properties. Many people have trouble with standard "models" and pictures of hyperbolic geometry in textbooks, because the intrinsic meanings of geodesics, lengths, angles, and areas cannot be directly seen. However, it is a common misconception that you cannot have a surface in 3-space whose intrinsic geometry is hyperbolic geometry.

Mathematicians looked for surfaces whose intrinsic geometry is complete hyperbolic geometry in the same sense that the intrinsic geometry of a sphere has the complete spherical geometry. In 1868, Eugenio Beltrami (1835–1900) described a surface, called the *pseudosphere*, whose local intrinsic geometry is hyperbolic geometry but is not complete in the sense that some geodesics (intrinsically straight lines) cannot be continued indefinitely. (See [25], p. 218, for photos of the surfaces that Beltrami constructed and further discussion of the pseudosphere.) In 1901, David Hilbert (1862–1943) [36] proved that it is impossible to define by (real analytic) equations a complete hyperbolic surface. In those days, "surface" normally meant one defined by real analytic equations and so the search for a complete hyperbolic surface was abandoned and still today many works state that a complete hyperbolic surface is impossible. For popularly written examples, see [53] (p. 158) and [37] (p. 243). In 1964, N. V. Efimov [17] extended Hilbert's result by proving that there is no isometric embedding defined by functions whose first and second derivatives are continuous. However, in 1955, Nicolas Kuiper [48] proved, without giving an explicit construction, the existence of complete hyperbolic surfaces defined by continuously differentiable functions. Then in some workshops in the 1970's William Thurston described the construction of complete hyperbolic surfaces (that can be made out of paper), see [61] (p. 49 and p. 50).

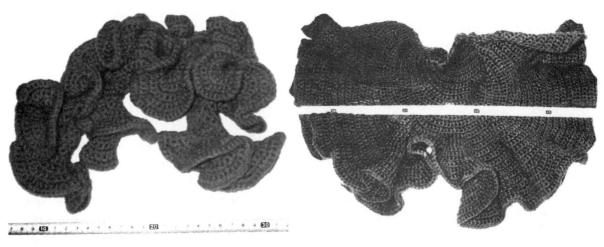

**Figure 1.** Crocheted hyperbolic planes of radius 4 cm and 8 cm. (crocheted by Daina Taimina, photo by David Henderson)

Directions for constructing Thurston's surface out of paper can be found in [32] (p. 48), [33], or [34], (p. 58). These references also contain a description of the method (invented by the first author of this article) for crocheting these surfaces. See Figure 1 for crocheted hyperbolic planes with different radii. In addition, there is in these references a description of an easily constructible polyhedral hyperbolic surface, called the "hyperbolic soccer ball," that consists of heptagons (7-sided regular polygons) each surrounded by 7 hexagons (the usual spherical soccer ball consists of pentagons each surrounded by 5 hexagons) — this construction was discovered recently by the second author's son, Keith Henderson. The intrinsic straight lines ("geodesics") on a hyperbolic surface can be found by folding the surface (in the same way that folding a sheet of paper will produce a straight line on the paper). This folding also determines a reflection about the intrinsic straight line.

# Projections, Coordinate Systems, "Models" of the Hyperbolic Plane

Question 4 asks: In what sense are the usual models of hyperbolic geometry "models"? To study the geometry of the sphere analytically we develop coordinate systems and projections (maps) of the sphere onto the plane. (Note that the usual spherical coordinates give a projection of portions of the sphere onto a rectangle in the plane.) For the same reason, coordinate systems and projections of the hyperbolic plane are useful for systematic analytic study. At the same time as he discovered the pseudosphere, Beltrami described (in 1868) a projection of the hyperbolic plane onto a disk in the plane, this projection was more fully developed in 1871 by Felix Klein (1849–1925). This projection, which can also be thought of as describing a coordinate system, is now called in the literature *projective disk model*, the *Beltrami–Klein model*, or *Klein model*.

The Beltrami–Klein model (and other models) was called a "model" instead of a "coordinate system" or "projection" because of the absence of a known surface whose intrinsic geometry was complete hyperbolic geometry. But after Thurston's construction we can now describe this model (and the ones described below) as projections or coordinate systems of the complete hyperbolic surface — for details, see [34], Chapter 17.

The Beltrami–Klein model was based on projective geometry and projective transformations, which had their origins in the Art/Pattern Strand. The Beltrami–Klein model (thought of as a transformation) takes straight lines in the hyperbolic plane to straight lines in the Euclidean plane in the same way that gnomic projection takes straight lines (great circles) on the sphere to straight lines on the Euclidean plane. As with gnomic projection, the measure of angles is not preserved in the Beltrami–Klein model. (See Chapters 14 and 17 of [34].)

Other well-known models of the hyperbolic plane were developed in 1882 by Henri Poincaré and were based on circles and inversions in circles. Both Poincaré models preserve the measure of angles but take straight lines in the hyperbolic plane to semicircles in the upper half-plane or disk. Reflection through a straight line on the hyperbolic plane corresponds in the Poincaré models to inversion in the semicircles. See [28] (Section 39), and [34] (Chapter 17).

Inversions in a circle are a necessary component of the analytic study of hyperbolic geometry. As we will see below they are also useful to understanding linkages that convert straight-line motion into circular motion. Our students ask questions about the origins of inversions, but there are not many sources on the history of inversions. We describe inversions as emerging from the Art/Pattern Strand and trace the ideas in the theory of inversions in circles back to Apollonius of Perga (225 B.C.–190 B.C.), who investigated one particular family of circles and straight lines. We know about this from a commentary of Pappus of Alexandria (290–350). Apollonius defined the curve $c_k(A, B)$ to be the locus of points $P$ such that $PA = k.PB$, where $A$ and $B$ are points in the Euclidean plane, and $k$ is a positive constant. Now this curve is a straight line if $k=1$ and a circle otherwise and is usually called an Apollonian Circle. Apollonius proved that a circle $c$ (with center $C$ and radius $r$) belongs to the Apollonian family $\{c_k(A, B)\}$ if and only if $BC.AC = r^2$ and $A$ and $B$ are on a same ray from $C$. [In modern terms, we say "if and only if $B$ is the inversion of $A$ with respect to $c$(and vice versa)".] See Figure 2. Theory of inversions in circles can be developed purely geometrically from Euclid's Book III but was not done so in ancient times. (We would say that this is because it was part of a separate strand.) For discussion of properties of circle inversions, see [6], [11], [28] (Section 37), [34] (Chapter 16), [39], and [51] (Chapter 15).

A systematic study of inversions in circles started only in the 19th century. Jakob Steiner (1796–1863) was among the first to use extensively the technique of inversions in circles. Steiner had no early schooling and did not learn to read or write until he was age 14. Against the wishes of his parents, at age 18 he went to the Pestalozzi School at Yverdon, Switzerland, where his extraordinary geometric intuition was discovered. By age 28 he was making many geometric discoveries using inversions. At age 38 he occupied

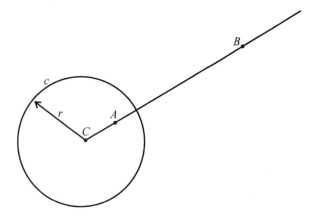

Figure 2. *B* is the inversion of *A* with respect to *C* (and vice versa).

the chair of geometry established for him at the University of Berlin, a post he held until his death. We find these facts inspiring to students. For more details, see [47]. A complete recent history of inversions can be found in the book [50].

Inversions were also used in the 19th century to solve a long-standing engineering problem (from the Motion/Machines Strand) we describe in the next section.

## What is "Straight"? How Can We Draw a Straight Line?

This is Question 5. When using a compass to draw a circle, we are not starting with a model of a circle; instead we are using a fundamental property of circles that the points on a circle are a fixed distance from a center. Or we can say we use Euclid's definition of a circle. So, now what about drawing a straight line: Is there a tool (serving the role of a compass) that will draw a straight line? One could say: We can use a straightedge for constructing a straight line. Well, how do you know that your straightedge is straight? How can you check that something is straight? What does "straight" mean? Think about it!

We can try to use Euclid's definition: "A straight line is a curve that lies symmetrically with the points on itself." (See [34], Appendix A, for a justification of this translation; and Chapter 1 for a discussion of "What is straight?") This leads to knowing that folding a flat piece of paper will produce a straight line. This use of symmetry, stretching, and folding can also be extended to spheres and hyperbolic planes (See [34], Chapters 2 and 5.) We can also use the usual high school definition, "A straight line is the shortest distance between two points." This leads to producing a straight line by stretching a string. Students are confused when reading in the literature that "straight line" is an undefined term, or that straight lines on the sphere are "defined to be arcs of great circles". We find that putting it in the context of the four strands helps clarify this: Symmetry comes from the Art/Pattern Strand, "undefined terms" come from the Building Structures Strand, and "shortest distance" from the Navigation/Star Gazing Strand.

But there is still left unanswered the question of whether there are mechanisms analogous to a compass that will draw an accurate straight line. Discussions about straightness in class had a different perspective after we learned ourselves and then presented to our students the following not widely known history of the problem. The students responded that it was important to learn this connection between a purely geometrical problem and a mechanical problem. It also shows that you can find a history of mathematical problems in the history of other sciences, in this case mechanics, which leads us into the Motion/Machines Strand.

Turning circular motion into straight line motion has been a practical engineering problem since at least the 13th century. As we can see in some 13th century drawings of a sawmill (see Figure 3) four

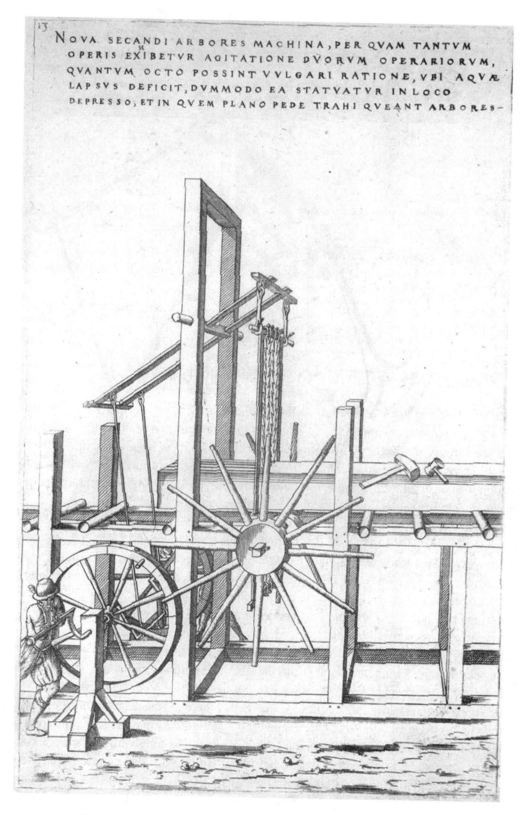

**Figure 3**. Hand cranked saw mill from before 1578, Plate 13 in [5].

bar linkage (rigid bars constrained to be near a plane and joined at their ends by rivets) was in use and probably was originated much earlier. In 1588, Agostino Ramelli published his book [58] on machines where linkages were widely used. In the late 18th century people started turning to steam engines for power. James Watt (1736–1819), a highly gifted designer of machines, worked on improving the efficiency and power of steam engines. In a steam engine the steam pressure pushes a piston down a straight cylinder. Watt's problem was how to turn this linear motion into the circular motion of a wheel (such as on steam locomotives).

It took Watt several years to design the straight-line linkage that would change straight-line motion to circular motion. Years later Watt told his son:

> Though I am not over anxious after fame, yet I am more proud of the parallel motion than of any other mechanical invention I have ever made. (quoted in [19], pp. 197–198)

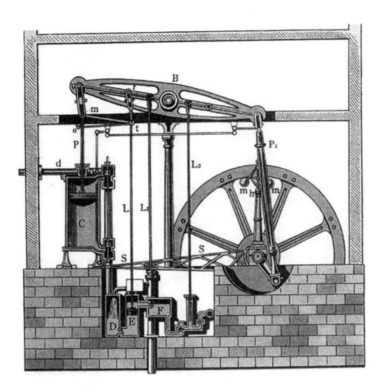

Figure 4. A steam engine with Watt's "parallel motion" linkage.

"Parallel motion" is a name Watt used for his linkage, which was included in an extensive patent of 1784. Watt's linkage was a good solution to the practical engineering problem. See Figure 4 where the linkage is the parallelogram *murt* connecting the piston *P* to the beam *B*. But Watt's solution did not satisfied mathematicians who knew that it can draw only approximate straight lines. In 1853, Pierre-Frederic Sarrus (1798–1861), a French professor of mathematics at Strassbourg, devised an accordion-like spatial linkage that traced exact straight lines (see Figure 5) if one end is held fixed.

Mathematics continued to look for a planar straight-line linkage. An exact straight-line linkage in a plane was not known until 1864–1871 when a French army officer, Charles Nicolas Peaucellier (1832–1913), and a Russian graduate student, Lipmann I. Lipkin (1851–1875), independently developed a linkage that draws an exact straight line. See Figure 6. (There is not much known about Lipkin. Some sources mentioned that he was born in Lithuania and was Chebyshev's student but died before completing his doctoral dissertation.)

Figure 5. Sarrus Spatial Linkage for drawing a straight line. (Photo by Francis C. Moon, Cornell University, courtesy of the Cambridge University Engineering Department.)

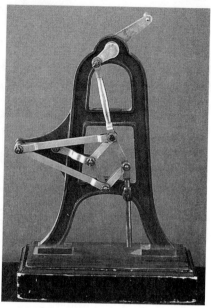

Figure 6. Peaucellier-Lipkin linkage from Cornell University's F. Reuleaux kinematic model collection. (Photo by Francis C. Moon, Cornell University.)

The drawing in Figure 7 depicts the working parts of the Peaucellier–Lipkin linkage. The linkage works because the point $P$ is inverted to the point $Q$ through a circle with center at $C$ and a radius squared equal to $s^2 - d^2$. The point $P$ is constrained to move on a circle that has center at $D$ and that passes through $C$ and thus $Q$ must move along a straight line. If the distance between $C$ and $D$ is changed to $g$ (not equal to $f$), then $Q$ instead of moving along a straight line will move along an arc of a circle of radius $(s^2 - d^2)f/(g^2 - f^2)$. This allows one to draw an arc of a large circle without using its center. Note that the definition of a straight line used here is: "A straight line is a circle of infinite radius." For a learning module on these topics, see [46].

In the late 19th century Franz Reuleaux (1829–1905), an engineer with a strong mathematical background, collected Watt's and 38 other linkages that had been constructed to turn circular motion into

Figure 7. Working parts of the Peaucellier–Lipkin linkage.

straight-line motion or vice versa. These linkages were a part of a larger collection of about 800 models of mechanisms that he considered to be a basis of a theory of machines. (For Reuleaux, "mechanisms" are the simple component parts that are put together to form a machine that does useful work.) The first president of Cornell University, A.D. White, purchased in 1882 a large number of these models and there are today more than 220 remaining. A picture of this new collection was on a cover of *Scientific American* in 1885. See Figure 8. This Reuleaux model collection at Cornell University is now being digitized and put on the web as part of the National Digital Science Library complete with viewer controlled motion, simulations, learning modules, and amazingly the possibility for printing out 3-D working models of many of the mechanisms. See [52]. It is viewable at [46]. Among mathematicians the most popular of these mechanisms is the Peaucellier–Lipkin linkage which is described above.

**Figure 8**. Reuleaux kinematic model collection in Cornell University Sibley School of Mechanical Engineering. (Reproduced from Scientific American, 1885 cover, Cornell University Library.)

We add here a few remarks that excited our students. In January 1874, James Joseph Sylvester (1814–1897) delivered a lecture "Recent Discoveries in Mechanical Conversion of Motion." Sylvester's aim was to bring the Peaucellier–Lipkin linkage to the notice of the English-speaking world. Sylvester learned about this problem from Chebyshev — during a visit of the Russian to England. He observed:

> The perfect parallel motion of Peaucellier looks so simple, and moves so easily that people who see it at work almost universally express astonishment that it waited so long to be discovered. (quoted in [19], p 206)

Later, Mr. Prim, "engineer to the Houses" (the Houses of Parliament in London), was pleased to show his adaptation of the Peaucellier linkage in his new "blowing engines" for the ventilation and filtration of the Houses. Those engines proved to be exceptionally quiet in their operation. See [44]. Sylvester recalled his experience with a little mechanical model of the Peaucellier linkage at a dinner meeting of the Philosophical Club of the Royal Society. The Peaucellier model had been greeted by the members with lively expressions of admiration, when it was brought in with the dessert, to be seen by them after dinner, as is the laudable custom among members of that eminent body in making known to each other the latest scientific novelties. (quoted in [19], p. 207)

And Sylvester would never forget the reaction of his brilliant friend Sir William Thomson (later Lord Kelvin) upon being handed the same model in the Athenaeum Club. After Sir William had operated it for a time, Sylvester reached for the model, but he was rebuffed by the exclamation:

No! I have not had nearly enough of it — it is the most beautiful thing I have ever seen in my life. (quoted in [19], p. 207)

In summer of 1876 Alfred Bray Kempe (1849–1922), a barrister who pursued mathematics as a hobby, delivered at London's South Kensington Museum (now Science Museum) a lecture with the provocative title "How to Draw a Straight Line" which in the next year was published in a small book [44]. In this book, which is viewable on the web as part of the Cornell University Library's online book collection, you can find pictures of the linkages we have mentioned here. Kempe essentially knew that linkages are capable of drawing any algebraic curve. Other authors provided more complete proofs of this fact during the period 1877–1902. The Peaucellier-Lipkin linkage is also used in computer science to prove theorems about workspace topology in robotics [38].

## Conclusion

We described four strands in the history of geometry. We used an understanding of these strands as part of our search for answers to the five questions in the Introduction and for clarifications of the issues surrounding these questions. In summary:

1. *What is the first non-Euclidean geometry?* Within the Building Structures Strand, hyperbolic geometry was the first non-Euclidean geometry. However, spherical geometry has an ancient history that goes back more than 2000 years within the Navigation/Stargazing Strand. In addition, other geometries (for example, projective geometry) developed out of the Art/Patterns Strand before the discovery of hyperbolic geometry.

2. *Does Euclid's parallel postulate distinguish the non-Euclidean geometries from Euclidean geometry?* Looking at it only from within the Building Structures Strand the answer is, Yes. However, in any context that includes spherical geometry the answer is clearly, No.

3. *Is there a potentially infinite surface in 3-space whose intrinsic geometry is hyperbolic?* As Hilbert proved there is no such surface that can be described by analytic equations. However, there are such surfaces that can be made from paper or can be crocheted.

4. *In what sense are the usual models of hyperbolic geometry "models"?* In light of the answers to Question 3, we can say that the Beltrami–Klein and Poincaré Models can be called "analytic models" because there is no analytic surface with complete hyperbolic geometry. However, in the context of the existence of a hyperbolic surface these analytic models serve as projections or coordinate systems.

5. *What does "straight" mean in geometry? How can we draw a straight line?* The meaning of "straight" seems to depend on the strand: In the Building Structures Strand, "straight" is an undefined term. "Straight" is described in terms of symmetries in the Art/Patterns Strand. "Shortest path" is the appropriate notion of "straightness" in the Navigation/Stargazing Strand. In the Motion/Machines Strand we obtain a straight line as a "circle of infinite radius."

Using history in a mathematics class shows the human side of mathematics and explores connections within mathematics and with other sciences. There are questions, which students and mathematicians have, that can only be clarified by reference to an appropriate history.

Partially supported by National Science Foundation's National Science, Technology, Engineering, and Mathematics Education Digital Library (NSDL) Program under grant DUE-0226238.

## References

1. Keith Albarn, Jenny Mial Smith, Stanford Steele, and Dinah Walker, *The Language of Pattern*, Harper & Row, New York, 1974.

2. L. Bagrow, *A History of Cartography*, Harvard University Press, Cambridge, 1964.

3. George Bain, *Celtic Arts: The Methods of Construction*, Constable, London, 1977.

4. J. Lennart Berggren and Alexander Jones, *Ptolemy's Geography: An Annotated Translation of the Theoretical Chapters*, Princeton University Press, Princeton, NJ, 2000.

5. Jacques Besson, *Theatrum instrumentorum er machinarum* (1578 text), Smithsonian Institution Libraries, Digital Edition, 1999. http://www.sil.si.edu/DigitalCollections/HST/Besson/besson.htm

6. David E. Blair, *Inversion Theory and Conformal Mapping*, American Mathematical Society, Providence, RI, 2000.

7. William Blackwell, *Geometry in Architecture*, John Wiley & Sons, New York, 1984.

8. Karol Borsuk and Wanda Szmielew, *Foundations of geometry: Euclidean and Bolyai-Lobachevskian geometry. Projective geometry*, Interscience Publishers, New York, 1960.

9. Abraham Bosse, *Traite des practiques geometrales,* Chez l'Auteur, Paris, 1665.

10. Walter Burkert, *Lore and Science in Ancient Pythagoreanism*, Harward University Press, Cambridge, 1972.

11. J.L. Coolidge, *A treatise on the circle and the sphere*, Oxford University Press, 1916.

12. Datta, *The Science of the Sulba*, University of Calcutta, Calcutta, 1932.

13. Phillip Davis and Reuben Hersh, *The Mathematical Experience*, Birkhäuser Boston, 1981.

14. L. Sprague DeCamp, *The Ancient Engineers*, Ballantine, New York, 1974.

15. Keith Devlin, *Mathematics: The Science of Patterns*, Scientific American Library, New York, 1994.

16. George B. Dyson, *Darwin among the Machines: The Evolution of Global Intelligence*, Perseus Books, Reading, MA, 1997.

17. N. V. Efimov, "Generation of singularities on surfaces of negative curvature" [Russian], *Mat. Sb.* (N.S.) 64 (106), pp. 286–320, 1964.

18. Ron Eglash, *African Fractals: Modern Computing and Indigenous Design*, Rutgers University Press, New Brunswick, 1999.

19. Eugene S. Fergusson, "Kinematics of Mechanisms from the Time of Watt", *United States National Museum Bulletin* **228**, Smithsonian Institute, Washington D.C., 1962, pp. 185–230.

20. Kitty Ferguson, *Measuring the Universe*, Walker & Co, New York, 1999.

21. Mike Field and Martin Golubitsky, *Symmetry in Chaos: A Search for Pattern in Mathematics, Art, and Nature*, Oxford University Press, New York, 1992.

22. Galileo Galilei, "Trattato della Sphaera," 1586–87, in *Galilei Opere*, ed. Antonio Favaro, G. Barbera, Florence, 1953.

23. Paulus Gerdes, *Geometrical Recreations of Africa*, African Mathematical Union and Higher Pedagogical Institute's Faculty of Science, Maputo, Mozambique, 1991.

24. Matila Ghyka, *The Geometry of Art and Life*, Dover Publications, New York, 1977.

25. Livia Giacardi, "Scientific Research and Teaching Problems in Beltrami's Letters to Houel" in Victor Katz (editor), *Using History to Teach Mathematics: An International Perspective*, MAA Notes #51, MAA, Washington, DC, 2000.

26. Ernst Gombrich, *The Sense of Order: A study in the Psychology of Decorative Art*, Cornell University Press, Ithaca, NY, 1978.

27. Istvan Hargittai and Magdolna Hargittai, *Symmetry: A Unifying Concept*, Shelter Publications, Bolinas, 1994.

28. Robin Hartshorne, *Geometry: Euclid and Beyond*, Springer, New York, 2000.

29. T.L. Heath, *Euclid: The Thirteen Books of The Elements*, Vol. 1, Dover Publications, New York, 1956.

30. ——, *Mathematics in Aristotle*, Clarendon Press, Oxford, 1949.

31. Torkil Heiede, "The History of Non-Euclidean Geometry" in *Using History to Teach Mathematics: An International Perspective*, Victor Katz, editor, MAA Notes #51, MAA, Washington, DC, 2000.

32. David W. Henderson (with Daina Taimina), *Experiencing Geometry in Euclidean, Spherical, and Hyperbolic Spaces*, 2nd Edition, Prentice Hall, Upper Saddle River, NJ, 2001.

33. David W. Henderson and Daina Taimina, "Crocheting the Hyperbolic Plane," *The Mathematical Intelligencer*, vol 23, no. 2, 2001, pp 17–28.

34. ——, *Experiencing Geometry: Euclidean and Non-Euclidean with Strands of History*, 3rd Edition, Prentice Hall, Upper Saddle River, NJ, 2005.

35. David Hilbert, *Foundations of geometry* (*Grundlagen der Geometrie*), translated by Leo Unger, Open Court, La Salle, IL, 1971.

36. ——, "Über Flächen von konstanter gausscher Krümmung," *Transactions of the A.M.S.*, 1901, pp. 87–99.

37. David Hilbert and S. Cohn-Vossen, *Geometry and the Imagination*, Chelsea Publishing Co., New York, 1983.

38. J. Hopkroft, D. Joseph, and S. Whitesides, "Movement problems for 2-Dimensional Linkages," *SIAM J. Comput.* Vol. 13, No.3, August 1984.

39. A.J. Hoffman, "On the foundations of inversion geometry," *Notices of the American Mathematical Society* 71, September, 1951, 218–242.

40. William Ivins, Jr., *Art & Geometry: A Study in Space Intuitions*, Dover Publications, New York, 1946.

41. Jay Kappraf, *Connections: The Geometric Bridge between Art and Science*, McGraw-Hill, New York, 1991.

42. Victor J. Katz, *A History of Mathematics: An Introduction*, 2nd Edition, Addison-Wesley Longman, Reading, MA, 1998.

43. —— (editor), *Using History to Teach Mathematics: An International Perspective*, MAA Notes # 51, Mathematical Association of America, Washington, DC, 2000.

44. A. B. Kempe, *How to Draw a Straight Line,* Macmillan and Co., London, 1877. Available from the Cornell University Mathematics Library at: http://historical.library.cornell.edu/math/index.html

45. Morris Kline, *Mathematical Thought from Ancient to Modern Times*, Oxford University Press, Oxford, 1972.

46. <http://kmoddl.library.cornell.edu>, *Reuleaux Kinematic Model Collection*, part of National Digital Science Library <www.nsdl.org>.

47. Louis Kollros, *Jacob Steiner*, Elemente der Mathematik, Beihefte Nr. 2, Verlag Birkhäuser, Basel, Dezember 1947.

48. Nicolas Kuiper, "On $C^1$-isometric embeddings ii," *Nederl. Akad. Wetensch.* Proc. Ser. A, 1955, pp. 683–689.

49. George E. Martin, *Geometric Constructions*, Springer, New York, 1998.

50. Elena Anne Marchisotto and James T. Smith, *The Legacy of Mario Pieri in Arithmetic and Geometry*, Birkhäuser, Boston, 2003.

51. John McCleary, *Geometry from a Differential Viewpoint*, Cambridge University Press, Cambridge, UK, 1994.

52. Francis C. Moon, "Franz Reuleaux: Contributions to 19th Century Kinematics and the Theory of Machines," *Applied Mechanics Reviews*, vol 56, no. 2, March 2003, pp. 1–25.

53. Robert Osserman, *Poetry of the Universe: A Mathematical Exploration of the Cosmos*, Anchor Books, New York, 1995.

54. John Pennethorne, *The Geometry and Optics of Ancient Architecture*, Williams & Norgate, London, 1878.

55. John Playfair, *The Elements of Geometry: Containing the First Six Books of Euclid...*, from the last London Edition, W.E. Dean, New York, 1839.

56. Polybius, *The Histories*, 9.26a, English translation by W.R. Paton, Harvard University Press, Cambridge, MA, 1922–1927.

57. V.V. Prasolov and V.M. Tikhomirov, *Geometry*, Translations of Mathematical Monographs, Vol 200, American Mathematical Society, Providence, RI, 2001.

58. Agostino Ramelli, *The Various and Ingenious Machines of Agostino Ramelli: A Classic Sixteenth-Century Illustrated Treatise on Technology*, trans. Martha Teach Gnudi, Dover Publications, New York, 1976.

59. A. Seidenberg, "The Ritual Origin of Geometry," *Archive for the History of the Exact Sciences*, 1(1961), pp. 488–527.

60. Baudhayana, *Sulbasutram*, G. Thibaut, trans., S. Prakash & R. M. Sharma, ed., Ram Swarup Sharma, Bombay, 1968.

61. William Thurston, *Three-Dimensional Geometry and Topology*, Vol. 1, Princeton University Press, Princeton, NJ, 1997.

62. Isaac Todhunter, *Spherical Trigonometry*, Macmillan, London, 1886. (Available online from the Cornell University Mathematics Library at: http://historical.library.cornell.edu/math/index.html)

63. Imre Toth, "Non-Euclidean Geometry before Euclid," *Scientific American*, 1969.

64. Jeffrey Weeks, *The Shape of Space*, Second Edition, Marcel Dekker, New York, 2002.

65. Trevor Williams, *A History of Invention: From Stone Axes to Silicon Chips*, Revised Edition, Checkmark Books, New York, 2000.

# 7

# Euler on Cevians

**Eisso J. Atzema**    and    **Homer White**
*University of Maine*          *Georgetown College*

## Introduction

For mathematicians looking to incorporate historical sources into their classroom teaching, Leonhard Euler is of special interest. His mathematical notation closely approximates contemporary use,[1] and his prose style is clear and not forbiddingly concise. Furthermore, he was active just before the era of increasing abstraction in mathematics, so that almost any problem he takes up is of immediate interest to a contemporary undergraduate. The motivation for the problem is usually clear, and it is seldom difficult for the student to come up with lines of further research, either by proving his results using modern techniques, generalizing them, or investigating related questions. Quite often, too, Euler's problem lies somewhat off the beaten track of contemporary mathematics, so that a student's further work stands some chance of resulting in new mathematics.

The present article is a historical examination of a little-known contribution of Euler to classical Euclidean geometry, combined with a free-floating elaboration on some of Euler's results. We firmly believe that this two-pronged approach provides a feasible method for the introduction into the mathematics classroom of historical material that might be stimulating to students and teachers alike (as opposed to only the former or, worse, just the latter). The main body of the paper may be adapted to classroom use, either in the form of lectures, in the form of exploratory activities, or as a combination of the two. Alternatively, parts of the text may be assigned as outside-class reading. With the latter use in mind, we maintain a style of exposition suitable for advanced undergraduates, and we have relegated pedagogical comments to an appendix, where the reader will find discussion of how the material in this paper has featured in previous geometry classes, and how it may be profitably used in other ways. Accordingly, the appendix also suggests various exercises and research projects for students. All diagrams for this paper have been drawn with Geometers' SketchPad®.The reader is strongly encouraged to explore the various findings in this paper using SketchPad or a similar dynamical geometry package as well.

## Euler's Theorem

One of the cornerstones of the study of the triangle is the concept of the so-called cevian. A cevian of a triangle is defined as a line that passes through any of the vertices of that triangle, but that is different from

---

[1]Or perhaps one should say, since Euler's notational choices were so influential, that modern notation closely approximates that of Leonhard Euler!

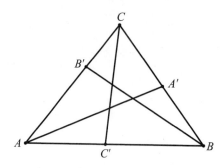

Figure 1. Three (non-concurrent) cevians in a triangle

its sides. This term is used in honor of the Italian mathematician Giovanni Ceva (1647–1734), who was one of the first to systematically study configurations of such lines. Students of classical triangle geometry will be familiar with what is now called Ceva's Theorem.[2] This theorem states that for any triangle $ABC$, if a cevian passing through $A$ intersects $BC$ at $A'$ with similarly defined cevians passing through $B$ and $C$ (see Fig. 1), then these three cevians are concurrent if and only if

$$AB' \cdot BC' \cdot CA' = AC' \cdot CB' \cdot BA',$$

where any of the lengths $AB'$, etc. is taken to be negative if the corresponding angle $BAB'$, etc. is an exterior angle of the triangle. Ceva also formulated what is now known as Menelaus' Theorem for the case of plane geometry.[3] The latter theorem states that for any three cevians $AA'$, $BB'$ and $CC'$ as above, their feet $A'$, $B'$ and $C'$ will be collinear if and only if

$$AB' \cdot BC' \cdot CA' = -AC' \cdot CB' \cdot BA',$$

where again any of the lengths $AB'$, etc. is taken to be negative if the corresponding angle $BAB'$, etc. is an exterior angle of the triangle. Until recently, both theorems got rather short shrift in many of the standard geometry textbooks. Of late, they have been given the more central place that they deserve.[4]

Note that both these results are expressed in terms of the ratios in which cevians cut their *opposite sides*, rather than in terms of the ratios in which the cevians cut *one another*. For this reason we may refer to these conditions for concurrency and collinearity as external. One may wonder then if the same conditions cannot be formulated in an internal way, i.e. purely in terms of the cevians and the ratios in which they intersect one another. Particularly, in the case of concurrency, suppose we have three line segments of given length (but not position) and intersecting one another in given ratios. Under what conditions might such a set of line segments be considered as a triple of concurrent cevians to some triangle and if so, can one 'solve' that triangle? It was precisely this question that the great Swiss mathematician Leonhard Euler (1707–1783) sought to answer in a paper written toward the end of his life.[5] What exactly inspired Euler

---

[2]Unlike many other theorems, this particular theorem was actually discovered by the person it is named after. It was published by Ceva in his *De lineis rectis* of 1678. For most of the 18th and 19th century, however, this result was attributed to John Bernoulli (1667–1748) who included a proof in his *Opera Omnia* of 1748 without indicating when he found this result. See [1, p. 59].

[3]Menelaus of Alexandria (fl. 100) was the author of a treatise on spherical triangles, which has only been preserved in Arabic translation. Until the 19th century, the theorem was ascribed to Ptolemy (ca 85-165) in whose *Almagest* it appears. Both Menelaus himself and Ptolemy were only interested in the spherical case, but in the 16th and 17th century others had considered the plane case as well. Ceva's investigations into cevians may have lead him to independently rediscover the theorem.

[4]See, for instance, [9, pp. 27–35 and pp. 45–50] and [2, pp. 53–58]. It will not come as a surprise that work for this paper was at least partly inspired by this renewed interest in the two principal cevian theorems.

[5]Although Euler presented his results to the Petersburg Academy on May 1, 1780, the actual paper entitled "Geometria et spherica quaedam" only appeared in 1815. See *Mémoires de l'Académie des Sciences de St-Pétersbourg* 5 (1812), pp. 96–114 (= *Opera Omnia*, T. 26, pp. 344–358). An English translation, prepared by the second author, will be posted online at the Euler Archive (http://www.dartmouth.edu/ euler).

to look into this question is not clear, but the kind of question was not completely original. Most notably, about half a century earlier, Philippe Naudé (1684–1745), the Berlin Academy of Sciences' president before its reform under Euler and Maupertuis, had published two papers on a closely related topic. In these papers he proposed nothing less than the creation of a new field of mathematical investigation he called *trigonoscopia*. Naudé envisioned this field as a branch of pure mathematics, the principal goal of which was to 'solve' triangles from given data. In that sense, it was to be an extension of the usual techniques of solving triangles from given angles and sides.[6] In addition to these time-honored techniques, trigonoscopia was to study the solution of triangles from combinations of sides and given points on unknown sides and other data. At the end of his first paper, Naudé distinguishes 16 categories of problems, totaling 678 configurations of givens. More relevant to us, among the configurations that he explicitly discussed we find the solution of a triangle from a triple of medians of given length as well as its solution from a triple of altitudes. It seems unlikely that the Naudé paper had much impact, but it is highly probable that Euler was aware of it. When Euler was still in St. Petersburg, he had been in correspondence with Naudé on various issues, including a problem that seems to have originated in Naudé's investigations into trigonoscopia. When Naudé's second paper was published in 1743, Euler had already been working at the Berlin Academy for two years.[7]

The point of departure for Euler's own investigations into trigonoscopia was the following observation:

**Theorem 1 (Euler, 1780)** *In a triangle $ABC$ with concurrent cevians $AA'$, $BB'$, and $CC'$ as defined above with $O$ their point of concurrency, the following property holds:*

$$\frac{OA'}{AA'} + \frac{OB'}{BB'} + \frac{OC'}{CC'} = 1,$$

*or, equivalently,*

$$\frac{AO}{OA'} \cdot \frac{BO}{OB'} \cdot \frac{CO}{OC'} = \frac{AO}{OA'} + \frac{BO}{OB'} + \frac{CO}{OC'} + 2.$$

*Proof.* In his paper Euler, not usually given to concise exposition anyway, set forth no less than three proofs of this result. Here, we will follow the most direct of these.[8] In Fig. 2, let segment $OF$ be parallel to $AB$, and let $OG$ be parallel to $AC$. The similarity of the three pairs of triangles, $BOG$ and $BB'C$, $COF$ and $CC'B$, $FOG$ and $BAC$ respectively yields the equations

$$\frac{OB'}{BB'} = \frac{GC}{BC}, \quad \frac{OC'}{CC'} = \frac{BF}{BC}, \quad \frac{OA'}{AA'} = \frac{FG}{BC}.$$

Summing these equations, we get

$$\frac{OA'}{AA'} + \frac{OB}{BB'} + \frac{OC'}{CC'} = 1,$$

---

[6]To be sure, Naudé did not use the expression 'solving triangles', nor was this topic as rigorously standardized as it is now. The techniques existed, but both the expression and the formal classication of the various cases (SSS, SSA, etc.) only came about in the 19th century.

[7]See Ph. Naudé "Trigonoscopia cuiusdam novae conspectus," *Miscellanea Berolinensia ad incrementum scientiarum* V (1737), pp. 10–32 and *Ibidem* VII (1743), pp. 243–270. In the correspondence between Euler and Naudé there is a reference to the problem of the determination of the diagonals of a cyclic quadrilateral. This problem was tackled by Naudé in an attempt to extend his trigonoscopia to (cyclic) quadrilaterals. See Ph. Naudé, "Problema geometricum de maximis in figuris planis," *Miscellanea Berolinensia ad incrementum scientiarum* VI (1740) , pp. 217–235. This paper may also have been the source of inspiration for Euler's own work on quadrilaterals around 1748.

[8]This proof actually appeared in an appendix to the paper and not in the main text. This might be an indication that it was added by Euler after he finished his paper. The other two proofs are in the main body of the text. The main text also contains two proofs for the analogous result in spherical geometry.

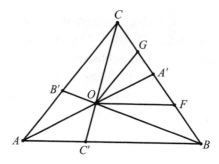

Figure 2. Euler's Proof of Theorem 1

the first of the two relations to be proved. The other is obtained with a bit of algebraic rearrangement. Let us define

$$\alpha = \frac{OA'}{AA'}, \quad \beta = \frac{OB'}{BB'}, \quad \gamma = \frac{OC'}{CC'},$$

for the ratios in which the cevians cut one another.[9] It is now easy to see that

$$\alpha = \frac{1}{\frac{AO}{OA'} + 1}, \quad \beta = \frac{1}{\frac{BO}{OB'} + 1}, \quad \gamma = \frac{1}{\frac{CO}{OC'} + 1},$$

so we have

$$\frac{1}{\frac{AO}{OA'} + 1} + \frac{1}{\frac{BO}{OB'} + 1} + \frac{1}{\frac{CO}{OC'} + 1} = 1,$$

from which a little more algebra gives

$$\frac{AO}{OA'} \cdot \frac{BO}{OB'} \cdot \frac{CO}{OC'} = \frac{AO}{OA'} + \frac{BO}{OB'} + \frac{CO}{OC'} + 2$$

which is the second relation. Note that Euler implicitly assumes that $O$ lies inside of the triangle, although later he notes that if $O$ lies outside the triangle the relationship will continue to hold, provided that one allows for directed line segments of negative length, much as in Ceva's Theorem. Euler goes on to prove this fact, but this part of the paper seems to be badly botched. In the printed text, his proof does not make any sense.

Euler does not try to generalize his theorem to tetrahedra. Indeed, it is not easy to see how his proof would allow for such a generalization. Since we will want to generalize to the case of tetrahedra, an alternative proof might be in order. Note that the cevians divide $ABC$ into three triangles $AOB$, $BOC$, and $COA$. Let us denote the areas of these triangles by $|ABC|$, etc. Since $AOB$ shares its base with $ABC$ and its height is $\gamma$ times the height of $ABC$, it follows that $|AOB| = \gamma|ABC|$ and similarly for the other triangles. Summing now immediately gives Euler's first result. From this proof, it is also obvious that similar relations hold in dimensions higher than 3. We will have occasion to resort to this kind of argument later on in this paper as well.

After having obtained his core theorem, Euler goes on to show that the condition of Theorem 1 is not only necessary but also sufficient to determine any triangle to a given set of cevians. Following a somewhat different approach, we will arrive at the same conclusion. Just like Euler, we first need to find the exact relations between the internal and external ratios and expressions for the theorems of Ceva and Menelaus in terms of the cevians themselves.

---

[9]In the main body of his paper, Euler uses the same three characters to denote the ratios $OA/OA$, etc. In the appendix, however, he switches to our notation. The advantage of this choice will be obvious from what follows.

## Ceva and Menelaus Revisited

Let us start with expressions for $\alpha$, $\beta$, and $\gamma$ in terms of the external ratios. We find:

**Proposition 2 (Inner-Outer Lemma A)** *Consider a triangle $ABC$, with concurrent cevians $AA'$, $BB'$, $CC'$ meeting at point $O$. Then, with ratios $\alpha$, $\beta$, and $\gamma$ defined as in the proof of Theorem 1,*

$$\frac{1-\alpha}{\alpha} = \frac{AC'}{C'B} + \frac{AB'}{B'C}, \quad \frac{1-\beta}{\beta} = \frac{BC'}{C'A} + \frac{BA'}{A'C}, \quad \frac{1-\gamma}{\gamma} = \frac{CB'}{B'A} + \frac{CA'}{A'B}.$$

*Proof.* Consider the triangle $BOC$ in Fig. 3. Note that the three line segments $BC'$, $OA'$, and $CB'$ are concurrent cevians to this triangle meeting at point $A$. Therefore, by Theorem 1 we have

$$-\frac{AC'}{BC'} + \frac{AA'}{OA'} - \frac{AB'}{CB'} = 1,$$

or

$$\frac{AA'}{OA'} - 1 = \frac{AC'}{BC'} + \frac{AB'}{CB'},$$

which becomes

$$\frac{1-\alpha}{\alpha} = \frac{AO}{OA'} = \frac{AC'}{C'B} + \frac{AB'}{B'C},$$

the first desired equation. The other equations are obtained similarly.

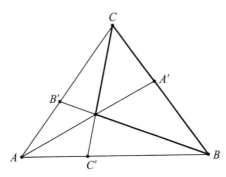

Figure 3. From External to Internal

**Corollary 3** *In triangle $ABC$ let cevians $AA'$ and $BB'$ meet at point $P$ inside $ABC$. With $\alpha$, $\beta$, and $\gamma$ defined as in the proof of Theorem 1, we have*

$$\frac{1-\alpha}{\alpha} = \frac{CA' \cdot AB'}{BA' \cdot CB'} + \frac{AB'}{B'C} = \frac{AB'}{B'C}\left(\frac{CA'}{BA'} + 1\right), \quad \frac{1-\beta}{\beta} = \frac{AB' \cdot BC'}{CB' \cdot AC'} + \frac{BC'}{C'A} = \frac{BC'}{C'A}\left(\frac{AB'}{CB'} + 1\right).$$

*Proof.* Draw cevian $CC'$ passing through $P$ and use Ceva's Theorem to eliminate $AC'/BC'$.

**Proposition 4 (The Inner-Outer Lemma B)** *Suppose that in triangle $ABC$, cevians $AA'$, $BB'$, and $CC'$ meet at a point $O$. Then, with ratios $\alpha$, $\beta$, and $\gamma$ still defined as in the proof of Theorem 1, we have*

$$\frac{BA'}{A'C} = \frac{\gamma}{\beta}, \quad \frac{CB'}{B'A} = \frac{\alpha}{\gamma}, \quad \frac{AC'}{C'B} = \frac{\beta}{\alpha}.$$

*Proof.* Consider the triangles $AOC$ and $AOB$. If, as before, $|...|$ denotes the area of a triangle, we have $|AOC| = \beta|ABC|$. Likewise we have $|AOB| = \gamma|ABC|$. Also, $|AOC| : |AA'C| = |AOB| : |AA'B|$. It follows that $|AA'C|/|AA'B| = \beta/\gamma$. But the left-hand side of this expression also equals $CA'/BA'$. We conclude that $CA'/BA' = \beta/\gamma$, which is the first of the expressions above.

**Corollary 5** *Suppose that in triangle $ABC$, two cevians, $AA'$ and $BB'$ intersect at a point $P$. Then, with ratios $\alpha$, $\beta$, and $\gamma$ defined as above, we have*

$$\frac{BA'}{A'C} = \frac{1-\beta-\alpha}{\beta}, \quad \frac{CB'}{B'A} = \frac{\alpha}{1-\beta-\alpha}.$$

*Proof.* Draw the cevian $CC'$ passing through $P$. By Theorem 1, $\gamma = 1 - \alpha - \beta$. Apply Proposition 4.

We are now ready to formulate the theorems of Ceva and Menelaus in terms of internal ratios. Consider a triangle $ABC$ in which arbitrary cevians $AA'$, $BB'$, $CC'$ are drawn. Say that $AA'$ and $BB'$ meet at $C''$, $AA'$ and $CC'$ meet at $B''$ and $BB'$ and $CC''$ meet at $A''$. We define ratios which describe how $AA'$ is cut by cevians $BB'$ and $CC'$ respectively, as follows:

$$\alpha_B = \frac{C''A'}{AA'}, \quad \alpha_C = \frac{B''A'}{AA'}.$$

Ratios $\beta_A, \beta_C$ and $\gamma_B, \gamma_C$ are defined similarly. With these notations, we now have

**Theorem 6 (Inner-Outer Theorem)** *In any triangle $ABC$, we have*

$$\frac{AC'}{C'B} \frac{BA'}{A'C} \frac{C'B}{BA'} = \frac{(1-\alpha_C-\gamma_A)}{\alpha_C} \frac{(1-\beta_A-\alpha_B)}{\beta_A} \frac{(1-\gamma_B-\beta_C)}{\gamma_B}$$

$$= \frac{\beta_C}{(1-\beta_C-\gamma_B)} \frac{\gamma_A}{(1-\gamma_A-\alpha_C)} \frac{\alpha_B}{(1-\alpha_B-\beta_A)}.$$

*Proof.* The first equation follows from cycling through all of the letters of the first equation of Proposition 3, the second from cycling through the letters of the second equation.

By combining the last two expressions, we obtain

**Corollary 7 (Inner Ceva and Menelaus I)** *Consider a triangle $ABC$ with cevians $AA'$, $BB'$, $CC'$ and $\alpha_B$, etc. defined as above. Then $AA'$, $BB'$ and $CC'$ are concurrent or $A'$, $B'$, and $C'$ are collinear if and only if*

$$\alpha_B \beta_C \gamma_A = \alpha_C \gamma_B \beta_A,$$

*with concurrency occurring if and only if $\alpha_B = \alpha_C$, $\beta_A = \beta_C$, and $\gamma_A = \gamma_B$.*

*Proof.* In case the cevians are concurrent or their feet collinear, the left-most expression of Theorem 6 equals $\pm 1$, so the product of the two other expressions equals 1, which gives rise to the equation above. If the equation above is satisfied, then the left-most expression of Theorem 6 equals $\pm 1$, so either the cevians are concurrent or their feet collinear. If concurrency occurs, then obviously $\alpha_B = \alpha_C$, etc. Conversely, if $\alpha_B = \alpha_C$, then clearly the cevians are concurrent.

## Solving a Triangle to Given Cevians

Now that we have laid the groundwork, we are ready to address Euler's main question: Suppose one has a triple of line segments $AA'$, $BB'$ and $CC'$ that have a point $O$ in common. Is it possible to find a triangle to which these segments are cevians? In other words, assuming that the line segments are cevians, can we solve the corresponding triangle? As we have seen before, for the segments to be cevians, it is a necessary condition that $OA'/AA' + OB'/BB' + OC'/CC' = 1$. Conversely, will the latter condition be also sufficient for the line segments to be the cevians to a triangle? Euler answers this question in the affirmative by giving a nice way to properly position the segments. Since his approach is computationally intensive, we will follow a somewhat different tack to arrive at the same result. Essentially, all that we

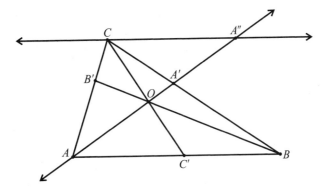

Figure 4. Extending a cevian

need is a slight generalization of a construction given by Hoehn in [5] for medians so as to make it apply to any triple of cevians. Consider a triangle $ABC$ with three concurrent cevians $AA'$, $BB'$ and $CC'$ and $\alpha$, $\beta$, $\gamma$ defined as before. Now construct a line through $C$ and parallel to the side $AB$ (see Fig. 4). Let $A''$ be the point of intersection of this line with the line $AA'$. Since $AA'B \sim A''AC$ and $CA'/A'B = \beta/\gamma$, it follows that $CA'' = \frac{\beta}{\gamma}c$ and $AA'' = \frac{(\beta+\gamma)}{\gamma}a'$, where $c = |AB|$ and $a' = |AA'|$. A triangle congruent to $AA''C$, but then rotated by $90^o$, we find by erecting a square $ACB_AB_C$ on $AC$ and a rectangle $ABC_AC_B$ on $AB$ with height $\frac{\beta}{\gamma}c$. It follows that $AA'$ and $B_CC_B$ are perpendicular. Clearly the same still applies if the heights of the rectangles on the side are scaled by the same factor. Because of this, the construction can also be extended to all three cevians at the same time. One choice would be to take $\alpha\beta c$ for the height of the rectangle on $AB$ and similarly for the other two rectangles. For this particular choice $B_CC_B$ is still perpendicular to $AA'$ and has length $\alpha(1-\alpha)a'$ and similarly for $B_CB_A$ and $C_AC_B$ (see Fig. 5).[10] Since these three line segments can be joined to form a triangle by simple translation, it follows that the three

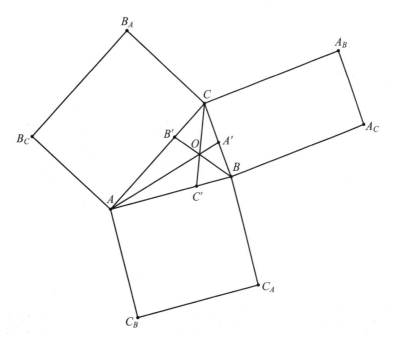

Figure 5. Hoehn's Construction Generalized

[10]For the sake of convenience, the actual heights of the rectangles in the figure are $\alpha\beta c/3$, etc. This yields much nicer pictures. Besides, for the case of medians the rectangles now reduce to the squares that [5] has.

cevians are perpendicular to a triangle $A\bar{B}C$ with sides of lengths $\alpha(1-\alpha)a'$, $\beta(1-\beta)b'$, and $\gamma(1-\gamma)c'$. This property suffices to show that for any three line segments with given lengths $a'$, $b'$, and $c'$ as well as the ratios $\alpha$, $\beta$, and $\gamma$, at most one triangle can exist to which these segments are cevians and from which one can construct this "mother" triangle in case it exists.[11] It remains to be shown under what conditions such a triangle will not exist.

Clearly, no mother triangle will exist if $\alpha(1-\alpha)a'$, $\beta(1-\beta)b'$ and $\gamma(1-\gamma)c'$ do not satisfy the triangle inequality. If this inequality is satisfied, however, we can easily compute the lengths of the sides of the mother triangle. In fact, if the triangle exists, then $\angle AOB$ and $\angle B\bar{C}A$ are supplements. It follows that

$$c^2 = (1-\alpha)^2(a')^2 + (1-\beta)^2(b')^2 + 2(1-\alpha)(1-\beta)a'b'\cos(\angle B\bar{C}A).$$

But using the law of cosines on $A\bar{B}C$, we find

$$\cos(\angle B\bar{C}A) = \frac{\alpha^2(1-\alpha)^2(a')^2 + \beta^2(1-\beta)^2(b')^2 - \gamma(1-\gamma)^2(c')2}{2\alpha\beta(1-\alpha)(1-\beta)(a')(b')}.$$

It follows that

$$c^2 = \left(\frac{1-\gamma}{\beta\alpha}\right)\left(\alpha(1-\beta)^2(b')^2 + \beta(1-\alpha)^2(a')^2 - \gamma^2(1-\gamma)(c')^2\right),$$

with similar expressions for $a^2$ and $b^2$. All we need to check now is whether for a triangle with sides thus determined, its three cevians through $A$, $B$ and $C$, dividing the opposite sides in ratios $\beta/\gamma$, $\gamma/\alpha$ and $\alpha/\beta$ respectively, have the same lengths as the given line segments. One way to do so is to make use of the following result:

**Proposition 8 (Stewart's Theorem)** [12] *For any triangle $ABC$ and a cevian $AA'$ as above, we have*

$$(AA')^2 = \frac{CA'}{CB}\cdot(AB)^2 + \frac{BA'}{CB}\cdot(AC)^2 - \frac{CA'}{CB}\frac{A'C}{CB}\cdot(CB)^2.$$

*Proof.* Find $AA'$ using the law of cosines on $ABA'$ and replace $\cos\beta$ by its expression in terms of sides of $ABC$.

A dreary computation shows that the cevians of the candidate mother triangle indeed equal the given line segments and we can now state

**Theorem 9** *Suppose we have three line segments of given lengths $a'$, $b'$ and $c'$ and given ratios $\alpha$, $\beta$, $\gamma$ as defined above. Then the line segments are the cevians to a triangle if and only if*

$$\alpha + \beta + \gamma = 1$$

*and the triangle inequality is satisfied for $\alpha(1-\alpha)a'$, $\beta(1-\beta)b'$ and $\gamma(1-\gamma)c'$.*

---

[11] Indeed, the construction given here allows us to solve Euler's problem as follows: given concurrent cevians $AOA$, $BOB'$, $COC'$ with sides of length $d'$, $b'$, $c'$ respectively and cutting ratios $\alpha$, $\beta$, $\gamma$ defined as usual (i.e., $\frac{OA'}{AA'} = \alpha$, etc.), construct a triangle with sides $\alpha(1-\alpha)d'$, $\beta(1-\beta)b'$ and $\gamma(1-\gamma)c'$. Now rotate the three given cevians around $O$ so that cevian-to-be $AA'$ is perpendicular to the side with length $\alpha(1-\alpha)d'$. Do the same for the other prospective cevian segments, making sure that no two of the segments $AO$, $BO$, and $CO$ are adjacent. The points $A$, $B$, $C$ will now be the vertices of the desired 'mother' triangle.

[12] The theorem is usually traced back to the Scottish mathematician Matthew Stewart (1717–1785) who published it in his *General Theorems* of 1746. It appears, however, that the theorem was first proved a few years earlier by his teacher Robert Simson (1687–1768). Euler himself is among the many who have rediscovered the result; see his "Problematis cuiusdam Pappi Alexandrini constructio" (*Opera Omnia* I, vol. 26, pp. 237–241). Euler may have been inspired by a special case of Stewart's Theorem, in which the point $A$ lies on the line $BC$, for the result in this form appears as Proposition 125 in Book Seven of the *Collectiones* of Pappus ca 300 AD. Indeed, as Proclus was at that point discussing a work of Apollonius (born ca 262 BCE), the result probably goes back much further.

*Proof.* See construction above. Strictly speaking the proof only applies to $\alpha$, $\beta$, $\gamma$ between 0 and 1. For the other cases, slight adjustments such as erecting rectangles into the triangle rather than away from it are necessary.

To really grasp this theorem, it might be worthwhile for the reader to recreate the diagram of Figure 5 using SketchPad. We have found that the actual construction in SketchPad of the triangle perpendicular to a triple of cevians of given lengths poses quite a challenge, but it is certainly helpful to give this a try as well.

Following a purely algebraic approach involving some rather thorny computations, Euler obtained the dimensions of the perpendicular triangle as well and could thus prove that Theorem 1 was both necessary and sufficient. For Euler this settled the question and all he had left to do was to extend this result to spherical geometry. We will not discuss this part of his paper. Instead, we will present some of investigations of our own that were inspired by Euler's work.

## The Case of Non-Concurrent Cevians

In the preceding we assumed that the given line segments were concurrent, which was the only case that Euler investigated. Let us now assume that the segments do not meet in one single point. If this is the case, the ratios in which the line segments intersect one another fully determine any triangle to which the segments might be cevians. All we need to verify is under what conditions $A'$, $B'$, $C'$ lie on $BC$, $CA$, $AB$ respectively. The easiest way to find such conditions is to note that $A$, $B'$ and $C$ lie on three different sides of the triangle $A''B''C''$. Applying Menelaus' theorem yields the condition

$$\frac{\beta_A}{\beta_C} \frac{1-\alpha_C}{1-\alpha_B} \frac{1-\gamma_B}{1-\gamma_A} = 1,$$

with two similar conditions for $A'$ and $C'$. Since Menelaus gives both necessary and sufficient conditions, these three equations give us necessary and sufficient conditions for the existence of a mother triangle. Actually, as each of these equations can be expressed in terms of the other two, only two of them suffice.

Somewhat more symmetric conditions can be obtained by noting that the points $B$, $A''$, $C''$, $B'$ and the points $C'$, $A''$, $B''$, $C$ lie in perspective with respect to $A$. It follows that their cross ratios $(BB'; A''C'')$ and $(C'C; A''B'')$ are equal or $\beta_A(1-\beta_C)/\beta_C(1-\beta_A) = \gamma_B(1-\gamma_A)/\gamma_A(1-\gamma_B)$.[13] Applying the same reasoning to the other two pairs of cevians, we find the necessary conditions

$$\frac{\beta_A(1-\beta_C)}{\beta_C(1-\beta_A)} = \frac{\gamma_B(1-\gamma_A)}{\gamma_A(1-\gamma_B)} = \frac{\alpha_C(1-\alpha_B)}{\alpha_B(1-\alpha_C)}.$$

Comparison with the previous set of conditions shows that this new set of conditions is equivalent and therefore sufficient as well.

The first set of conditions also can be used to give an alternative internal formulation of Ceva's and Menelaus' theorems. Indeed, from these equations it follows that

$$\frac{\beta_C}{\beta_A} \frac{\gamma_A}{\gamma_B} \frac{\alpha_B}{\alpha_C} = \frac{(1-\gamma_B)(1-\alpha_C)}{(1-\gamma_A)(1-\alpha_B)} \frac{(1-\alpha_C)(1-\beta_A)}{(1-\alpha_B)(1-\beta_C)} \frac{(1-\beta_A)(1-\gamma_B)}{(1-\beta_C)(1-\gamma_A)}$$
$$= \left(\frac{(1-\alpha_C)(1-\gamma_B)(1-\beta_A)}{(1-\alpha_B)(1-\beta_C)(1-\gamma_A)}\right)^2.$$

In other words, we have

---

[13] On cross ratios and their use, see any book on projective geometry.

**Theorem 10 (Inner Ceva and Menelaus II)** *Consider a triangle $ABC$ with cevians $AA'$, $BB'$, $CC'$ and $\alpha_B$, etc. defined as above. Then $A'$, $B'$ and $C'$ are collinear if and only if*

$$(1 - \alpha_B)(1 - \beta_C)(1 - \gamma_A) = -(1 - \alpha_C)(1 - \gamma_B)(1 - \beta_A).$$

*Concurrency of the cevians will occur if and only if*

$$(1 - \alpha_B)(1 - \beta_C)(1 - \gamma_A) = (1 - \alpha_C)(1 - \gamma_B)(1 - \beta_A),$$

*in which case $\alpha_B = \alpha_C$, $\beta_C = \beta_A$, and $\gamma_A = \gamma_B$.*

*Proof.* See above. The correct sign can be determined upon inspection.

Again, we have found it helpful in understanding this theorem to reconstruct the situation above in SketchPad and to verify the theorem by simple measurements.

## Beyond the Plane

As we will show in this section, Euler's Theorem 1 can be fairly easily generalized to higher dimensions as can many of the other properties that we derived. Specifically, the triangle $A\bar{B}C$ with each side perpendicular to one of the cevians of a triangle $ABC$ has a direct analog in three-space. To see this, it is probably easiest to verify the properties of $A\bar{B}C$ using vector algebra and then generalize our approach to tetrahedra. Perhaps a warning is in order that most of this section goes somewhat beyond the kind of elementary geometry of the preceding sections.

Let $\vec{a}$, $\vec{b}$, $\vec{c}$ be vectors in the $xy$-plane in three-space with their initial points at the origin. The endpoints of these vectors are the vertices $A$, $B$, and $C$ of a triangle. Without loss of generality, we may assume that three concurrent cevians to this triangle meet in the origin. Thus, the vectors $\vec{a}$, $\vec{b}$, $\vec{c}$ also indicate the directions of the cevians and the directions of the sides of any triangle $A\bar{B}C$ "perpendicular" to these cevians (see the Solving a Triangle (...) section) will be given by the vectors $\vec{a} \times \mathbf{e}_3$, $\vec{b} \times \mathbf{e}_3$, and $\vec{c} \times \mathbf{e}_3$. If we fix the length of the side $\bar{B}C$ at $|\vec{a} \times \mathbf{e}_3|$, then the length of the other two sides will be of the form $\lambda_B |\vec{b} \times \mathbf{e}_3|$ and $\lambda_C |\vec{c} \times \mathbf{e}_3|$. Now, note that because of the area formula for a triangle, we have

$$(\vec{a} \times \mathbf{e}_3) \times \lambda_B (\vec{b} \times \mathbf{e}_3) = \lambda_B (\vec{b} \times \mathbf{e}_3) \times \lambda_C (\vec{c} \times \mathbf{e}_3),$$

from which it follows that $\lambda_C = |\vec{a} \times \vec{b}| / |\vec{b} \times c|$. Likewise, we have $\lambda_B = |\vec{a} \times \vec{c}| / |\vec{b} \times c|$. Since triangles perpendicular to the cevians are determined up to a scalar, it follows that the cevians can be taken to be perpendicular to a triangle with sides of lengths $|\vec{a}||\vec{b} \times \vec{c}|$, $|\vec{b}||\vec{c} \times \vec{a}|$, and $|\vec{c}||\vec{a} \times \vec{b}|$. Now, the length of $\vec{a}$ equals the length of that part of the cevian of $ABC$ on the side $BC$ that lies between $A$ and the point of concurrency of the three cevians. In other words, if $\alpha$ is defined as in the proof of Theorem 1 and $a'$ denotes the total length of the cevian departing from $A$, we have that $|\vec{a}| = (1 - \alpha)a'$ and $|\vec{b} \times \vec{c}|$ equals $|BOC| = \alpha|ABC|$. Scaling again, it follows that we can take the lengths of the sides of $A\bar{B}C$ as $\alpha(1 - \alpha)a'$ and so on, which lengths completely define $A\bar{B}C$. Note that for this choice of the sides of $A\bar{B}C$, their lengths are conveniently measured in the same units as the lengths of the sides of the original triangle. We therefore can meaningfully compare the perpendicular triangle to the original triangle.

In the case of a tetrahedron in three-space, let $\vec{a}$, $\vec{b}$, $\vec{c}$, $\vec{d}$ be vectors with their initial points at the origin. The endpoints of these vectors are the vertices $A$, $B$, $C$, and $D$ of a tetrahedron. Without loss of generality, we may assume that four concurrent cevians to this tetrahedron meet in the origin. Thus, the vectors $\vec{a}$, $\vec{b}$, $\vec{c}$, $\vec{d}$ also indicate the directions of the cevians. Let now $AB\bar{C}D$ be a tetrahedron "perpendicular" to these cevians, i.e., a tetrahedron such that the face $B\bar{C}D$ is perpendicular to $\vec{a}$ and so on. By construction, its edge $A\bar{B}$ has to be perpendicular to both $\vec{c}$ and $\vec{d}$. In other words, the vector $\vec{v}_{A\bar{B}}$ with length and direction of $A\bar{B}$ has to have the direction of $\vec{c} \times \vec{d}$ and similarly for the other edges of the tetrahedron. Using the same trick as in the two-dimensional case on the faces of the tetrahedron and symmetrizing, we find that we may take

$$\vec{v}_{A\bar{B}} = |\vec{a}, \vec{b}, \vec{c}||\vec{a}, \vec{b}, \vec{d}| \cdot (\vec{c} \times \vec{d}),$$

where $|., ., .|$ is the absolute value of a determinant. Or, by the same argument as used in the two-dimensional case, we could take $\vec{v}_{A\bar{B}} = \gamma\delta(\vec{c} \times \vec{d})$. Finally, if we want the lengths of the edges of $A\bar{B}CD$ to be comparable to those of the edges of the original tetrahedron, i.e. if we want both sets to be measured in the same units, we will have to scale the expression for $\vec{v}_{A\bar{B}}$ once again. One expression that would do the job is

$$\vec{v}_{A\bar{B}} = \frac{\gamma\delta}{\sqrt[3]{6|ABCD|}}(\vec{c} \times \vec{d}).$$

Similar expressions apply for the other edges of this choice for $A\bar{B}CD$ and together these expressions completely determine a perpendicular tetrahedron $A\bar{B}CD$. For the area $|A\bar{B}C|$ of the face $A\bar{B}C$ of this particular perpendicular tetrahedron we find

$$2|A\bar{B}C| = |\vec{v}_{A\bar{B}} \times \vec{v}_{AC}| = \frac{\gamma\delta}{\sqrt[3]{6|ABCD|}}\frac{\delta\beta}{\sqrt[3]{6|ABCD|}}|(\vec{c} \times \vec{d}) \times (\vec{d} \times \vec{b})|$$

$$= \frac{\gamma\delta \cdot \delta\beta \cdot 6\alpha|ABCD| \cdot (1-\delta)d'}{\sqrt[3]{36|ABCD|}^2} = \alpha\beta\gamma\delta \sqrt[3]{6|ABCD|} \cdot \delta(1-\delta)d',$$

where $d'$ is the length of the cevian departing from $D$. Similar expressions apply for the three other faces of $A\bar{B}CD$ and it follows that

$$\frac{|B\bar{C}D|}{\alpha(1-\alpha)a'} = \frac{|A\bar{C}D|}{\beta(1-\beta)b'} = \frac{|A\bar{B}D|}{\gamma(1-\gamma)c'} = \frac{|A\bar{B}C|}{\delta(1-\delta)d'}.$$

These relations are analogous to what we found in the two-dimensional case. It should be pointed out, however, that in contrast to the two-dimensional case, this property alone does not uniquely determine a similarity class of perpendicular tetrahedra for a given set of line segments and their intersection ratios. Therefore it does not define a unique 'solution' of the tetrahedron and an additional property will be needed. To see this, it suffices to have a look at the medians of the so-called equifacial or isosceles tetrahedra, i.e. the class of tetrahedra with congruent faces. For these tetrahedra, all medians are of the same length and $\alpha = \beta = \gamma = \delta = 1/4$. Therefore, any tetrahedron perpendicular to the medians of a given equifacial tetrahedron has to be equifacial as well. Since there are infinitely many non-similar equifacial tetrahedra, this property does not allow for the reconstruction of the original set of medians.

## Aftermath

To the best of our knowledge, the results of the two preceding sections have not been published before. In fact, the internal properties of cevians appear to have been hardly studied at all. As we mentioned before, even Euler's Theorem 1 received very little notice in the mathematical community. It appears though that from time to time this specific result was rediscovered. It shows up as an exercise in at least two modern geometry books.[14] Before Euler's paper was published in 1812, cevians (or transversals, as they were

---

[14]See [4, p.119] and [8].

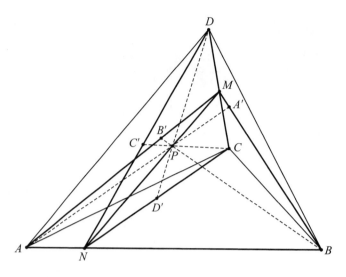

Figure 6. Vallès' construction

then called) were studied by, among others, Lazare Carnot (1753-1823) in his *Essai sur la théorie des transversales* of 1806. All of these studies, however, used an external ratio approach only. Early in the nineteenth century, shortly after Euler's paper was finally published, Theorem 1 and its generalization to tetrahedra were posed as a problem in the *Annales de mathématiques pures et appliquées*, edited by Joseph-Diez Gergonne (1771–1859). Most likely Gergonne saw Euler's paper and thought the problem was suitable for his journal. It is equally likely that some of the solvers of the problem saw Euler's paper. Although Euler is never referred to, three of the four submitted solutions for the plane case were essentially identical to Euler's proof as given above. A rather elegant and surely original proof of the generalization of Theorem 1 to tetrahedra was given by one Vallès, a student at the Collège Royale de Montpellier, a Jesuit boarding school in the same town where Gergonne taught.[15] Following Euler's reasoning closely, Vallès assumes that he has a tetrahedron $ABCD$ with a point $P$ inside (see Fig. 6). For this tetrahedron, he considers the plane containing both $P$ and the edge $AB$ of the tetrahedron. This plane intersects the edge $CD$ in some point $M$. Similarly, the plane containing edge $CD$ and $P$ meets $AB$ at some point $N$. Considering triangle $ABM$, he notices that its three concurrent cevians $AP$, $BP$ and $MP$ meet the sides of the triangle at $A'$, $B'$ and $N$ respectively. Hence,

$$\frac{PA'}{AA'} + \frac{PB'}{BB'} + \frac{PN}{MN} = 1$$

by Euler's Theorem 1. Likewise, for triangle $CND$, its concurrent cevians $CP$, $NP$ and $DP$ meet opposite sides at $C'$, $M$ and $D'$, yielding, by Theorem 1,

$$\frac{PC'}{CC'} + \frac{PD'}{DD'} + \frac{PM}{MN} = 1.$$

Summing these two equations, and noting that $MP + NP = MN$, Vallès obtains the desired result.

It is interesting to note that neither Euler nor Vallès connect Theorem 1 with the determination of the areas of the triangles into which the cevians divide the mother triangle. Today, this would be the standard approach, and indeed this is the proof we find in [4]. Taking this tack would seem all the more obvious to anyone versed in geometry since most proofs of Ceva's theorem use a similar approach as well. But

---

[15]The original problem appeared on p.116 of vol. 9 (1818–19). A first batch of solutions by Vecten, Durrande, Frégier, Fabry, and Gergonne himself was published on pp. 277–284 of the same volume. For Vallès' proof, see vol. 12, pp. 178–9.

then again, using the area approach to prove Ceva's Theorem is relatively recent as well. Working in the tradition of Torricelli, Ceva's main proof was based on mechanical considerations and of the two other purely geometrical proofs that he gives, one is the usual area comparison proof, but the other does not use areas either.[16] John Bernoulli's proof uses similar triangles. Might it be that at least in the 18th and early 19th century, mathematicians were reluctant to use comparison of areas to prove theorems?

August Crelle (1780–1855) might have been the first to give a modern proof of Ceva's theorem using area considerations in his study on transversals of 1816 (see [1, p. 61]). The first mathematician to explicitly formulate a theoretical framework for these kinds of techniques was August Ferdinand Moebius (1790–1764) in his *Barycentrische Calcul* of 1827. Using mechanical considerations, essentially what Moebius noted was that to every point $P$ within a triangle $ABC$ one can associate a unique triple formed by the areas of the triangles $AOB$, $BOC$ and $COA$, thereby coordinatizing the plane in an alternative way. In many ways, this new approach was very well suited to study the kind of problems in geometry on which Euler had mostly published. Indeed, Moebius does mention Euler at various places, albeit not in connection with Theorem 1. Not surprisingly, however, Theorem 1, together with its three-dimensional analog, does show up. We find it at p. 212, at the very end of a long discussion on what Moebius calls affinity.[17] Both are just provided as an example and no reference to Euler or Gergonne is made. Euler's paper seems to have gone completely unnoticed for practically all of the 19th century. There is no mention of it in [3]. In his very thorough survey on elementary geometry of 1905, Max Simon does have a section on the property of Theorem 1, but there is no mention of Euler (see [10, p. 181]). Even if Euler's paper had been more widely known, it certainly was published too late to exercise any influence on the study of the problem that he had first tackled. When it finally appeared in 1815, the study of elementary plane geometry had developed along different lines from those suggested by Naudé's trigonoscopia. Rather than focusing on the study of other objects associated with the triangle, attention had shifted to the study of quadrilaterals and polygons in general. This field of study was known as *polygonometry* and it seemed a highly promising area of research around 1800 and for some time afterwards. Ironically, perhaps, one of the two principal contributors to this new field, Anders Lexell (1740–1784), was a close collaborator of Euler. The other, Simon l'Huilier (1750–1840), received his mathematical training from a student of Euler.

## In Conclusion

In the preceding we discussed Euler's 1780 paper on the transversals of a triangle, as well as some ramifications of the results it contains. We still have to address in more detail how all this mathematics might be used in the classroom. Indeed, although Euler's paper may never have had many readers, we do think his results merit attention. To be sure, we do not claim that his work breaks any new ground today. We do feel that the paper has great pedagogical value. The problems Euler studies are of an elementary nature, and the fact that his solutions have never become part of main-stream geometry allows for some fresh explorations for a reasonably well-prepared college geometry class.

In fact, early on during the Euclidean portion of an undergraduate geometry course, the second author was prompted to inject some elements of Euler's paper into his treatment of cevian concurrence for that class when one of his students, James Campbell, reported that he had hit upon an interesting result. Using Geometer's SketchPad, Campbell noted that given an arbitrary triangle, it is possible to translate its sides so that they are the concurrent medians of a new triangle. It is then readily shown that this triangle has

---

[16]See *De lineis rectis*, pp. 15–17. According to Ceva, the area proof was supplied by one Pietro Paulo Caravaggio Jr. For a modern version of this proof see [4, p. 123].

[17]Roughly a correspondence between mathematical objects defined by what we now call affine transformations, i.e. linear transformations preserving area.

an area equal to four thirds of the area of the original triangle.[18] Quite naturally, Campbell wondered how this result might be generalized and a perfect opportunity to introduce Euler's work in this particular class had arisen. Examples of exercises that resulted from this use of Euler's original paper are given below (Exercises 3, 4, 6, and 8).

This leaves the question how Euler's work might be used in other classes. Even though Euler in translation is quite accessible to the average undergraduate, it is our view that in this case students need not consult Euler's original paper. Some of his demonstrations are much more involved than they need to be. Rather we suggest one of the following two approaches:

- Assign this article in full as independent reading, and have the student solve several exercises as a "final project" and/or present some of Euler's results to the rest of the class. At the second author's home institution, the Honors program has an "honors increment" component, in which for several regular courses a student selects, in consultation with the instructor, an extra project related to the course. This article along with some of the more involved exercises would make a suitable Honors Increment.

- Intersperse the Euler material at several different points in the course:

  1. Euler's theorem should be introduced fairly early, as soon as students have encountered a few basic concurrency results (medians, altitudes, etc.). It is also good to discuss the higher-dimensional versions of Euler's theorem at this time, as well as the Inner-Outer Lemmas. Exercises 13 and 14 are now appropriate to assign.

  2. Later on, after Ceva's theorem has been introduced, have students do Exercise 8. Optionally, after Ceva and Menelaus are covered, introduce the Internal versions of these theorems.

  3. Most treatments of geometric constructions include the construction of a triangle given its three medians, or its three altitudes. These are two of the *trigonoscopia* problems that so interested Naudé. This would be an appropriate time to introduce the Hoehn-like version of Euler's construction of a triangle given the segments into which three concurrent cevians cut one another, or else have a student, one who has read the entire article, make this presentation with the help of some dynamic geometry software. Exercises 11 and 12 would serve to motivate this material.

## Appendix: Exercises and Projects

The following exercises and project suggestions only serve as an example of the kind of questions that might be asked to deepen student understanding of the mathematical ideas discussed in the main text. None of the exercises would count as a routine exercise, but no additional theory is needed to solve any of them either. Their arrangement roughly follows the discussion of the ideas in the main text. No ranking according to difficulty has been attempted. Solutions are available from the authors upon request.

**Exercise 1** In his second paper on trigonoscopia, Philippe Naudé discusses the construction of a triangle to a given set of altitudes. Euler's Theorem 1 would not have been of much help to Naudé in this case. The problem is easier solved by using the fact that in a triangle $ABC$ with sides $a$, $b$, and $c$ and corresponding altitudes $a'$, $b'$, and $c'$ we have $a' : b' : c' = bc : ca : ab$. Prove this property and explain why this suffices to construct the mother triangle.

---

[18]This particular result is well-known, but in this case has been rediscovered. Being a physics major himself, Campbell proved it using vector methods. He spoke on this result at a state MAA meeting; for an abstract see http://web.centre.edu/mat/kymaa/meetings/abstracts02.doc. See also Exercise 11 below.

**Exercise 2** Show that application of Euler's Theorem 1 to the altitudes of a triangle $ABC$ with angles $\alpha$, $\beta$ and $\gamma$ leads to the triangle identity

$$\cot\alpha\cot\beta + \cot\beta\cot\gamma + \cot\gamma\cot\alpha = 1$$

**Exercise 3** In Euler's original paper, his proof of Theorem 1 for the case in which the point of concurrence $O$ lies outside the triangle $ABC$ was badly botched, either by Euler himself or by persons involved in transcribing or printing the article. Set things right by proving the result mentioned in the note to Theorem 1 above. You can employ a similarity argument very like the one Euler gave.[19]

**Exercise 4** A result that can be easily handled using "internal" cevian-cutting ratios is the following. Trisect each side of the triangle $ABC$ and denote the trisection points as in Fig. 7. Now let $A''$ be the point of intersection of $BB_A$ and $CC_A$, etc. Show that the triangle $A''B''C''$ is a dilation of $ABC$ with respect to the centroid of the triangle by a factor $-1/5$, where the negative sign means that in addition to a scaling, there is a point reflection around the centroid involved.

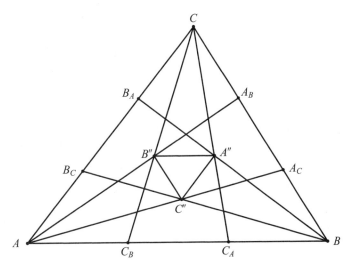

Figure 7. A similar triangle

**Exercise 5** For a triangle $ABC$ and angles $A$, $B$, $C$ erect similar isosceles triangles $ABC''$, $BCA''$, and $CAB''$ on each of its sides. Let the base angles of the isosceles triangles be called $\zeta$ and assume that $AA''$ intersect $BC$ in $A'$, etc. Use area considerations and Ceva's theorem to prove that $AA''$, $BB''$ and $CC''$ are concurrent in a point $O$ with

$$\frac{OA'}{A'A} : \frac{OB'}{B'B} : \frac{OC'}{C'C} = \frac{a}{\sin(A+\zeta)} : \frac{b}{\sin(B+\zeta)} : \frac{c}{\sin(C+\zeta)}.$$

**Exercise 6** (After [7]) Suppose that we have a triangle $ABC$ with concurrent cevians $AA'$, $BB'$, $CC'$ meeting at a point $O$ inside $ABC$. Let $\alpha$, $\beta$, and $\gamma$ be as defined before. Prove that

$$\frac{1}{\alpha} + \frac{1}{\beta} + \frac{1}{\gamma} \geq 9,$$

with equality if and only if $O$ is the centroid of the triangle. Hint: Use Theorem 1 and Lagrange multipliers.

**Exercise 7** Generalize the statement of the previous exercise to higher dimensions.

---

[19]Euler himself appears to have tried to use the case in which $O$ lies inside $ABC$ to derive the case in which $O$ lies outside.

**Exercise 8** (Ceva's Theorem–The necessary part) Suppose we have a triangle $ABC$ with cevians $AA'$, $BB'$, and $CC'$. Use Inner-Outer Lemma B to prove that the equality

$$AB' \cdot BC' \cdot CA' = AC' \cdot CB' \cdot BA'$$

is a necessary condition for concurrency of the three cevians.

**Exercise 9** (Routh's Theorem–Internal) In the case of non-concurrent cevians, with notations as before, show that

$$|A''B''C''| = |\alpha_B + \alpha_C + \beta_C + \beta_A + \gamma_A + \gamma_B - 2| \cdot |ABC|.$$

**Exercise 10** (Routh's Theorem–External) Show that the relation of the previous exercise can be rewritten in external form as

$$\frac{|A''B''C''|}{|ABC|} = \frac{\left(AC' \cdot CB' \cdot BA' - AB' \cdot BC' \cdot CA'\right)^2}{(BC \cdot CA - A'B \cdot B'A)(AC \cdot AB - B'C \cdot C'B)(AB \cdot BC - C'A \cdot A'C)}.$$

Hint: Use Inner-Outer Lemma A and a decent computer algebra package.

**Exercise 11** (See [5]) For any given triangle, its sides can be translated (that is, they can be moved without changing their slopes) so that they form three concurrent cevians of a new triangle. In fact, they are the medians of the new triangle, the area of which is equal to four-thirds of the original triangle. Prove this result using one of the following three approaches: (1) the rectangle construction of this paper; (2) vector methods; (3) Given the original triangle, $ABC$, draw its medians $AA'$, etc., meeting at the centroid $G$. Next extend one of these medians, say $AA'$, to a point $A''$ outside the triangle so that $A'A''$ equals $GA'$. Next draw $A''B$, and extend it past $B$ to a point $B''$ so that $BB''$ equals $A''B$. Then $A''B''C$ is the desired mother triangle, as can be seen by playing around with parallelograms.

**Exercise 12** (Area of a mother triangle) The previous exercise gives the area of the mother triangle to three medians. In his paper, Euler derived a more general formula for the area of a mother triangle $ABC$ to any three concurring cevians: Let the perpendicular triangle $A\bar{B}C$ be defined as in the Solving a Triangle (...) section. Prove that

$$|ABC| = \frac{1}{\alpha\beta\gamma}|A\bar{B}C|,$$

with $\alpha$, $\beta$, and $\gamma$ as in the definition of $A\bar{B}C$. Hint: Compute the area of the mother triangle by determining each of the areas of triangles $AOB$, $BOC$, $COA$ and then comparing each of these with the area of the triangle $A\bar{B}C$. Alternatively, you may use the vector approach explained in the Beyond the Plane section.

**Exercise 13** Any relation between the volume of a mother tetrahedron $ABCD$ and that of a perpendicular tetrahedron $A\bar{B}CD$ is only meaningful in case these volumes can be compared, i.e. in case they are measured in the same units. For the specific perpendicular tetrahedron suggested in the Beyond the Plane section for which this is the case, prove that

$$|ABCD| = \frac{1}{(\alpha\beta\gamma\delta)^2}|A\bar{B}CD|,$$

with $\alpha$, $\beta$, $\gamma$, and $\delta$ as in the definition of $A\bar{B}CD$.

**Exercise 14** Without using Euler's area formula, show that in a triangle $ABC$ for any set of concurrent cevians with lengths $a'$, $b'$, $c'$ meeting in a point $O$ and $\alpha$, $\beta$, and $\gamma$ defined as in the proof of Theorem 1, one has

$$|ABC|^3 = \frac{1}{8\alpha\beta\gamma}\left((1-\alpha)a' \cdot (1-\beta)b' \cdot (1-\gamma)c'\right)^2 \sin p \sin q \sin r,$$

where $p = \angle BOC' = \angle B'OC$, $q = \angle COA' = \angle C'OA$, and $r = \angle AOB' = \angle A'OB$.

**Exercise 15** Suppose that in a triangle $ABC$ three given concurrent cevians $AA'$, $BB'$, $CC'$ (with $A'$ on $BC$ etc.) each have the same length. Using the area formula of the previous exercise, show that the area of the mother triangle is maximal for $\alpha = \beta = \gamma = 1/3$, in which case the cevians are the medians to an equilateral triangle. You may want to optimize seperately for the variables $\alpha$, $\beta$, $\gamma$ and the variables $p$, $q$, $r$ and show that the values that you find for these six variables correspond to the same configuration.

## Acknowledgments

The authors wish to thank James Campbell and the other students of the second author's Advanced Geometry class for inspiring the work that led to this paper.

## References

1. Peter Baptist, *Die Entwicklung der neueren Dreiecksgeometrie*, Wissenschaftsverlag, Mannheim, 1992.

2. Arthur Baragar, *A Survey of Classical and Modern Geometries*, Prentice Hall, Upper Saddle River, 2001.

3. Michel Chasles, *Aperçu Historique sur l'Origine et le Développement des Méthodes en Géométrie*, Gauthier-Villars, Paris, 1887 (= third edition).

4. John L. Heilbron, *Geometry Civilized. History, Culture, and Technique*, Clarendon Press, Oxford, 2000 (= second edition).

5. Larry Hoehn, "Extriangles and Excevians," *Mathematics Magazine*, 74 (2001) 384–388.

6. Norbert Hungerbühler, "Proof Without Words: The Triangle of Medians is Three-Fourths the Area of the Original Triangle", *Mathematics Magazine*, 72 (1999) 142.

7. Geoffrey A. Kandall e.a., "A Centroidal Equality," *Mathematics Magazine*, 75 (2002) 320–1 (= Problem 1630 with solutions).

8. Z.A. Melzak, *Invitation to Geometry*, Wiley, New York, 1983.

9. Alfred S. Posamentier, *Advanced Euclidean Geometry*, Key College Publishing, Emeryville, 2002.

10. Max Simon, *Ueber die Entwicklung der Elementar-Geometrie im XIX. Jahrhundert*, Teubner, Leipzig, 1906 (= *Jahresberichte der Deutschen Mathematiker-Vereinigung. Ergänzungsbände*, B. I).

# 8

# Modern Geometry after the End of Mathematics

**Jeff Johannes**
*State University of New York College at Geneseo*

It appears to me also that the mine [of mathematics] is already very deep and that unless one discovers new veins it will be necessary sooner or later to abandon it. Physics and chemistry now offer the most brilliant riches and easier exploitation; also our century's taste appears to be entirely in this direction and it is not impossible that the chairs of geometry in the Academy will one day become what the chairs of Arabic presently are in the universities.

—Lagrange, September 21, 1781 [5]

## Introduction

In *Mathematical Thought from Ancient to Modern Times*, Morris Kline writes "By the end of the eighteenth century ... the mathematicians began to feel blocked." [5] Several prominent mathematicians expressed concern regarding the future of mathematics. Lagrange fears the end of mathematical evolution, writing that perhaps "it will be necessary sooner or later to abandon it." Kline continues: "Euler and d'Alembert agreed with Lagrange that mathematics had almost exhausted its ideas, and they saw no new great minds on the horizon." As final evidence of his assertions, he includes thoughts from both Diderot and Delambre.

I dare say that in less than a century we shall not have three great geometers left in Europe. This science will very soon come to a standstill where the Bernoullis, Maupertuis, Clairauts, Fon-taines, d'Alemberts and Lagranges will have left it.... We shall not go beyond this point.

—Diderot, 1754 [5]

It would be difficult and rash to analyze the changes which the future offers to the advancement of mathematics; in almost all its branches one is blocked by insurmountable difficulties; perfection of detail seems to be the only thing which remains to be done. All these difficulties appear to announce that the power of our analysis is practically exhausted.

—Delambre (permanent secretary of the mathematics and physics section of the Institut de France), 1810 [5]

At the very least, some mathematicians were not confident that mathematics would develop beyond the eighteenth century. It seems that today's students have a similar perspective on the scope of mathematics. Specifically, they hold that the evolution of mathematics ended with the completion of calculus.

93

As mathematicians today, it is our duty to dispel this myth by giving students at all levels (especially those who will not study more advanced mathematics) a view into the expansive growth of mathematics since the close of the eighteenth century. In particular, geometry and topology were revolutionized during this time. This yielded not only new disciplines and more formal methods of study, but also changes in the philosophical nature of mathematics. Rather than positing mathematics as the absolute truth of the universe, the discipline became focused on the study of potential reality.

The nineteenth century saw many changes in the perception of mathematics. The work of Lobachevsky, Bolyai and Gauss in non-Euclidean geometry led to questions of the meaning of mathematical truth. Monge and Riemann's contributions to differential geometry continued to apply ideas from calculus to new and surprising geometric spaces. Further study by Monge and Poncelet in projective geometry analyzed the way the world is seen. One of the most popular ideas of this time was the geometric realization of higher dimensions as introduced by Cayley and Grassman. Finally, the creation of geometric topology by Poincaré, Möbius and Klein provided an entirely new view on the significance of geometric properties independent of metric.

Each of these areas demonstrates the vitality of mathematics in the past two centuries and leads to major research efforts in modern mathematics. Sharing these topics with students gives them hope for the future of mathematics and helps them to appreciate the view that mathematics is unbounded in its possibilities. Without focusing on details as much as some mathematicians believe is necessary, I give the students in my classes a picture of these and other modern developments in mathematics.[1] Hereafter I connect nineteenth and twentieth century advances in geometry and topology to the mathematical content of undergraduate courses. This paper also discusses my comments upon these advances in my classes. (For more information on the history or mathematics of any of these topics, the reader is recommended to consult [4] or [5].)

## Non-Euclidean Geometry

The insights gained from the exploration of non-Euclidean geometry not only greatly expanded the objects of study within geometry; they were a major step in changing the very nature of mathematics. Gauss was uneasy about publicizing his work in non-Euclidean geometry because he was concerned that society was not yet prepared for the idea that mathematics does not necessarily represent *true reality*, but rather only *possibility*. Uneasiness with these consequences of work of Lobachevsky, Bolyai, Beltrami and Gauss led to a backlash against intuitive reasoning. Subsequent attempts to place a rigorous foundation beneath mathematics led to the successes of Russell and Whitehead, but also to Gödel's proof of the doomed nature of the project from the beginning. This leads, as stated above, to the thought that mathematics presents possibility, not absolute truth. At best, mathematics provides truth about relationships among abstract concepts, not physical entities.

Similarly, my students are anxious about this new vision of mathematics. While I was teaching a senior geometry course, a student revisited this historical unrest, saying to me that she found hyperbolic geometry offensive because it is not the Euclidean geometry given by God. In each of my classes, particularly in the senior geometry course, but also in the geometry course for preservice elementary teachers, I emphasize that Euclidean geometry is only one option, one possible geometry. I stress that as we live on a sphere, there is another geometry relevant on a global scale that is clearly not Euclidean. In the larger scope, there is the open astronomical question of the shape of space: whether we live in a hyperbolic, spherical, or Euclidean universe.

---

[1]Curiously, this tension within the mathematical community itself developed from concerns around 1900 about the need for rigor in mathematics.

Even in my proofs course, I emphasize the danger of moving completely toward formalism, of which Russell and Whitehead's work was the high point. Now, their work stands as a relic in light of Gödel's famous proof on incompleteness: any logical system complicated enough to contain exponentiation on the natural numbers must necessarily have either unprovable statements or contradictions. Rather, I encourage my students to struggle with the difficulties that they have with their intuition and to develop it into a reliable guide for the possible worlds of mathematics. Furthermore, I emphasize that mathematical results depend more upon human assumptions than on the nature of the world. Mathematics is a human creation much more than the sciences that describe the world.

## Differential Geometry

In Calculus I students focus on the geometric information of graphs of single variable functions revealed by the calculus. In Calculus II there is some consideration of plane curves given by equations in two variables and plane curves given by parametric equations. These ideas are the beginnings of differential geometry and were extended by Clairaut and Lambert to consider the geometric properties of parametric space curves. In Calculus III students begin to consider geometric properties of surfaces, examining not only maxima and minima, but saddle points and both cylindrical and flat sections with zero curvature. As elaborated by Gauss and Riemann, these are elementary topics in differential geometry of surfaces.

Also, Calculus III is an important time to mention that there are constraints on how maxima, minima and saddle points balance. These are the beginnings of Morse theory and lead to the Gauss–Bonnet theorem, which goes beyond vague ideas of curvature and Euler characteristic to address angle sum of triangles. Calculus is an ideal place to point out that detailed graphing of functions is not the end of a graphing program that they began in middle school, but rather the beginning of graphing more intricate curves and surfaces than they previously had imagined.

## Projective Geometry

The world that we see is very different from the idealized three-dimensional world that we imagine. In the visual world, objects pass over one another and so-called parallel lines appear to converge. For the most part, vision is two-dimensional with limited depth perception. This is one reason that the two-dimensional entertainment of films and television seems so realistic. When we see a cubic box, we do not see all six faces simultaneously, and certainly we do not see the inside of the box. These same ideas are at the heart of projective geometry as expanded upon by such nineteenth century mathematicians as Monge, Poncelet, Plücker, and Chasles. Since we rely so heavily upon vision to maneuver in the world, it seems that projective geometry is more descriptive of worldly experiences than is Euclidean geometry.

I mention these ideas in the elementary geometry course for preservice teachers, where they are explored in terms of perspective not only for painting and drawing, but also for photography. In the senior geometry course, we explore these ideas in more detail, considering that all "parallel" lines meet at a line at infinity.

## Geometric Topology

The field of geometric topology was first proposed in the late nineteenth century in the seminal work of Poincaré and others. While many results in this field are abstract and sophisticated, there are facets that can be understood by students of all levels. The curious properties of orientation that are evident in the Möbius strip can be seen even by elementary school students. This generalizes to the peculiarities of the surface of a Klein bottle, which is a surface without holes that has no well-defined interior or exterior.

Also, my topology students learn an elementary classification of orientable surfaces that merely involves cutting and gluing polygons, which gives some insight into the significant position that the torus holds as the first non-spherical surface.

Another topic in geometric topology that is broadly accessible is the study of knots and links. This exploration began after Lord Kelvin proposed a theory of classifying chemical bonds in terms of knotting between the atoms. At that time Tait and Little began a table of knots, which was extended in the late twentieth century to include over a million knots. Manipulating string and diagrams is a tangible activity that makes this topic particularly appealing to students. Knot theory also features many simply stated questions that are surprisingly difficult to answer. The most important unsolved question is simply: how does one tell if a piece of string is knotted? [2]

I mention these studies to my preservice elementary school teachers, emphasizing the wide variety of topics within mathematics. Many students are quite surprised that mathematics extends to such diverse realms.

## Higher Dimensions

The contemplation of higher dimensions is perhaps the most broadly disseminated new (in the past two centuries) idea of mathematics. It is one of the primary themes of science fiction, and most people in Western culture have had some exposure to (and some confusion with) the fourth dimension. Abbott and Hinton set the stage in the late nineteenth century for a mystical view of higher dimensions as representing a deeper reality. [1, 3] Mixing this with modern astrophysics that contemplates a ten- or twenty-four-dimensional universe, it becomes clear how it has become a science fiction staple. As there is so much public curiosity about this topic, I am eager to mention it to anyone who will listen.

Students are always interested to hear about the fourth dimension. Usually they have been exposed to an idea of it as time, or some idea that the fourth dimension is a mystical arena. It is rewarding to explain to them that the fourth dimension is no different than any of the first three, but that there are things that can happen with four dimensions that are not possible with three, just as there are possibilities with three dimensions that are not there with two.

The fourth dimension arises when modeling numbers in my elementary education classes. We use a new dimension to model each of the place values: the units place is modeled as zero-dimensional, the tens place as one-dimensional, the hundreds place as two dimensional, and the thousands place as three-dimensional. There are higher places than the thousands place, so the ten-thousands place is modeled as four-dimensional and the hundred-thousands place as five-dimensional, with no end to the number of dimensions that we may consider.

Each of the aspects of geometry mentioned in this chapter opens up new possibilities as we move into higher dimensions. The models of hyperbolic spaces are at times difficult to work with because they run out of room in our limited three dimensions. The most intriguing surfaces that are discussed in differential geometry are those that do not fit in three dimensions. Furthermore, the Klein bottle of topology does not fit in three dimensions without intersecting itself, but easily can be placed into four dimensions without this concern. Many topological theorems are more coherent when viewed as special cases of higher dimensional results, much in the same way that properties of functions of the real numbers are more consistent when viewed as a subset of the higher dimensional complex numbers. Contrarily, knotting of string vanishes as we move to four dimensions. In this case there is so much extra space that the string can pass "over" itself, thus unknotting. However, not all knots are eliminated in higher dimensions; there are knotted two-dimensional spheres in four dimensions, something that is completely foreign to our three-dimensional perspective.

## Conclusion

Most of the aforementioned topics are not central subject areas in my courses. I do not teach a course on projective or differential geometry. I do not evaluate students on their understanding of these comments, but I make them nonetheless in order to give students the view that there is always more to mathematics. It is our responsibility as active mathematicians to show students that mathematics did not end in 1800 and that, in fact, much has been and continues to be done.

## References

1. E. A. Abbott, *Flatland: A Romance of Many Dimensions,* Dover, New York, 1952.

2. C. Adams, *The Knot Book: An Elementary Introduction to the Mathematical Theory of Knots,* W. H. Freeman, New York, 1994.

3. C. H. Hinton, *Speculations on the Fourth Dimension,* Dover, New York, 1980.

4. V. Katz, *A History of Mathematics: An Introduction,* Addison-Wesley, New York, 1998.

5. M. Kline, *Mathematical Thought from Ancient to Modern Times,* Oxford University Press, New York, 1972.

# III

# Discrete Mathematics, Computer Science, Numerical Methods, Logic, and Statistics

*The purpose of computing is insight, not numbers.*
—Richard W. Hamming

# 9

# Using 20th Century History in a Combinatorics and Graph Theory Class

**Linda E. McGuire**
*Muhlenberg College*

## Introduction

A few years ago, during the first week of the fall semester, I assigned a writing project in an upper-division undergraduate mathematics course. The class was charged with pondering the often-asked question "Was mathematics discovered or invented?" and then crafting a three to four page essay establishing and supporting their personal position. Students were instructed to cite specific and appropriate historical happenings in mathematics to argue their points.

I anticipated that the assignment would serve as an ice-breaker for the class as a whole and that I would gain some early insight into each student's expository writing ability. While these goals were (to some extent) achieved, I received an unexpected revelation during the course of reading these assignments.

Almost without exception, students aligned themselves firmly with the camp whose opinion is that mathematics is a man-made invention. It was after making this assertion that the quality of most essays deteriorated, and it became clear why almost immediately. These genuinely skilled junior and senior mathematics majors had no working knowledge of mathematical history and, as a result, possessed no sense of a historical context in which to search for evidence to strengthen their cases. Most bandied about names like Euclid, Euler, Newton, Cauchy, and Riemann, but virtually no one conveyed the idea that they truly understood how the work of these mathematicians fit into the mosaic of mathematics. Nor did they give any indication that individuals other than the likes of the aforementioned giants practiced the art of mathematics at all.

I was overwhelmed by the feeling that we, as an institution, had failed these students. They were bright, hardworking, and inquisitive young people. There was no other reason for this massive gap in their mathematical foundation except for the obvious one. Either we did not stimulate within them an interest in the historical setting in which all of the mathematics they had been studying took place or we did not encourage that curiosity if it did arise. Their classes, for the most part, had focused exclusively on content and only superficially on context.

This was a painful revelation, and one that I resolved to work hard to avoid encountering again. As I began reading papers devoted to introducing the history of mathematics into one's pedagogy, I was heartened to find that many others in the mathematics community were hard at work trying to address this very problem. The opinion expressed by V. Frederick Rickey in his article *The Necessity of History in Teaching Mathematics* [12] closely mirrored my own feelings:

It is not sufficient that we simply present the mathematical details. Our responsibility is much greater. We would be derelict in our responsibilities as teachers if we presented mathematics as a fully developed discipline, a discipline that seemingly appeared millennia ago in perfect form.

Teaching by means of utilizing historical sources has become a more prevalent instructional technique in mathematics classes over the last several years. I teach a substantial number of calculus and analysis-based courses and the opportunities for adopting a rich, historically based pedagogy in such classes are ample. The passage of time has allowed a library of well-written, accessible history texts and papers to be produced. From ancient to pre-modern times; be it Greek, Hindu, Islamic, Chinese, European, or early American mathematics, an instructor can find any number of excellent historical sources to employ in the classroom.

However as an applied mathematician by training and interest, I spend much of my time teaching courses in discrete mathematics, combinatorics, graph theory, network analysis, modeling, numerical analysis and the like. A historically-driven pedagogical approach can be quite challenging when one focuses on the more modern branches of mathematics mentioned above, where the "newness" of the discipline renders traditionally consulted sources scarce or non-existent. Yet with the interest in applied mathematics that students of mathematics, computer science, economics, and the physical sciences demonstrate, designing activities through which students can gain a historical perspective of such applied topics should become a high priority.

This treatise will describe and analyze a semester-long, historically-based research project used in an upper division undergraduate course in combinatorics and graph theory. The assignment was constructed to provide all students in the class with historical background pertaining to these fields, the origins of which are primarily in the twentieth century. This description will include discussions of the nature of the project, the quality of the research produced by the students, the innovative research techniques they employed, the amount of intricate mathematical reading and writing that the task demanded, and the motivating element that this assignment proved to be for the entire class. Issues involving assessment, time constraints, material coverage, and writing difficulties will also be addressed.

## Motivating Factors

The course in question was billed as a special topics course in combinatorics and graph theory. One of the nicest aspects of offering this course in the liberal arts setting that Muhlenberg College provides is that the class draws a varied audience. Students come to this course from major fields of study such as mathematics, computer sciences, economics, physics, chemistry, history, and philosophy. This diversity does not come at the expense of a certain level of mathematical sophistication as each student had successfully completed a prerequisite course in proof technique and logical reasoning.

Early in the planning stages of this course it was decided that students would be required to complete an extensive, semester-long research project. Several motivating factors influenced the design of that project and, ultimately, the entire course. Chief among them was the desire on my part to increase the historical content of the course and to test the suspicion that such a project would appeal to the mixed audience that this class enrolls. Other issues strengthened my resolve to require this research assignment.

Contrary to published reports proclaiming that our society is becoming increasingly ahistoric, most students when asked indicated that they wanted to see more historical information woven throughout their mathematical course work. They bemoaned the fact that they had no sense of history where mathematics was concerned. Most knew the names of the major players like Newton, Gauss and von Neumann, yet these students had no sense of the context in which these men lived and worked. One of our strongest majors summed it up best when she said that she always felt as though she had "arrived at a play halfway through Act I; by the time you figure out who is who and who did what, you are almost through Act III."

As is the case in many colleges and universities throughout the United States, Muhlenberg College is dedicated to encouraging and emphasizing the "Writing Across the Curriculum" ethos. Integrating an historically themed research project seemed like a natural way to introduce a relevant writing assignment. Such an exercise had the additional appeal of requiring not only solid English exposition, but also necessitating clear and concise mathematical descriptions and proofs of pertinent results.

Another key element in the decision to create this component of the course was the nature of the subject matter itself. The modern, visual, and hands-on nature of subjects like combinatorics, graph theory and computer science would stimulate students' interest in twentieth century mathematics. Also, from the point of view of both student and professor, focusing on mathematics from the last century appeared to be slightly less daunting than opening the flood gates and allowing topics to be taken from the entire spectrum of mathematics.

## Course Structure

Before describing the project and the results it produced in detail, some general information about the structure of the course is in order. The set physical parameters of the course proved to be ideally conducive to discussing mathematical history. The class had two, seventy-five minute meetings per week in a classroom that possesses excellent audio-visual equipment including fast-reacting computer systems, video capabilities, and well maintained overhead projection units.

The course itself would focus on problem solving within a framework that emphasizes the primary themes of any combinatorics class: examining problems of existence, enumeration and optimization. Major topics included counting techniques, game theory, recurrence relations, generating functions, basic graph theory, matchings, planarity, traversal theory, chromatic numbers, and computer science applications. With these general topics on the agenda, I was confident that the mathematical content of the course was such that students would be able to find some topic they wished to explore in depth.

## Project Description

The research component of the course was briefly described on the syllabus. On the second day of the term students were presented with an outline of what the project would entail.

Each student would research the life and work of a significant figure in (predominantly) applied mathematics with both an oral presentation and a term paper being the major elements of their required work. At the end of the semester we would assemble a portfolio containing the work that the entire class had produced. The names of possible subjects, partially replicated below, was distributed.

| | | |
|---|---|---|
| Fan Chung Graham | Frank Harary | Edouard Lucas |
| Ingrid Daubechies | D. A. Huffman | John McCarthy |
| Edsger Dijkstra | Richard M. Karp | John Nash |
| G. A. Dirac | Andrei Kolmogorov | John von Neumann |
| Paul Erdős | Donald Knuth | Frank Ramsey |
| Ronald Graham | Joseph Kruskal | Bertrand Russell |
| Richard Hamming | Kasimir Kuratowski | Alan Turing |

While these names were mentioned to offer students ideas, they were encouraged to put forward their own suggestions if they felt strongly about wanting to examine the work of an appropriate person not included on this list.

For example, one young woman came to see me to ask if she might do her project on René Descartes. When pressed for a rationale she said that since he appeared to have solved many mathematics problems primarily to apply those results to questions in physics, music, meteorology, and optics, he had an "applied math" perspective. Also, he lived at a time in history that was of particular interest to her. As a philosophy minor she was already familiar with a variety of Descartes' writings. In particular she expressed a curiosity about his essay on the hydrostatic paradox. This paradox, stated as "the pressure exerted by a fluid does not depend on the area at the bottom of the container but only on the height of the liquid" [13], was where she wanted to begin her reading.

Her interest was genuine and her informed argument was well thought out, so I consented to grant her request under two conditions. First, given the historical parameters of Descartes' lifetime, that she give the first oral presentation of the term. Second, that she make a concerted effort to ascertain and establish how Descartes influenced twentieth century mathematics.

Students were then given a week and a half to scout around, read about each person and formulate a list of their top three choices. By the end of the second week of class, assignments were made allowing for no duplicates.

The selections that the students submitted were such that it became clear that their oral presentations could fulfill an important pedagogical purpose. I scheduled the oral presentations so that each would serve as either an introduction to the next section of course material or as a special topic that somehow related to current discussions. For example, students who were studying about Richard Hamming and D.A. Huffman would give their talks immediately preceding the sections on coding. Presentations on Alan Turing, Fan Chung Graham, and Joseph Kruskal would be sprinkled throughout the sections on graph theory. Those topics that did not quite fit any category of course material, like reports on Ingrid Daubechies and wavelet theory, would be inserted into the schedule between major units.

The class was informed that each of their presentations should be 25 to 30 minutes in length. The presentation should be designed in such a way that they introduce the individual in question to the rest of the class by means of a biographical sketch. Then they should attempt to address, to the degree possible for each subject, the following aspects of each person's work:

1. the context, both mathematical and physical, in which their work takes/took place;
2. the implications and significance of their work;
3. how their work continues to influence other researchers;
4. when applicable, how the work of their subject tied into our course topics;
5. complete mathematical descriptions of problems whose solutions their researcher is responsible for solving;
6. a description of at least one open problem whose origins can be traced to the subject under investigation.

This "wish list" of information gave rise to a variety of intriguing problems that the students encountered in their research process. These issues will be discussed shortly.

The 20 to 25 page term paper associated with each project would not be due, in finalized form, until the end of the semester. However two drafts were required at the one-third and two-third marks of the term. This requirement ensured that no one left the project to the last minute.

## Assessment

In that combinatorics contains a great deal of rich mathematical content without such a research component, the question of appropriate assessment had to be considered early in the course planning process. Given the often innovative reasoning and intricate problem-solving techniques that combinatorics requires, one could easily restrict all evaluative measures to exams and problem sets. So how should the project be woven into an assessment paradigm? Clearly it had to carry significant weight in that the mathematical and expository scope of the paper was extensive, and the presentation was to be an intrinsic part of the course material. Exams would include questions based upon these presentations. On the other hand, combinatorics and graph theory are all about problem-solving, so the exercise sets and exams were also crucial evaluative devices. After much deliberation and revision, the following grading scheme was employed:

| two in-class exams | 15% each |
|---|---|
| final exam | 20% |
| weekly problem sets | 25% |
| term paper | 15% |
| oral presentation | 10% |

This scheme proved to yield a balanced, fair assessment. No crises of conscience on my part arose. Of note is how much of the grade was determined by the research project. Post-semester analysis showed that, in most cases, the quality of the project closely matched the level of the student's mathematical work in other areas of the class. In a few instances students who struggled with the problem-solving content of the course did benefit once the project grade was included. However this did not inflate their grades in an inappropriate or exaggerated way. The reverse scenario also materialized. There were students who demonstrated solid mathematical skills but were not as proficient at (in particular) the writing component of the research project. It appears that the project served to reward the student who had cultivated not only their mathematical abilities, but also their abilities in the fields of oral and written communication and information analysis. On the whole the projects allowed students to display and utilize all of the elements of their education obtained within a liberal arts environment.

## Pros and Cons

As is the case when implementing any new approach in the classroom, this term-long research experience had both positive and negative consequences. Since, in my opinion, the pros far outweigh the cons, we will consider the bad news first.

Combinatorics and graph theory are packed with excellent mathematical material. Giving over class time to the oral presentations did reduce the amount of material we covered in class. The result was that students had to do careful preliminary reading before class and also had to grapple with some difficult material on their own or in groups outside of class. While this was initially a cause for concern, it proved to be an important lesson in self-reliance for these upperclassmen. For the most part, students performed admirably. From an instructional point of view, only two or three smaller topics had to be jettisoned due to time constraints.

Despite my best efforts, there was often an end-of-term pile up of presentations. This can be avoided by carefully scheduling the special topics presentations throughout the term, even if they are an apparent digression from current discussions.

Significant research issues arose. For many students, commonly consulted sources and standard investigatory techniques were of little use since the "history" they were examining was so recent. They frequently came to my office complaining that they could not find sufficient information. While our excellent science and history librarians proved to be invaluable resources at this juncture, the problem of inadequate or nonexistent source material dogged many students throughout this experience. A suggestion of mine that proved to be useful was that students try to determine if their subject had any co-authors with whom they frequently worked. Through analyzing available information about the co-author, students often uncovered important facts and research leads regarding their project. To their credit, diligent students unearthed a wealth of interesting material through clever methods of their own. Many of these approaches will be discussed later.

In an effort to honestly acknowledge the time crunch that all academics face, I must confess that the work involving this project took up a significant amount of my time. Reasons for this included my own need to read up on project topics, scheduling extra office hours and consultation times to help students interpret some of the weighty mathematics they encountered, reading and commenting on drafts, dealing with plagiarism issues both large and small, and assembling and editing the final portfolio. I did not have many spare moments that term. I feel that many refinements and alterations to my approach to the course would reduce this extra workload substantially. For example, I would reduce the number of required drafts due and, enrollment numbers permitting, allow students to work in pairs on the project.

Negative student comments on the course were of the "this class required too much work" variety. While this was, to some degree, true this is a student comment that is difficult to interpret. It was a common but by no means universal complaint. Frankly, had at least a handful of such comments *not* arisen I would be likely to conclude that I did not challenge the class adequately.

While the negative elements described above were in evidence, they were not deterrents to the prospect of utilizing this instructional device again. On the contrary, those obstacles were minor in the overall scheme of things and are now merely viewed as pitfalls with regard to weaving the project into the academic calendar that need to be anticipated and avoided. The far-reaching rewards that this project granted to everyone in the class were easily identifiable.

Students were motivated, involved and, by necessity, well-prepared for each class meeting. Having worked with several of these students in other courses, it became clear that their writing skills had improved dramatically. While many struggled with designing the "script" for their presentations, most overcame their difficulties with little intervention on my part. In particular, it was evident that they were taking great care to understand and carefully explain the often complex mathematics they were analyzing.

The topic choices that students made generated several interesting match-ups. Some did focus on subjects in their own areas of interest . . . a physics major researched Ingrid Daubechies and wavelets and a computer science major studied the life and work of Donald Knuth. For each example like these, one could find a companion case of a student studying someone seemingly well outside of their fields of interest. There were students one would categorize as pure mathematics majors looking at figures like Hamming, Turing, and Dijsktra, a philosophy major examining the graph theory work of Paul Erdős, a chemistry major looking into the logical inference work of Bertrand Russell, and an economics major studying John McCarthy and the beginnings of artificial intelligence. These pairings led to a lot of extracurricular communication amongst the students. They helped get each other up to speed on topics in their own majors and many worked together to analyze and interpret the mathematics being considered. This contributed mightily to the overwhelmingly positive classroom atmosphere. Students were more supportive of, instead of competitive with, one another. Almost everyone felt as though they had something relevant to bring to class discussions.

It was also the case that one presentation topic often led beautifully into the next. A talk on John von Neumann and self-replicating automata was immediately followed by a presentation on John McCarthy. The introduction to the life and work of Frank Ramsey preceded the presentation about Paul Erdős. Students got a sense of how these researchers influenced and inspired each other, as well as future scholars.

The research resource difficulties many students encountered were overcome in some innovative, bold and clever ways. In that many of these researchers are still very much alive, students (on their own) attempted to contact the people whose work they were investigating. One student included a transcript of a conversation/interview with Joseph Kruskal as part of her report. She inquired about his young life, inspirations, and work philosophy and he was quite gracious in his willingness to respond. Another student corresponded with Donald Knuth and included the letters in both his presentation and paper. Others found relevant vitae on the web and established contact by using information found there.

While these "plusses" are worthy of mention in and of themselves, the mathematics education that students received from this project is the primary positive result that emerged after completing this assignment. Students had to read not only historical sources, but mathematically dense articles as well. This was a new and difficult experience for most of them. However, the nature and quality of the conversations I had with these students in my office as we tried to analyze the mathematics at hand were the kind of moments all academics hope to have with their students. They were given the chance to see good open problems, to witness how original proofs were improved upon by subsequent mathematicians and, perhaps most significantly, they were made privy to the problem solving techniques that distinguish mathematicians from other types of researchers.

Students were excited by the prospect of being "historian for a day," introducing their classmates to new information. In general, students walked away from this class with the knowledge that mathematics is something that is actively occurring, instead of feeling as though it is something that has occurred.

## Conclusion

How successful was this project from the student's perspective? Students were asked to comment extensively on the project at the end of the semester and the resulting course evaluations contained comments like:

> When I started I thought it was unfair to assign a paper on topics that had few references in book form. By the end I felt like a cross between a private investigator and a stalker. I was determined to find out about this guy.

> I'll admit that "Oh God, not another paper...." was my first reaction to the research assignment. Looking back I really think that the project was the most complete learning experience I had this year.

While comments such as the last one are gratifying, I believe that the "complete experience" alluded to can be attributed to a single pedagogical device. The historical presentations served to provide the framework upon which the related mathematics the students examined was constructed. With varying degrees of success, students obtained some context in which to place the theories being discussed. I am convinced that this component of the course brought about the sense of "completeness" to which many class members, in one form or another, referred. I can say in all honesty that this class was one of the most rewarding experiences of my teaching career to date.

The first implementation of a historically-inspired pedagogy presents the instructor with numerous new challenges. While many of these revolve around issues of front-end organization, time management and research strategies, none of these elements are so crippling that they should negatively influence one's decision to assign such a project. The overall results are too good, the class experience too valuable to reject the use of such an approach. An appropriately modified model of this course will require a different type of work on an instructors part, not necessarily more work.

I have assigned this type of project on two occasions subsequent to the one described here. In both instances the quality of each students' research and presentation experience was excellent. However, in these cases I put considerable effort into devising a schedule to balance material coverage and presentation time well before the semester began. Such information was included in course syllabi to better inform students as to when they could expect to give their presentations. This brought better structure to the entire course and the oral reports fit more naturally into the flow of the course. The end-of-term rush of presentations was virtually eliminated.

Knowing what I do now about using such an assignment as a classroom construct, I would not hesitate for a moment to continue to employ it in future classroom settings. I would encourage any of my colleagues to do the same.

## References

1. Eric T. Bell, *Men of Mathematics*, Two Volumes, Penguin Publishing, London, 1953.

2. John Conway, *On Numbers and Games*, Academic Press, London, 1976.

3. William Dunham, *Journey Through Genius: The Great Theorems of Mathematics*, John Wiley and Sons, New York, 1990.

4. ——, *The Mathematical Universe*, John Wiley and Sons, New York, 1994.

5. Jekuthiel Ginsburg, "Rabbi Ebn Ezra on Permutations and Combinations" in *The Mathematics Teacher* **15**, 1922, 347–356.

6. Robert L. Hayes, "History—A Way Back to Mathematics" in *History and Pedagogy of Mathematics* Newsletter, **22**, 1991, 10–12.

7. Torkil Heiede, "Why Teach History of Mathematics?" in *The Mathematical Gazette* **76**, 1992, 151–157.

8. Victor J. Katz, "Combinatorics and Induction in Medieval Hebrew and Islamic Mathematics" in *MAA Notes 40: Vita Mathematica*, 1996, 99–106.

9. Israel Kleiner, "A History-of-Mathematics Course for Teachers, Based on Great Quotations" in *MAA Notes 40: Vita Mathematica*, 1996, 261–268.

10. ——, "Famous Problems in Mathematics: An Outline of a Course" in *For the Learning of Mathematics*, **6**(1), 1986, 31–38.

11. Reinhard Laubenbacher and David Pengelley, "Great Problems of Mathematics: A Course Based on Original Sources" in *American Mathematical Monthly*, **99**, 1992, 313–317.

12. V. Frederick Rickey, "The Necessity of History in Teaching Mathematics" in *MAA Notes 40: Vita Mathematica*, 1996, 251–256.

13. William R. Shea, "The Magic of Numbers and Motion: The Scientific Career of Rene Descartes" in *Science History Publications*, 1991, 28.

14. John Stillwell, *Mathematics and Its History*, Springer-Verlag, New York, 1989.

# 10

# Public Key Cryptography

**Shai Simonson**
*Stonehill College*

## Introduction

When teaching mathematics to computer science students, it is natural to emphasize constructive proofs, algorithms, and experimentation. Most computer science students do not have the experience with abstraction nor the appreciation of it that mathematics students do. They do, on the other hand, think constructively and algorithmically. Moreover, they have the programming tools to experiment with their algorithmic intuitions.

Public-key cryptographic methods are a part of every computer scientist's education. In public-key cryptography, also called trapdoor or one-way cryptography, the encoding scheme is public, yet the decoding scheme remains secret. This allows the secure transmission of information over the internet, which is necessary for e-commerce. Although the mathematics is abstract, the methods are constructive and lend themselves to understanding through programming.

The mathematics behind public-key cryptography follows a journey through number theory that starts with Euclid, then Fermat, and continues into the late 20th century with the work of computer scientists and mathematicians. Public-key cryptography serves as a striking example of the unexpected practical applicability of even the purest and most abstract of mathematical subjects.

We describe the history and mathematics of cryptography in sufficient detail so the material can be readily used in the classroom. The mathematics may be review for a professional, but it is meant as an outline of how it might be presented to the students. "In the Classroom" notes are interspersed throughout in order to highlight exactly what we personally have tried in the classroom, and how well it worked. The use of history is primarily for context and for interest. There are great stories to tell, and the results are better appreciated in context. Plenty of references are given for those who would like to extend our work and design their own labs.

The material presented here was taught in a *learning-community* course at Stonehill College [5]. The course is a three-way collaborative of three courses: Discrete Mathematics, Data Structures, and Mathematical Experiments in Computer Science. Students register for all three courses. The first two courses are standard lecture and discussion. The third course is a *closed* lab: a scheduled time when students work on specially prepared laboratory assignments and interact with the computer in order to discover solutions and principles. The course has five 3-week units, one of which is Cryptography. In each unit: the first week is spent analyzing, proving theorems, and writing programs; the second week is spent using the programs and theorems of the first week for experimenting and exploring; and the third week is used for "enrichment", which usually means videos, stories, and related material of a lighter nature. The course is interactive, with only short impromptu lectures during which we can react on the spot to the student's questions and experiments. The labs appear at the Stonehill College computer science department's website [24].

## A Motivating Puzzle

In the 1995 movie *Die Hard: With a Vengeance* (*aka Die Hard III*), Bruce Willis and Samuel L. Jackson play a bomber's deadly game as they race around New York trying to prevent explosions. In order to stop one explosion, they need to solve the following puzzle:

> Provided with an unlimited water supply, a 5-gallon jug, and a 3-gallon jug, measure out precisely 4 gallons, by filling and emptying the jugs. (See photo in Figure 1).

© 1995 Twentieth Century Fox

Figure 1. Bruce Willis and Samuel L. Jackson solving a mathematical puzzle.

This puzzle is useful for studying public-key cryptography, because the solution embodies the two major related number-theoretic results: Euclid's algorithm and Fermat's Little Theorem. Of course this puzzle was not invented for this movie. Most references attribute the puzzle and its variations to Tartaglia [29], but the earliest known version of the problem occurs in the *Annales Stadenses* compiled by Abbot Albert of the convent of the Blessed Virgin Mary in Stade. Stade is a small city on the west side of the Elbe estuary a bit downriver from Hamburg. The date of compilation is uncertain, but seems to be 1240 [27]. The puzzle is also discussed in detail at the wonderful educational mathematics site created by Alex Bogolmony [4]. Bogolmony retells the story of how the famous French mathematician Simeon Poisson (1781-1840) was pressured against his will by his father to study medicine. Poisson stumbled upon the puzzle, solved it, and decided to abandon medicine in favor of mathematics.

In the Classroom: The Poisson story allows us to digress and discuss what kinds of mathematics are inspiring to what kinds of people. We wonder whether there are any budding mathematicians who are inspired by viewing *Die Hard III*.

## A Solution to the Puzzle

When solving this puzzle, it doesn't take the students long to realize that the only thing worth doing is to repeatedly pour one container into the other, emptying the second container when it gets filled. For example, to solve the case presented in the movie, where 3-gallon and 5-gallon jugs must be used to measure out four gallons, we fill the 3-gallon jug, pour it into the 5-gallon jug, fill 3 again, pour it into 5, empty 5, finish pouring 3 into 5, yielding 1 gallon in the 5-gallon jug, fill 3 once again, and pour into 5, yielding four gallons. In this way, we could cycle through the following values in the 5-gallon jug: 3, 1, 4, 2, and 0. One could also pour in the opposite direction. That is, from the 5-gallon into the 3-gallon jug, giving the cycle of values: 2, 4, 1, 3, and 0. The students discover that every value modulo 5 is obtainable and that the two cycles are the reverse of each other.

> In the Classroom: We screen the scene in the movie and compare it to the solution the class discovered on its own. Afterwards, we try to solve the problem in its general form given jugs of sizes $a$ and $b$. We do this with an informal class discussion, allowing the class to follow dead ends up to a point, and eventually coming to some conclusions.

## Using Programs to Experiment and Understand

Gerry Sussman, one of the authors of the award winning text *Structure and Interpretation of Computer Programs,* once explained [28] that one grasps a mathematical idea if and only if one is able to write a program to illustrate it.

> In the Classroom: In our closed labs, the students write a program to generate possible values for the quantities in the jugs. The programming is used to reveal and reinforce an understanding of this puzzle. It is an exercise that supports Sussman's hypothesis.

When the students write a program to generate the cycles, it becomes clear that the sequence of values obtained by repeatedly pouring 3 into 5, is simply the successive multiples of 3 modulo 5:

$$1 \times 3 \quad \mod 5 = 3$$
$$2 \times 3 \quad \mod 5 = 1$$
$$3 \times 3 \quad \mod 5 = 4$$
$$4 \times 3 \quad \mod 5 = 2$$
$$5 \times 3 \quad \mod 5 = 0$$

The students notice that if $x$ appears in this cycle then $x = 3u - 5v$, for some $u$ and $v$. For example, when measuring four gallons with 3-gallon and 5-gallon jugs, we calculate: $4 = 3 \times 3 - 5 \times 1$.

The students look at another example. If the jugs held 17 and 7 gallons then the cycle would be: 7, 14, 4, 11, 1, 8, 15, 5, 12, 2, 9, 16, 6, 13, 3, 10, 0. Because 1 is the fifth number in the list, $7 \times 5 = 1$ mod 17, and $1 = 7 \times 5 - 17 \times 2$. Naturally then:

$$2 = 7 \times 10 - 17 \times 4$$
$$3 = 7 \times 15 - 17 \times 6$$
$$4 = 7 \times 20 - 17 \times 8$$
$$\vdots$$

Therefore, after they can calculate 1, they can also calculate the rest of the numbers by using multiples of the values used for 1. Looking at the cycle, the numbers 1 through 16 can be counted in order, by starting with 1 and moving five to the right each time, cycling around when you hit the end of the list. In general, given two jugs of sizes $a$ and $b$, $a > b$, the values that can be obtained are the values between 0 and $a$ that are multiples of the smallest positive value that can be obtained.

What is this smallest number that can be obtained from pouring water back and forth between jugs of sizes $a$ and $b$? In other words, given $a$ and $b$, $a > b > 0$, how can we find the smallest positive integer $x$ such that $au + bv = x$?, where $u$ and $v$ are integers.

Euclid's algorithm gives the answer: $x = \gcd(a, b)$, the greatest common divisor of $a$ and $b$. Hence, a complete solution to the two jug puzzle for sizes $a$ and $b$, $a > b$, is that we can measure all multiples of the greatest common divisor of $a$ and $b$. In particular, when $a$ and $b$ are relatively prime, all values less than or equal to $a$ can be obtained.

## Euclid's Algorithm

The students' first stop in mathematical history is Euclid. Euclid's method provides a fast recursive algorithm, given $a$ and $b$, to calculate $u$ and $v$, such that $au + bv = \gcd(a, b)$. Euclid's algorithm is described in The Elements, Book VII, Proposition 2 [13].

*To find the greatest common measure of two given numbers... Let AB and CD be the two given numbers not relatively prime. It is required to find the greatest common measure of AB and CD. If now CD measures AB, since it also measures itself, then CD is a common measure of CD and AB. And it is manifest that it is also the greatest, for no greater number than CD measures CD. But, if CD does not measure AB, then, when the less of the numbers AB and CD being continually subtracted from the greater, some number is left which measures the one before it...*

> In the Classroom: We take this opportunity to briefly discuss Euclid's Elements and its place in the history of mathematics. We also spend some time discussing Euclid's style of writing and in particular the absence of any algebra or modern notation. Eventually, we interpret the paragraph above well enough to deduce the famous algorithm, that given $a > b$, $\gcd(a, b)$ can be computed recursively as follows: **$\gcd(a, b)$: if $b$ divides $a$ then return $(b)$ else return $(\gcd(b, a \mod b))$.**

The students try some examples. For 28 and 123, the tail recursive algorithm (tail recursion means that the recursion does nothing on the way back but pass the result upwards) computes: $\gcd(123, 28) = \gcd(28, 11) = \gcd(11, 6) = \gcd(6, 5) = \gcd(5, 1) = 1$.

Working backwards the students recursively calculate $u$ and $v$ such that $123u + 28v = 1$. We start at the penultimate recursive call by calculating a linear combination of 6 and 5 that equals the greatest common divisor:

    (A)   $1 = 6 - 1 \times 5$.

Now, using $5 = 11 - 1 \times 6$, and substituting for 5 in equation (A) gives:

    (B)   $1 = 6 - 1 \times (11 - 1 \times 6) = 2 \times 6 - 1 \times 11$.

Continuing backwards, using $6 = 28 - 2 \times 11$, and substituting for 6 in equation (B), we get:

    (C)   $1 = 2(28 - 2 \times 11) - 1 \times 11 = 2 \times 28 - 5 \times 11$.

Similarly, using $11 = 123 - 4 \times 28$, and substituting for 11 in equation (C), we get:

    (D)   $1 = 2 \times 28 - 5 \times (123 - 4 \times 28) = 22 \times 28 - 5 \times 123$.

## Euclid's Algorithm Extended

This idea can be used to extend Euclid's algorithm, so that given $a > b$, the algorithm below, CalculateUV(a,b), calculates $u$ and $v$, such that $au + bv = \gcd(a, b)$. This algorithm is not tail recursive, and the results are changed as they are passed back up. Note that the division is integer division, and truncates the remainder.

**CalculateUV($a, b$):**
**if $a$ mod $b = \gcd(a, b)$ then return $(u, v) = (1, -a/b)$.**
**else let $(u', v') = $ CalculateUV($b, a$ mod $b$), and return $(v', u' - (a/b)v')$.**

In the last example, when $a = 123$ and $b = 28$: CalculateUV(123, 28) calls CalculateUV(28, 11), which calls CalculateUV(11, 6), which calls CalculateUV(6,5), which returns $(1, -1)$.

Winding back from the recursion: CalculateUV(11, 6) returns $(-1, 1 - 1 \times (-1)) = (-1, 2)$. CalculateUV(28, 11) returns $(2, -1 - 2 \times 2) = (2, -5)$. CalculateUV(123, 28) returns $(-5, 2 - 4 \times (-5)) = (-5, 22)$. This means that $\gcd(123, 28) = 123 \times (-5) + 28 \times 22 = 1$.

In the Classroom: As short as this algorithm may be, we find that students always have some trouble remembering exactly how to do the calculations. In order to make the idea more concrete, they are asked to code the algorithm, and to print out and interpret the values at each level of recursion.

## The Complexity of Euclid's Algorithm

The following brief discussion regarding the time complexity of Euclid's algorithm will be important later when we talk about public key cryptography. Euclid's algorithm is usually not taught to middle school children, perhaps because the reason why it works is not obvious, or perhaps because for small numbers other methods seem more intuitive, simpler, and faster. Here are two intuitive methods for calculating greatest common divisors that *are* usually found in middle school textbooks.

Given $a > b$, $\gcd(a, b)$ can also be computed by:

1. Factoring: Factor $a$ and $b$ into prime factors, and take the intersection of the prime factors.

2. Brute Force: Try all the numbers from $b$ down to 1, and return the first one that divides both $a$ and $b$ evenly.

Of course, middle school children should learn how to factor, and should understand why these two methods work. From that point of view, these algorithms are worth teaching; however, both algorithms have a horribly large time complexity. The time complexity of the first algorithm is proportional to the size of $a + b$, and that of the second is proportional to the size of $b$.

In contrast, Euclid's algorithm is very fast. Gabrielle Lamé (1795–1870), using Fibonacci numbers, proved that the complexity of Euclid's algorithm, including the extended version CalculateUV, is proportional to the number of digits in $b$ [23]. This time complexity is exponentially faster than the middle school algorithms described in the last paragraph! On 100-digit numbers, the two slow algorithms will take hundreds of centuries even on the world's fastest computer, while Euclid's algorithm will appear to work instantaneously even on the hand-me-down PCs often donated to middle schools!

In the Classroom: Students write programs for the greatest common divisor in three different ways. Sure, one way would be enough, and the *right* way is Euclid's simple recursive formula. However, by doing it the other two ways, we introduce a side lesson about computational complexity, which parallels the discussion of the complexity for factoring that is so crucial later on. Simple measurements are done to illustrate the speed differences between these algorithms. We focus on the preoccupation that theoretical computer scientists have had with exponential versus polynomial time complexity since the beginnings of computational complexity in the 1960s [10].

# Fermat's Little Theorem

The next stop through mathematical history that relates to modern cryptography is Fermat's Little Theorem, not to be confused with Fermat's Last Theorem. Fermat's Little Theorem was stated in a letter from Fermat to the amateur mathematician Frenicle de Bessy (1605–1675) dated October 18, 1640.

**Fermat's Little Theorem**   *Let $p$ be a prime that does not divide the integer $a$, then $a^{p-1} = 1 \mod p$.*

Fermat wrote "I would send you the demonstration, if I did not fear its being too long"[19]. Frenicle de Bessy was an excellent amateur mathematician but not the equal of Fermat, and he was not able to provide a proof. He wrote angrily to Fermat and although Fermat gave more details in his reply, Frenicle de Bessy felt that Fermat was teasing him. Bessy was able to solve many other problems posed to him by Fermat, but it was Euler who first published a proof of Fermat's Little Theorem in 1736.

In the Classroom: As with Euclid, we spend a brief time reviewing the life and influence of Fermat. We emphasize how Fermat published almost nothing in his lifetime, and gave no systematic explanation of his methods. Instead, he left his results on the margins of works that he had read and annotated, or in letters to mathematicians of his day. He almost always described his results without proof, leaving the details of the proofs to other mathematicians, who usually supplied them. We make sure the class knows about the famous Last Theorem, and contrast it with Fermat's Little Theorem which is less famous but more applicable. One interesting note is that the same Gabrielle Lamé, who proved the logarithmic complexity of Euclid's algorithm, also proved the special case of Fermat's Last Theorem when $n = 7$. For both Fermat and Euclid, there are dozens of good references, and the reader can follow the lists in [16].

Euler's proof, essentially identical to an unpublished version by Leibniz around 1680, can be understood and motivated by the same water jug puzzle that helped us with Euclid. Consider again the case of a 17-gallon jug and a 7-gallon jug. The sequence of quantities that end up in the 17-gallon jug as we repeatedly fill and empty the 7-gallon jug into it is: 7, 14, 4, 11, 1, 8, 15, 5, 12, 2, 9, 16, 6, 13, 3, 10, 0.

In the Classroom: Students use the program they wrote earlier that generates the successive possible values for the quantities in the jugs, and experiment to see what happens when the sizes of the two jugs are relatively prime versus when they are not. Examples and special cases are always helpful when trying to motivate the statement and proof of a theorem.

The students notice how the 16 multiples of 7 are all distinct modulo 17. This would be true for any pair of relatively prime numbers $a$ and $p$. If any two of the $p - 1$ multiples of $a$ were equal modulo $p$, then their difference, call it $ax$, would be divisible by $p$. However, $p$ does not divide $ax$. By assumption, $p$ does not divide $a$, and $p$ does not divide $x$ because $x$ is less than $p$. Hence $p$ does not divide $ax$, because:

Euclid (VII.30) : *If two numbers, multiplied by one another make some number, and any prime number measures the product, then it also measures one of the original numbers.*

Now if the $p - 1$ multiples of $a$ are all distinct modulo $p$ then their product, $a \times 2a \times 3a \times \cdots \times (p-1)a$ equals the product $1 \times 2 \times \cdots \times (p-1)$, modulo $p$. Fermat's Little Theorem follows by dividing both sides of the equality by $1 \times 2 \times \cdots \times (p-1)$. With the proof of Fermat's Little Theorem in hand, we are ready to study public-key cryptography.

## Cryptography: A Brief History

The history of cryptography covers thousands of years with dozens of interesting stories [14, 25], including:

- The story of Lysander of Sparta in 404 B.C. who decoded a message written on a belt by winding the belt around a wooden staff, and reading the letters that appeared adjacent to each other, (see Figure 2). He learned that Persia was planning an attack, and he was able to thwart the attack.

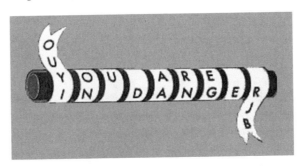

Figure 2. Lysander Decoding the Transposition Code

- The story of Mary, Queen of Scots, who was executed after her encoded correspondence regarding her plan to murder Queen Elizabeth I was intercepted and decoded.
- The strange story of the Beale Ciphers, describing the hidden location of a fortune in gold, buried somewhere in Virginia in the nineteenth century and still not found.
- The cracking of German codes by Alan Turing in Bletchley Park, England, during World War II. This story is brought to life by Sir Derek Jacobi in the Broadway play and later PBS TV special *Breaking the Code*.

A complete history of cryptography is beyond our scope here, and the reader is referred to [14, 25]. We will review only enough of cryptographic history to give context to the work of Rivest, Shamir, and Adelman [21].

In the Classroom: These stories are all very colorful and fun. They catch the interest of the class, who are about to get hit with some much less colorful and more difficult mathematics.

The code of Lysander was a *transposition* code where the letters transmitted were correct but scrambled out of order. The winding of the belt put them back in order. Modern cryptographic methods are all *substitution* ciphers in which letters or numbers are replaced by other letters or numbers.

One of the oldest substitution ciphers is called the Caesar cipher, invented by the Roman Emperor Julius Caesar. Caesar's method is similar to the decoder rings given away in cereal boxes in the 1960s (see Figure 3).

The alphabet is rotated by some number of letters, so for example if we rotate five places, "a" becomes "f," "b" becomes "g",..., and finally "z" becomes "e," wrapping all the way around. The substitutions for this rotation (call it the "f-shift") are shown below:

a b c d e f g h i j k l m n o p q r s t u v w x y z
f g h i j k l m n o p q r s t u v w x y z a b c d e

For example, Caesar's famous phrase "veni vidi vici" ("I came I saw I conquered") would read "ajsn anin anhn" using the f-shift. Actually, Caesar himself was partial to the c-shift. A codebreaker only has to try 25 possibilities to break an encoding like this. By translating the message 25 times using the reverse

Figure 3. Ovaltine Nostalgia Decoder Ring

of each possible letter shift, until one translation looks like a real message, he will eventually find the real message.

A better substitution method that is much more difficult to decode is called the Vigenère Cipher. Blaise de Vigenère (1523–1596), from the court of Henry III described his method in a book titled A Treatise on Secret Writing. Vigenère was also known for his 1578 rationalist treatise debunking superstitions about comets. "In this little book various statements regarding comets as signs of wrath or causes of evils are given, and then followed by a very gentle and quiet discussion, usually tending to develop that healthful skepticism which is the parent of investigation." [31]

The Vigenère cipher is an example of a polyalphabetic substitution cipher, in which each subsequent letter uses a different shift. After a while these shifts cycle. The sequence of shifts is represented by a codeword consisting of letter shifts. For example, if the codeword is "remarkable", then the first letter of the message would shift by 17, the second letter by 4,..., the 7th letter by 0, etc. The 11th letter would shift by 17 again like the first and the cycle would repeat.

For example, using the codeword "remarkable," the sentence "we meet in new york for a rendezvous" would be encoded as "ni yevd io yin carb pos 1 vvrpeqfovd." Note that spaces between words are not actually transmitted or encoded. For a codeword like "remarkable" of length ten, there are $26^{10}$ different possible encodings. The longer the codeword, the more possibilities there are.

In the Classroom: Students write a program that encodes and decodes a Vigenère cipher. The program takes plaintext along with a codeword and produces encrypted text. In this exercise as well as others that follow, we have the students compete against each other by creating messages for each other to decode. This is a lot of fun, and the students love it.

## Breaking the Vigenère Cipher: Determining the Length of the Codeword

The first step in breaking a Vigenère cipher is to determine the length of the codeword. There are a number of ad hoc ways of doing this including the work of Kasiski (1863) and Kerckhoff (1883). The former searches for commonly recurring *bigrams* (two letter combinations) that might encode words like "me" or "it," hoping that the distances between repeated occurrences will be the length of the codeword. The latter tries different lengths, groups the letters into disjoint sets each of which contains letters that use the same substitution, and sees if the sorted frequencies of the encoded letters match the sorted expected English frequencies. This latter method also suggests guesses of the actual key word. However, neither of these methods is as reliable as the more systematic method of Philip Friedman published in 1922 [8].

Friedman's ingenious method to determine the length of the codeword is described by David Kahn [14] as "the most important single publication in cryptology." Friedman defines the *index of coincidence,* a statistical measure of text, that for normal English is about 6.6%, but for a random collection of letters is only about 3.8%. He uses this measurement to calculate the length of the codeword.

To determine the index of coincidence of a particular piece of text, take the text, rotate it by some random number of places, and write the rotated text underneath the original text. For example, the sentence below is rotated by 58 characters and the rotated version appears beneath it.

alanturingbreakscodeslikenobodysbusinessbuthispersonallifesadlybecameeverybodysbusiness

dysbusinessbuthispersonallifesadlybecameeverybodysbusinessalanturingbreakscodeslikenobo

The index of coincidence is the number of places in which the same letter occurs in both strings of text. In this example, there are 6 coincidences for 87 characters of text, a rate of 6.9% and close to the 6.6% of normal English text. An important point to notice is that when a text is encrypted with a Caesar cipher, where every letter is shifted by the same value, the index of coincidence remains constant.

For a Vigenère encryption, we would expect to see an index of coincidence more like the 3.8% of random letters, *unless* we happen to have rotated by a multiple of the Vigenère cycle length. In the case where the rotation is a multiple of the cycle length, the letters that are lined up underneath each other are encoded using the same shift, and the usual English index of coincidence would be expected, because the index of coincidence is invariant under a Caesar cipher. Hence the way to determine the Vigenère cycle length is to rotate the encrypted text by 1, 2, 3, etc. symbols, until we see an index of coincidence that looks more like 6.6% than 3.8%.

The Vigenère cipher itself did not catch on, but variations of it did, including the Gronsfeld cipher, (which is essentially a Vigenère cipher with a codeword of digits), and the more complex ciphers of the Germans during World War II. Until the work of Kasiski, Kerkehoff, and later Friedman, the Vigenère cipher and its variants, were for 300 years considered unbreakable, especially if long codewords were used and short messages were encoded.

In the Classroom: Students write a program using Friedman's technique to determine the cycle lengths of messages encrypted with different codewords. Seeing the correct percentage pop up is perhaps the best way to appreciate Friedman's amazing contribution to code breaking.

## Breaking the Vigenère Cipher Given the Length of the Codeword

Once one knows the length of the cycle, 10 for the "remarkable" example, there are a number of techniques for breaking the code without having to try all the possibilities. One idea is to divide the letters of the encoded message up into 10 groups, one for each shift in the cycle. The 1st, 11th, 21st, etc. make the first group. The 2nd, 22nd, 32nd, etc. make the second group. And so on. The letters in each group are encoded using the same shift. This method is very much like Kerckhoff's method but without having to guess the length of the codeword.

We try to decode the letters a group at a time. For each group, we try all 26 possible shifts, but since the letters in each group are scattered throughout the message, we are unlikely to learn anything by doing this. The chance of the partially decoded message looking familiar will be very small.

Nevertheless, there is a way to learn something about the message. Every language has a characteristic statistical frequency for each letter. For example, in English the letter "e" is the most frequently used letter.

For a given group, we compare the frequencies of the letters in each of the 26 decodings to the expected frequencies. Matching the two sets of frequencies helps identify the shift or at least narrows down the number of possibilities from 26 to just a few. Of course this method requires a large text sample.

In the Classroom: We have not yet designed a lab to experiment with this method. It requires a linear least squares regression and our students have not yet studied this in their math classes. Instead, we have the students design more ad hoc methods for guessing the letters, which allow for computer-human cooperation. These methods force the students to be creative and unique. The methods are also true to the history of breaking codes. The kind of programming that went on at Bletchley Park during World War II was exactly this ad hoc style of combining analytical methods with practical necessity. It was the combined power of machine and human problem solving that ultimately cracked the German codes.

In World War II, the Germans used an encoding scheme similar to the Vigenère cipher but more complex. They used a machine to generate the letter substitutions, and the cycles were extremely long. The machine the Germans used was called the Enigma. An online simulator of such a machine can be found at [6]. Breaking Nazi codes required the sophisticated computer-aided decoding effort led by Alan Turing.

The important thing to note about the Vigenère Cipher and all its variants is that *if* one knows the encoding method (via espionage for example), then the decoding is trivial. At first thought, a kind of encryption where the encoding method is not easily reversible might appear impossible. However, if such a scheme were possible, then even if the encoding method were to fall into the wrong hands, the enemy could still not easily decode messages! Now we are ready to jump ahead to the late twentieth century to the work of Rivest, Shamir and Adelman.

## Public Key Cryptography: The RSA Breakthrough of 1978

Cryptographic methods do not have to be reversible! Today a new kind of encoding is used which is called *public-key* cryptography (or *one-way*, or *trapdoor*). With this new method, the whole world is able to encode messages, but unless Sam Hacker has more information, he still cannot decode a message.

This new method is what allows e-commerce to flourish without fear of a security breach. Suppose I want to send my credit card number to Amazon.com. I encode my number with a publicly published method that anyone else could use called the *public key*, but only Amazon.com can decode it because only they have the *private key*!

### Authentication

What if I don't trust that I am actually talking to Amazon.com? That is, I suspect that Sam Hacker posing as Amazon, sent me a fake public key, and that he is planning to decode my reply with his own private key and get my credit card number! In that case, we do the process in reverse. I ask Amazon to send me a message encoded with their private key. If I decode their message with their public key and it states, "Hi I am Amazon.com", I know that the message had to come from Amazon.com, because nobody else would have known how to encode it correctly. This is called authentication and is described in nice detail by VeriSign, the largest digital signature provider [30].

The students often wonder what prevents Sam Hacker from using Amazon's private signature. After all, everyone who decodes "Hi I am Amazon.Com" knows what those characters look like before they are decoded. Sam could decode their signature just like anyone does, and start signing messages with it! This is a serious problem. The solution is to run the whole message sent by Amazon through a *hash* function—and then encode the result of the hash function with Amazon's private key. (A hash function $H$ is a function that takes a message $M$ of any size and computes $H(M)$ a fixed size output, with the property that it is computationally infeasible to find another message $N$, where $H(N) = H(M)$.) Then all Sam could do is retransmit the very same message Amazon meant to send, but he could not send his own messages with Amazon's signature.

Another objection often arises in class: How do we know that the public key of Amazon is correct? Perhaps an adversary has published fake Amazon.com public keys all over the Internet. The solution is to have a company like VeriSign guarantee the authenticity of Amazon's public key by the same process. Of course someone could pretend they are VeriSign, but that is hard to do since VeriSign's raison d'être is to make sure that nobody masquerades as them. Their business depends on it.

## The Mathematics Behind Public Key Cryptography

How do we construct and use these private and public keys? Interestingly, the method is based on number theory, one of the oldest branches of pure mathematics, more famous for its beauty and elegance than its practical applications.

We start by describing a simple version of the Rivest, Shamir, and Adelman (RSA) algorithm that is *not* public-key cryptography because the private key can be computed from the public key using the Extended Euclid's Algorithm described earlier. This simpler version isolates the main ideas from the public-key part, and helps one better appreciate the contribution of RSA.

Public key cryptographic methods encode integers into integers, so we assume that our messages are first converted somehow to a sequence of integers. The exact method of conversion uses hash functions and is not trivial but that won't concern us here.

To encode a number, we will need the *public key*. This consists of two integers, for example 5 and 17. The second integer must be prime, and the first must be relatively prime to the second integer minus 1. In this case, 17 is prime, and 5 is relatively prime to 16.

For example, to encode 6 using this key, we calculate $6^5 \bmod 17$. You can check that this equals 7. To decode 7 back into 6, a brute force approach requires trying all possible values from 0 to 16 to see which one would encode into 6. This is computationally prohibitive when the prime has 40 or more digits. However, we can decode quickly if we calculate the private key. The private key also consists of two numbers, one of which is part of the public key, namely the prime 17, and one of which is private, in this case 13. To decode, we calculate $7^{13} \bmod 17$, which you can verify equals 6.

How is the private key, 13 in our example, calculated? It is the solution to the equation $5u = 1 \bmod 16$. This solution can be computed efficiently with Euclid's Extended Algorithm, by finding $u$ and $v$ such that $5u + 16v = \gcd(5, 16) = 1$.

Why does the private key decode correctly? In our example, why does $6^{(13)} \bmod 17 = 6 \bmod 17$? It all comes down to Fermat's Little Theorem. Fermat's Little Theorem implies that $6^{16} = 1 \bmod 17$. Since $5(13) = 1 \bmod 16$, we can write $5(13) = 16(4) + 1$. Thus $6^{5(13)} = 6^{16(4)+1}$. Finally, since $6^{16} = 1 \bmod 17$, $6^{16(4)+1} = 6 \bmod 17$.

In the Classroom: We have found that isolating the RSA idea from the part that requires the factoring enables students to more easily understand the algorithm. Furthermore, there is some historical justification for this pedagogy because the results were discovered in this layered way.

## The RSA Algorithm

The real RSA algorithm is very similar to the algorithm described in the previous few paragraphs with one important difference: with the real RSA it is *not* easy to compute the private key from the public key.

This time we start with two prime numbers, $p$ and $q$, say 2 and 17, and compute their product $pq = 34$. We calculate $(p-1)(q-1) = 16$, and then choose a value that has no common factors with 16, let's try 5. The public key becomes the pair of numbers 34 and 5.

The encoding and decoding is done just like before. For example, to encode 6, we compute $6^5$ mod 34 = 24, and to decode 24 we compute $24^{13}$ mod 34 = 6. As before, 13 is the solution to the equation $5u = 1$ mod 16.

The difference between the real RSA idea and our first attempt is that previously 16 was calculated simply by subtracting 1 from the common public prime. The equation, $5u = 1$ mod 16, was then solved by Euclid's Extended Algorithm. But now the only way to calculate 16 is to factor the number 34 into 2 and 17 and compute $(2-1) \times (17-1)$. And the factoring part is hard! Nobody knows how to factor numbers quickly. The best methods are exponentially slower than the time complexity of the Extended Euclid's Algorithm.

Of course, anybody can factor 34, but in practice the two primes that are chosen for the public key are on the order of 100 digits each. This makes all currently known factoring algorithms take years. If you come up with an efficient algorithm that can factor numbers, you will be famous!

In the Classroom: Students write programs to encode integers using the RSA algorithm. As with the Vigenère cipher, they compete by trying to break each other's codes. The importance of choosing large enough prime numbers comes to life, as otherwise their messages are easily cracked. As the large prime factors defeat the cracking attempts, the students appreciate first hand just why it is safe to send a credit card number over the internet.

## The Problem of Factoring

What makes the RSA algorithm a one-way, or *trapdoor* method, is that decoding requires factoring, and encoding does not. There is currently no better way to factor an integer $n$ than to try all possible prime factors, up to $\sqrt{n}$. This is an exponential time algorithm. At first, a lot of students mistakenly think that this is a polynomial time algorithm, because the process takes about $\sqrt{n}$ operations for an integer $n$. However, computer scientists naturally measure time complexity as a function of the *size* of the input, and the *size* of an integer $n$ is the number of bits it takes to store the integer, which is proportional to $\log n$. Letting $m = \log n$, the best factoring algorithm takes time proportional to $2^{m/2}$ operations.

Is there a faster way to factor numbers? There is, after all, no proof that factoring is inherently hard. The problem of factoring is not even known to be NP-Complete. Perhaps one day someone will come up with a polynomial time factoring algorithm.

In 2002, a trio of Indian computer scientists, Agrawal, Kayal, and Saxena invented a polynomial time algorithm [1] to determine whether or not a number is prime. Let $m$ be the number of bits representing the number to be factored. Their algorithm runs in time proportional to $m^{12}$. This was a huge breakthrough, since up until then the best result was $m^{c \log \log m}$ discovered in 1983 [20]. This latter result is not polynomial in $m$ because of the $\log \log m$ factor in the exponent.

For determining whether or not an integer is prime, it is possible that the exponential time algorithm, $m^{c \log \log m}$, will run faster than the provably polynomial time algorithm, $m^{12}$, but that is a question of engineering. Even though $\log \log m$ is theoretically not a constant, it is less than 4 for all input sizes currently being used for secure message transmission, and less than 10 for all input sizes ever likely to be used. A joke from Carl Pomerance [20] summarizes "that although it has been *proved* that $\log \log m$ tends to infinity with $m$, it has never been observed doing so." Nevertheless, from the complexity theorists' point of view, the polynomial time breakthrough of Agrawal, Kayal, and Saxena [1] is a milestone in the way the 4-minute mile was. It opens possibilities, debunks impossible barriers, and confirms what most thought would be true and what some suspected would not.

This breakthrough in the complexity for determining whether a number is prime or composite is reminiscent historically of the 1980s breakthrough in linear programming. Linear programming is an optimization problem that is used in all sorts of practical situations, such as determining the best way to schedule airline flights so as to minimize costs and maximize profits. The simplex algorithm of Dantzig (1947) had been used for years to successfully solve linear programming problems, despite its theoretical worst-case exponential time complexity. In practice, the exponential time behavior was not observed, although theoretically it was not clear why this was the case. Then in 1979 [17], a polynomial time algorithm was invented. It took another breakthrough in 1984 [15] and years of software engineering and testing before this polynomial time algorithm was competitive with the exponential time simplex algorithm. Complexity theory, a 20th century field of study, has a long way to go before it completely models practical computer science problems.

In the Classroom: This historical parallel between linear programming and factoring makes an impression on the students. They are often suspicious of theory and its practical uses. Although there are plenty of examples to show them the applicability of theory in computer science, here we show them examples where the theory is weak! They learn that the critique of theory in science is a historical process that forces the theory to improve.

Can this new breakthrough for determining whether or not a number is prime be used to help factor numbers quickly? Not yet as far as anyone knows, and most people suspect it cannot. However, in many cases, including the well-known Traveling Salesman Problem, the *decision* version of a problem is polynomially related to the *optimization* version. [10]. That is, if you can decide whether or not something is possible, then you can also figure how to do it. If this were the case for factoring, then a polynomial time algorithm for deciding whether or not a number is prime would hint at a polynomial time algorithm for determining the factors.

In the Classroom: The students learn where the history of this research is heading and what problems are still open. Examples of problems where the decision and optimization versions are *not* polynomially related is rare. It seems that this *may* be the case for factoring. When will we know?

Research on new cryptographic methods continues simultaneously with research on breaking the current codes. There are long-term strategies, like quantum computing and DNA computing, and short-term strategies using conventional computing. In all cases, mathematics will no doubt play a central role. In the meantime, e-commerce is still on secure ground.

In the Classroom: Quantum computing and DNA computing are very new fields that try to deal with the fundamental intractability that is inherent in hundreds of NP-Complete problems. They represent two parallel paths of future work that may or may not bear fruit. They are difficult fields but could be used to design labs, although we have not yet attempted to do this. The interested reader should consult [9].

## Cryptographic Decoding Challenges for Practice and Review

In the Classroom: In our labs, we not only have the students compete against each other by designing and cracking each other's codes, but also by seeing who can first decode messages that we created. There are usually cash prizes where the amount is proportional to the level of difficulty. The Vigenère ciphers pay more when we do not tell the students the length of the codeword. Here are some examples:

## Historical Notes on RSA Encryption

Ron Rivest [22] was kind enough to share with me his own recollections of the RSA discovery and his thoughts about the future directions of cryptography. Rivest credits the original motivation for the RSA work to the 1976 seminal ground-breaking paper "New Directions in Cryptography" by Diffie and Hellman [7]. The paper explains how two users can exchange a secret key over an insecure medium without any prior secrets. The challenge of public-key cryptography was directly proposed in this paper, and was brought to the attention of Rivest by a graduate student. Rivest, Shamir, and Adelman were assistant professors at MIT at the time, Rivest in computer science, Shamir and Adelman in mathematics. The departments in MIT overlap through the many independent research labs, and the three colleagues collaborated at LCS, the lab for computer science.

The three tried a number of unsuccessful approaches, including a method using the knapsack problem as a way to thwart the *bad guys*. I suggested that it was ironic that the approach using knapsack, a known NP-Complete problem, failed, while the approach using factoring, a problem not known to be NP-Complete, succeeded. Rivest responded that NP-complete problems often have many different solutions, and for the purposes of public key cryptography, it is easier to work with factoring which has a unique solution.

Rivest is not currently working on the complexity of factoring, although he continues to work on new applications in cryptography related to internet voting and to radio frequency ID tags. When I asked him whether he believed that factoring was provably hard, he responded "I have a built-in bias in favor of hoping that factoring is hard, but to prove lower bounds for factoring—we'll need something a bit different than NP-completeness." Rivest does not expect that the new polynomial decision algorithm for prime numbers will yield a better factoring algorithm, but he is willing to be surprised. "I'm always in favor of the truth."

Did Rivest realize the broad applications and notoriety that their work would eventually yield? Rivest explained that RSA was discovered long before the days of the internet and just before the personal computer age. He had no inkling of the eventual supporting and crucial role his research would have on e-commerce. "It wasn't obvious, a priori, that our result would stand the test of time." Theoretically, there could be advances in factoring methods. Practically, the applications might not find enough uses.

After PCs came on the scene, Rivest et al started a company in 1983, called RSA Data Security, in an attempt to find commercial uses for the theory. By 1986, "we thought we would go bust," but then Lotus Notes signed a contract. The 1990s introduced the world-wide-web and RSA became ubiquitous. VeriSign was spun-off, in order to maintain its independence from the technology, and RSA Data Security was bought and renamed simply RSA Security (http://www.rsasecurity.com/).

Commercial success notwithstanding, Rivest seems most pleased with how cryptography as a field has evolved. "The rich interplay between theory and practice brought vitality to the subject." It has also brought controversy.

The National Security Agency (NSA) is a government agency with a heavy investment in secret cryptographic research. They do not publish their results, nor commercialize their implementations, but they do monitor what others are doing. At one time they threatened that the planned delivery of a paper related to RSA, at a Cornell IEEE conference, would violate the Export Control Act! The subject matter they claimed is classified. Eventually and sensibly, the government backed down. Rivest pointed out the absurdity of classifying research when one or more of the authors has no security clearance themselves. Today the NSA allows academic freedom in cryptographic research but still insists that software companies register their cryptographic products with the Bureau of Industry and Security (formerly the Bureau of Export Administration); see http://www.bxa.doc.gov/Encryption/Default.htm. The role of the NSA in monitoring cryptographic research and its commercial development will no doubt continue to evolve.

**Vigenère Ciphers I**    A two-cycle Vigenère encryption gives an encoding of a well-known cerebral song:

BQHIERPVBZXOPORHASACNFLQHBYSKFBBZKBHAHASYZIIKXFLQHBLIEHBBZKBHAHASKOBB
PWMVMVXHACNUAHLWWPXHAWGYBBBQHIERUSTBHHASKZBBVCEBBTBCGZRVTRTPKOBB.

What is the original text?

**Vigenère Ciphers II**    A five cycle Vigenère encryption gives an encoding of a Woody Allen math joke:

BOQWMNWOOQSSFHQKASRLAGOWRAADWRKAONCIOHKJKCYHVYWHLLC.

What is the original text?

**RSA Ciphers I**   An RSA encoding of nine ASCII values, using public key (10555, 21971), gives the name of a hunter: 16912 19531 20676 16912 6613 2348 17835 15770 15770. What is the original text?

**RSA Ciphers II**   An RSA encoding using public key (5555551, 118513313) gives: 80217189 107242213 96490860 79543571 25953566. What are the original numbers?

## Conclusion

Cryptography is an ancient yet vibrantly active field for mathematicians and computer scientists. It is fun to teach because it contains a beautiful combination of elegant theory and practical application, and lends itself to exploration and learning through programming and experiments. The history of cryptography, while providing stories of intrigue and excitement, is a mathematical metaphor for the delicate balance between theory and practice.

**Acknowledgments:**   Thanks to Ralph Bravaco, Tara Holm, Andrea Simonson, and Amy Shell-Gellasch for their suggestions on early versions of this article. A special thanks is due to Ron Rivest for sharing his personal thoughts and memories.

## References

1.  Agrawal, M., Kayal, N. and Saxena, N., PRIMES in P, August 2002.
    http://www.cse.iitk.ac.in/users/manindra/primality.ps
    http://www.cse.iitk.ac.in/news/primality.pdf

2.  Albers, Donald J. and Reid, Constance, "An Interview with George B. Dantzig: The Father of Linear Programming," College Mathematics Journal, 17:4, (1986) 293–313.

3.  Ball, W.W. Rouse and Coxeter, H.S.M., *Mathematical Recreations and Essays*, Dover, 1987.

4.  Bogolmony, Alex, Cut-the-Knot, http://www.cut-the-knot.com/water.shtml.

5.  Bravaco, R., Simonson, S., "Mathematics and Computer Science: Exploring a Symbiotic Relationship," to appear in *Mathematics and Computer Education*, 2004.

6.  Carlson, Andy, Enigma Simulator, http://homepages.tesco.net/~andycarlson/enigma/enigma_j.html

7.  Diffie, W. and Hellman, M.E., "New Directions in Cryptography," *IEEE Trans. Information Theory*, IT-22, 6, (1976) 644–654, http://citeseer.nj.nec.com/diffie76new.html.

8.  Friedman, W., *The Index of Coincidence and Its Applications in Cryptography*, Publication No. 22, Geneva, IL, Riverbank Publications, 1922.

9.  Gramss, T., Grob, M., Mitchell, M., et al., Eds. *Non-Standard Computation: Molecular Computation — Cellular Automata — Evolutionary Algorithms — Quantum Computers*, John Wiley & Sons, 1998.

10. Garey and Johnson, *Computers and Intractability, A Guide to the Theory of NP-Completeness*, Freeman and Co., San Francisco, 1979.

11. Hodges, Andrew, *Alan Turing — The Enigma*, Walker and Company, New York, 2000.

12. The Imperial War Museum Online Exhibitions, "The Story of Alan Turing and Bletchley Park":
    http://www.iwm.org.uk/online/enigma/eni-intro.htm

13. Joyce, David, Euclid's *Elements*: http://aleph0.clarku.edu/~djoyce/java/elements/elements.html

14. Kahn, D., *Codebreakers: The Story of Secret Writing*, Macmillan, 1967.

15. Karmarkar, N., "A New Polynomial-time Algorithm for Linear Programming," *Combinatorica*, 4, (1984) 373–395.

16. Katz, Victor, *A History of Mathematics*, Harper Collins, 1993.

17. Khachian, L., "A Polynomial Algorithm in Linear Programming," *Soviet Math. Dokl.*, 20, (1979) 191–194.

18. Menezes, Alfred J., van Oorschot, Paul C., Vanstone, Scott A., Handbook of Applied Cryptography, CRC Press, 2001.

19. O'Connor, J. J. and Robertson, E. F., "Biography of Pierre de Fermat": http://www-gap.dcs.st-and.ac.uk/ history/Mathematicians/Fermat.html

20. Pomerance, Carl, "A New Primal Screen," *FOCUS*, Vol. 22, No. 8, The Newsletter of the MAA, (2002) 4.

21. Rivest, Ron, Shamir, Adi and Adleman, Len, "A method for obtaining Digital Signatures and Public Key Cryptosystems," *Communications of the ACM*, 21(2), (1978) 120–126.

22. Rivest, Ron, personal communication.

23. Rosen, Kenneth H., *Discrete Mathematics and its Applications*, McGraw Hill, 4th edition, (1999) 206.

24. Simonson, Shai, Cryptography Labs for Mathematical Experiments in Computer Science—a Learning Community Course Exploring Discrete Math and Data Structures, http://www.stonehill.edu/compsci/LC/Cryptography.html

25. Singh, Simon, *The Code Book: The Science of Secrecy from Ancient Egypt to Quantum Cryptography*, Anchor Books, 2000.

26. ——, Cryptography Links by Simon Singh, http://www.simonsingh.net/owtasite/Crypto_Links.html

27. Singmaster, David, personal communication.

28. Sussman, Gerry, personal communication.

29. Tweedie, M.C.K., "A Graphical method of Solving Tartaglian Measuring Puzzles," *Mathematical Gazette*, 23, (1939) 278–282.

30. VeriSign, Authentication, http://www.verisign.com/docs/pk_intro.html.

31. White, Andrew Dickson, *A History of the Warfare of Science with Theology in Christendom*, D. Appleton and Co., 1896, (reprinted in the public domain at: http://www.santafe.edu /~shalizi/White/).

# 11

# Introducing Logic
# via Turing Machines

**Jerry M. Lodder**
*New Mexico State University*

## Introduction

A curious situation has arisen today in the undergraduate curriculum with many computer science majors learning the fundamentals of logic from a memorized list of truth tables and rules of inference, without regard to the original problems whose solutions involved the logic that would become part of the programmable computer. Current discrete mathematics textbooks, which often cover combinatorics, deductive reasoning and predicate logic, present the material as a fast-paced news reel of facts and formulae, with only passing mention of the original work and pioneering solutions that eventually found resolution through the modern concepts of induction, recursion and algorithm. Presented here are curricular materials, based on primary historical sources, designed for use in an introductory discrete mathematics course, particularly one with a significant number of computer science majors. The materials are organized into two-week written projects for students, and offer excerpts from Alan Turing's (1912–1954) original 1936 paper "On Computable Numbers with an Application to the Entscheidungsproblem"[1] [49], a paper which outlines a logical device, a *Turing Machine*, that is the forerunner of a modern computer program.

The two projects included here, "An Introduction to Turing Machines" and "Turing Machines, Induction and Recursion," were both assigned recently in a beginning discrete mathematics course at New Mexico State University, and build on the pedagogical idea of calculus projects [5]. For each project the students wrote a detailed paper, answering a sequence of guided questions designed to illuminate the ground-breaking ideas of Turing's work. Each project contributed significantly towards the student's course grade (about 25% per project), and each took the place of an in-class examination. Student reaction was overwhelmingly positive, with most becoming actively involved in both the historical significance of the piece as well as the details of the solutions to particular questions. Several students excelled beyond the specifics of the project, researched independently the impact of Turing's work, and discussed this in their papers. Particular advantages of the written historical project include providing context and direction to the subject matter, honing the students' verbal and deductive skills through reading the original work of some of the greatest minds in history, and the rediscovery of the roots common to discrete mathematics and computer science.

---

[1] *Das Entscheidungsproblem* is German for *the decision problem.*

## The Decision Problem

David Hilbert (1862–1943), an intellectual giant of the last century, is well known for his list of mathematical problems in 1900 [9, pp. 290–329] [34], whose study and (attempted) solutions would occupy many mathematicians for at least another hundred years. Among the problems were the foundational questions of whether the axioms of arithmetic are consistent, and whether there is a number system in cardinality between the rational numbers and the continuum of real numbers. In 1928 Hilbert posed another group of problems [11, 12], this time dealing with the consistency, completeness and independence of the axioms of a logical system in general, as well as the problem of deciding whether a given statement is valid within a logical system. The solutions to these problems, in particular Kurt Gödel's (1906–1978) demonstration of the incompleteness of arithmetic with the existence of statements that are not provable (as true or false) [7], had profound consequences for mathematics, and brought to the fore mathematical logic as a separate field of study [22]. In this exposition, however, we deal primarily with the decision problem:

> ... there emerges the fundamental importance of determining whether or not a given formula of the predicate calculus is universally valid. ... A formula ... is called *satisfiable* if the sentential variables can be replaced with the values truth and falsehood ... in such a way that the formula [becomes] a true sentence. ... It is customary to refer to the equivalent problems of *universal validity* and *satisfiability* by the common name of the *decision problem* [13, pp. 112–113].

Following Gödel's results, the decision problem remained, although it must be reinterpreted as meaning whether there is a procedure by which a given proposition can be determined to be either "provable" or "unprovable". In the text *Introduction to Mathematical Logic* [4, p. 99], Alonzo Church (1903–1995) formulated this problem as: "*The decision problem* of a logistic system is the problem to find an effective procedure or algorithm, a *decision procedure*, by which for an arbitrary well-formed formula of the system, it is possible to determine whether or not it is a theorem ... ." To be sure, Church proves that the decision problem has no solution [2, 3], although it is the algorithmic character of Turing's solution that is pivotal to the logical underpinnings of the programmable computer. Moreover, the simplicity of a Turing machine provides a degree of accessibility to the subject ideal for a first course in logic or discrete mathematics.

Briefly a Turing machine, $M$, is a device which prints a sequence of 0s and 1s on a tape based on (i) the figure currently being scanned on the tape, and (ii) a set of instructions. Moreover, Turing describes the logical construction of a universal computing machine, $U$, which accepts the set of instructions of a given machine $M$ in some standard form, along with the input to that machine, and then outputs the same sequence as $M$. Applying a machine $U$ to another machine $M$ is denoted as $U(M)$ for this exposition. It follows from Turing's paper [49] that if the decision problem has a solution, then there is a machine $D$ which accepts the set of instructions of another machine $M$ and decides whether $M$ prints a finite or an infinite number of symbols on the tape. Determining whether a machine terminates in a finite number of executable steps is today known as the halting problem in computer science. If the decision problem, and hence the halting problem has a solution, then a new machine $T$ can be defined so that $T(M)$ halts if $M$ does not halt, and $T(M)$ does not halt if $M$ halts. By considering the behavior of $T(T)$, we conclude that $T$ halts and $T$ does not halt, a contradiction, from which it follows that the decision problem has no solution.

## The Projects

What follows are descriptions and the actual text of the projects "An Introduction to Turing Machines" and "Turing Machines, Induction and Recursion," along with a few words of advice for the instructor.

The first project acquaints students with the workings of a Turing machine, asks the reader to compute the output of a particular machine, and then to design a machine with a given output. All background needed to answer these questions is gleaned from reading the excerpts from Turing's original paper included in the project, without a special in-class lecture to define a Turing machine. In fact, an entire class session was set aside to allow the students to work on their project and discuss their progress among themselves. During office hours or while students worked during class, I would answer questions about the output of the examples presented in Turing's paper, but only on the students' initiative. The project also asks the students to design a Turing machine to test the property of set containment (for finite sets), which builds on the topic of set theory, covered just prior to the assignment of the project. For that question, the operation of storing a character in memory is allowed, although Turing did not include this feature in his paper. The instructor is encouraged to tailor the projects to mesh with what has transpired in class, add some questions relevant to topics being covered, and delete or rephrase others. The details of a project should be clear before assignment.

The second project, "Turing Machines, Induction and Recursion," explores the machine's capacity for a limited type of memory, and its use in elementary arithmetic operations. In particular, the students are asked to describe a machine, which given two positive integers $n$ and $m$, produces the product $n \cdot m$. In doing so, the student witnesses the appearance of recursion when repeating steps in a Turing machine as well as the use of induction when verifying that the output of a machine is the desired result. For the second project, the students worked in groups of two or three, which often resulted in a group dynamic with one member proposing an algorithm for multiplication, and the other group members testing its accuracy. The first project, on the other hand, was an individual project, requiring that every student learn the rudiments of a Turing machine before tackling the more sophisticated concepts of induction and recursion needed for the design features of the second project. The second highlights the need of several abstract topics in discrete mathematics for programming needs.

Two projects appear to be ample material for a one-semester course, supplemented with a midterm and final, both of which could include questions about the projects. Two additional projects could easily be authored from Turing's paper. The third in this sequence would be to develop Turing's idea of a universal computing machine (today known as a compiler or an interpreter), and the fourth would outline Turing's negative solution to the decision problem in what has been called the halting problem in computer science. These last two projects, available in [1], are best suited for an intermediate or advanced undergraduate course in the subject. Finally, for a detailed history of related topics in logic, see [22]; for a leisurely account of the events leading up to the programmable computer, see [6]; and for a detailed description of Turing's life and times, see [14].

## Project I: An Introduction to Turing Machines

During the International Congress of Mathematicians in Paris in 1900, David Hilbert (1862–1943), one of the leading mathematicians of the last century, proposed a list of problems for following generations to ponder. On the list was whether the axioms of arithmetic are consistent, a question which would have profound consequences for the foundations of mathematics. Continuing in this direction, in 1928 Hilbert proposed the decision problem (das Entscheidungsproblem), which asked whether there was a standard procedure that can be applied to decide whether a given mathematical statement is true. Both Alonzo Church (1903–1995) and Alan Turing (1912–1954) published papers in 1936 demonstrating that the decision problem has no solution, although it is the algorithmic character of Turing's paper "On Computable Numbers, with an Application to the Entscheidungsproblem" [49] that forms the basis for the modern programmable computer. Today his construction is known as a *Turing machine*.

The goal of this project is to read a few excerpts from Turing's original paper [49] and outline a machine that would verify the condition of set containment, a topic discussed in class. After carefully reading the attached pages, answer the following:

(a) Describe the workings of a Turing machine (referred to as a "computing machine" in the original paper).

(b) What is the precise output of the machine in Example 1? Certain squares may be left blank. Be sure to justify your answer.

(c) Design a Turing machine which generates the following output. Be sure to justify your answer.

$$010010100101001 \ldots$$

(d) Describe the behavior of the following machine, which begins with a blank tape, with the machine in configuration $\alpha$.

| Configuration | | Behavior | |
|---|---|---|---|
| m-config. | symbol | operation | final m-config. |
| $\alpha$ | none | R P1 | $\beta$ |
| $\alpha$ | 1 | R P0 | $\beta$ |
| $\alpha$ | 0 | HALT | (none) |
| $\beta$ | 1 | R P1 | $\alpha$ |
| $\beta$ | 0 | R P0 | $\alpha$ |

(e) Given finite, non-empty, sets $A$ and $B$, design a Turing machine which tests whether $A \subseteq B$. Suppose that the first character on the tape is a 0, simply to indicate the beginning of the tape. To the right of 0 follow the (distinct, non-blank) elements of $A$, listed in consecutive positions, followed by the symbol &. To the right of & follow the (distinct, non-blank) elements of $B$, listed in consecutive positions, followed by the symbol Z to indicate the end of the tape:

| 0 | | | . . . | | & | | . . . | | Z |
|---|---|---|---|---|---|---|---|---|---|

The symbols 0, &, Z are neither elements of $A$ nor $B$. The machine starts reading the tape in the right-most position, at Z. If $A \subseteq B$, have the machine erase all the elements of $A$ and return a tape with blanks for every square which originally contained an element of $A$. You may use the following operations for the behavior of the machine:

- R: Move one position to the right.
- L: Move one position to the left.
- S: Store the scanned character in memory. Only one character can be stored at a time.
- C: Compare the currently scanned character with the character in memory. The only operation of C is to change the final configuration depending on whether the scanned square matches what is in memory.
- E: Erase the currently scanned square.
- P( ): Print whatever is in parentheses in the current square.

You may use multiple operations for the machine in response to a given configuration. Also, for a configuration $q_n$, you may use the word "other" to denote all symbols $S(r)$ not specifically identified for the given $q_n$. Be sure that your machine halts.

The following is an excerpt from Turing's original paper [49, pp. 231–234], reprinted with permission from the London Mathematical Society.

# On Computable Numbers, with an Application to the Entscheidungsproblem

*By* A. M. Turing

## 1. *Computing Machines*

We have said that the computable numbers are those whose decimals are calculable by finite means. This requires more explicit definition. No real attempt will be made to justify the definitions given until we reach §9. For present I shall only say that the justification lies in the fact that the human memory is necessarily limited.

We may compare a man in the process of computing a real number to a machine which is only capable of a finite number of conditions $q_1, q_2, \ldots, q_R$, which will be called the "$m$-configurations". The machine is supplied with a "tape" (the analogue of paper) running through it, and divided into sections (called "squares") each capable of bearing a "symbol". At any moment there is just one square, say the $r$-th, bearing the symbol $S(r)$ which is "in the machine". We may call this square the "scanned square". The symbol on the scanned square may be called the "scanned symbol". The "scanned symbol" is the only one of which the machine is, so to speak, "directly aware". However, by altering its $m$-configuration the machine can effectively remember some of the symbols it has "seen" (scanned) previously. The possible behaviour of the machine at any moment is determined by the $m$-configuration $q_n$ and the scanned symbol $S(r)$. This pair $q_n$, $S(r)$ will be called the "configuration"; thus the configuration determines the possible behaviour of the machine. In some of the configurations in which the scanned square is blank (i.e. bears no symbol) the machine writes down a new symbol on the scanned square; in other configurations it erases the scanned symbol. The machine may also change the square which is being scanned, but only by shifting it one place to right or left. In addition to any of these operations the $m$-configuration may be changed. Some of the symbols written down will form the sequence of figures which is the decimal of the real number which is being computed. The others are just rough notes to "assist the memory". It will only be these rough notes which will be liable to erasure.

It is my contention that these operations include all those which are used in the computation of a number. The defense of this contention will be easier when the theory of the machines is familiar to the reader. In the next section I therefore proceed with the development of the theory and assume that it is understood what is meant by "machine", "tape", "scanned", etc.

## 2. *Definitions.*

### *Automatic machines*

If at each stage the motion of a machine (in the sense of §1) is *completely* determined by the configuration, we shall call the machine an "automatic machine" (or $a$-machine).

For some purposes we might use machines (choice machines or $c$-machines) whose motion is only partially determined by the configuration (hence the use of the word "possible" in §1). When such a machine reaches one of these ambiguous configurations, it cannot go on until some arbitrary choice has been made by an external operator. This would be the case if we were using machines to deal with axiomatic systems. In this paper I deal only with automatic machines, and will therefore often omit the prefix $a$-.

### *Computing machines*

If an $a$-machine prints two kinds of symbols, of which the first kind (called figures) consists entirely of 0 and 1 (the others being called symbols of the second kind), then the machine will be called a computing machine. If the machine is supplied with a blank tape and set in motion, starting from the correct initial $m$-configuration the subsequence of the symbols printed by it which are of the first kind will be called the *sequence computed by the machine*. The real number whose expression as a binary decimal is obtained by prefacing this sequence by a decimal point is called the *number printed by the machine.*

At any stage of the motion of the machine, the number of the scanned square, the complete sequence of all symbols on the tape, and the *m*-configuration will be said to describe the *complete configuration* at that stage. The changes of the machine and tape between successive complete configurations will be called the *moves* of the machine.

3. *Examples of computing machines*

I. A machine can be constructed to compute the sequence 010101 . . .. The machine is to have the four *m*-configurations "*b*", "*c*", "*f*", "*e*" and is capable of printing "0" and "1". The behaviour of the machine is described in the following table in which "*R*" means "the machine moves so that it scans the square immediately on the right of the one it was scanning previously". Similarly for "*L*". "*E*" means "the scanned symbol is erased and "*P*" stands for "prints". This table (and all succeeding tables of the same kind) is to be understood to mean that for a configuration described in the first two columns the operations in the third column are carried out successively, and the machine then goes over into the *m*-configuration described in the last column. When the second column is blank, it is understood that the behaviour of the third and fourth columns applies for any symbol and for no symbol. The machine starts in the *m*-configuration *b* with a blank tape.

| Configuration | | Behaviour | |
|---|---|---|---|
| m-config. | symbol | operation | final m-config. |
| *b* | none | P0, R | *c* |
| *c* | none | R | *e* |
| *e* | none | P1, R | *f* |
| *f* | none | R | *b* |

If (contrary to the description §1) we allow the letters *L*, *R* to appear more than once in the operations column we can simplify the table considerably.

| Configuration | | Behaviour | |
|---|---|---|---|
| m-config. | symbol | operation | final m-config. |
| *b* | none | P0 | *b* |
| *b* | 0 | R, R, P1 | *b* |
| *b* | 1 | R, R, P0 | *b* |

# Project II: Turing Machines, Induction and Recursion

The logic behind the modern programmable computer owes much to Turing's "computing machines," discussed in the first project, which the reader should review. Since the state of the machine, or *m*-configuration as called by Turing, can be altered according to the symbol being scanned, the operation of the machine can be changed depending on what symbols have been written on the tape, and affords the machine a degree of programmability. The program consists of the list of configurations of the machine and its behavior for each configuration. Turing's description of his machine, however, did not include memory in its modern usage for computers, and symbols read on the tape could not be stored in any separate device. Using a brilliant design feature for the tape, Turing achieves a limited type of memory for the machine, which allows it to compute many arithmetic operations. The numbers needed for a calculation are printed on every other square of the tape, while the squares between these are used as "rough notes to 'assist the memory.' It will only be these rough notes which will be liable to erasure" [49, p. 232].

11. Introducing Logic via Turing Machines

Turing continues:

The convention of writing the figures only on alternate squares is very useful: I shall always make use of it. I shall call the one sequence of alternate squares $F$-squares, and the other sequence $E$-squares. The symbols on $E$-squares will be liable to erasure. The symbols on $F$-squares form a continuous sequence. ... There is no need to have more than one $E$-square between each pair of $F$-squares: an apparent need of more $E$-squares can be satisfied by having a sufficiently rich variety of symbols capable of being printed on $E$-squares [49, p. 235].

Let's examine the Englishman's use of these two types of squares. Determine the output of the following Turing machine, which begins in configuration $a$ with the tape

| $X$ | | | ... | | | |

and the scanner at the far left, reading the symbol $X$.

| Configuration | | Behavior | |
|---|---|---|---|
| m-config. | symbol | operation | final m-config. |
| $a$ | $X$ | R | $a$ |
| $a$ | 1 | R, R | $a$ |
| $a$ | blank | P(1), R, R, P(1), R, R, P(0) | $b$ |
| $b$ | $X$ | E, R | $c$ |
| $b$ | other | L | $b$ |
| $c$ | 0 | R, P($X$), R | $a$ |
| $c$ | 1 | R, P($X$), R | $d$ |
| $d$ | 0 | R, R | $e$ |
| $d$ | other | R, R | $d$ |
| $e$ | blank | P(1) | $b$ |
| $e$ | other | R, R | $e$ |

Recall the meaning of the following symbols used for operations.

- R: Move one position to the right.
- L: Move one position to the left.
- E: Erase the currently scanned square.
- P( ): Print whatever is in parentheses in the current square.

(a) What is the precise output of the machine as it just finishes configuration $a$ and enters configuration $b$ for the first time? Justify your answer.

(b) What is the precise output of the machine as it just finishes configuration $a$ and enters configuration $b$ for the second time? Justify your answer.

(c) What is the precise output of the machine as it just finishes configuration $a$ and enters configuration $b$ for the third time? Justify your answer.

(d) Guess what the output of the machine is as it just finishes configuration $a$ and enters configuration $b$ for the $n$th time. Use induction to prove that your guess is correct. Be sure to write carefully the details of this proof by induction.

(e) Design a Turing machine which, when given two arbitrary natural numbers, $n$ and $m$, will compute the product $n \cdot m$. Suppose that the machine begins with the tape

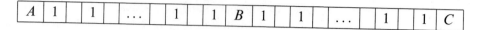

where the number of 1s between $A$ and $B$ is $n$, the number of ones between $B$ and $C$ is $m$, and the machine begins scanning the tape at the far left, reading the symbol $A$. The output of the machine should be:

| $A$ | 1 | | ... | 1 | $B$ | 1 | | ... | 1 | $C$ | 1 | 1 | | ... | | 1 | $D$ | |
|---|---|---|---|---|---|---|---|---|---|---|---|---|---|---|---|---|---|---|

where the number of 1s between $C$ and $D$ is $n \cdot m$. Use induction to verify that the machine produces the correct output.

Letting $T$ denote the Turing machine which multiplies $n$ and $m$ together, so that the value of $T(n, m)$ is $n \cdot m$, design $T$ so that for $n \in \mathbf{N}$,

$$T(n, 1) = n$$

and for $m \in \mathbf{N}$, $m \geq 2$, we have

$$T(n, m) = T(n, m - 1) + n.$$

Such an equation provides an example of a recursively defined function, an important topic in computer science. In our case, the algorithm for multiplication, $T$, is defined in terms of addition, a more elementary operation.

## Conclusion

The written historical projects presented here combine learning techniques from both a projects-based approach to instruction [5] and an historically informed teaching strategy [15]. For implementation in the classroom, introduction to the relevant historical material should begin early in the course, with a brief description of Hilbert's problem list, its effect on the development of mathematics, and the significance of the decision problem. Further commentary about the evolution of Turing's ideas into the modern programmable computer should be provided by the instructor before assigning the projects. If the projects are used verbatim, allow about two weeks for the completion of each. Monitor the students' progress on what for them is a lengthy assignment, provide in-class time for them to work on the projects, and offer guidance when they become perplexed (and they will). The projects should count for a significant percentage of the course grade, with each taking the place of an in-class examination, thus providing incentive to complete them. Two projects are included here, although just one could be used in a given course, or a third could be authored. Modify the projects to fit the course, with the course content and class size being factors which may well influence the questions asked. For example, if set containment has not been covered, this part of Project I (part(e)) may be deleted, or for a large class, deleting part (e) would much simplify the project, and reduce the time needed for grading. Project II could be shortened by asking the students to design a Turing machine to compute the sum of two positive integers, instead of the product. Nonetheless, pedagogical advantages are to be had by asking significant questions, ones which engage the students and reflect the actual development of the subject. Two additional projects in this direction would be the logical description of Turing's "universal computing machine," and Turing's (negative) solution to the decision problem, available at the web resource [1] along with other historical projects.

## Acknowledgments

The author gratefully acknowledges permission from the London Mathematical Society to reproduce excerpts from Alan Turing's paper "On Computable Numbers with an Application to the Entscheidungsproblem," *Proceedings of the London Mathematical Society* **42** (1936), 230–265. The development of curricular materials for discrete mathematics has been partially supported by the National Science Foundation's Course, Curriculum and Laboratory Improvement Program under grant DUE-0231113, for which the author is most appreciative. Any opinions, findings, and conclusions or recommendations expressed in this material are those of the author and do not necessarily reflect the views of the National Science Foundation.

## References

1. G. Bezhanishvili, H. Leung, J. Lodder, D. Pengelley, D. Ranjan, "Teaching Discrete Mathematics via Primary Historical Sources," www.math.nmsu.edu/hist_projects/

2. A. Church, An Unsolvable Problem of Elementary Number Theory, *American Journal of Math.*, **58**, (1936), 345–363.

3. A. Church, A Note on the Entscheidungsproblem, *Journal of Symbolic Logic*, **1**, (1936), 40–41.

4. A. Church, *Introduction to Mathematical Logic*, Princeton University Press, Princeton, New Jersey, 1996.

5. M. Cohen, E. Gaughan, A. Knoebel, D. Kurtz, D. Pengelley, *Student Research Projects in Calculus*, Mathematical Association of America, Washington, DC, 1992.

6. M. Davis, *The Universal Computer*, W. W. Norton & Co., New York, 2000.

7. K. Gödel, Über Formal Unentscheidbare Sätze der Principia Mathematica und Verwandter Systeme, *Monatshefte für Mathematik und Physik* **38**, 1931, 173–198. (English translation in *The Undecidable*, M. Davis, editor, Raven Press, New York, 1965.)

8. I. Grattan-Guiness, *The Search for Mathematical Roots, 1870–1940: Logics, Set Theories and the Foundations of Mathematics from Cantor through Russell to Gödel*, Princeton University Press, Princeton, New Jersey, 2000.

9. D. Hilbert, *Gesammelte Abhandlungen*, Vol. III, Chelsea Publishing Co., New York, 1965.

10. D. Hilbert, Mathematical Problems, Newson M., translator, *Bulletin of the American Mathematical Society*, **8** (1902), 437–439.

11. D. Hilbert, Probleme der Grundlegung der Mathematik, *Mathematische Annalen*, **102**, (1930), 1–9.

12. D. Hilbert, W. Ackermann, *Grundzüge der Theoretischen Logik*, Dover Publications, New York, 1946.

13. D. Hilbert, W. Ackermann, *Principles of Mathematical Logic*, L. Hammond, G. Leckie, F. Steinhardt, translators, Chelsea Publishing Co., New York, 1950.

14. A. Hodges, *Alan Turing: The Enigma*, Simon & Schuster, Inc., New York, 1983. *Breaking the Code*, video available at http://www.wgbh.org/shop/

15. F. Swetz, J. Fauvel, O. Bekken, B. Johansson, V. Katz, editors, *Learn From The Masters!*, Mathematical Association of America, Washington, DC, 1995.

16. A. Turing, On Computable Numbers with an Application to the Entscheidungsproblem, *Proceedings of the London Mathematical Society* **42** (1936), 230–265.

# 12

# From Hilbert's Program to Computer Programming

**William Calhoun**
*Bloomsburg University*

## Introduction

The impact of computers on our lives is obvious to everyone, while mathematical logic is an esoteric subject for most people. Yet the histories of mathematical logic and computers are tightly interwoven. I will discuss the connections between logic and computing, and how I use these connections in my teaching. I find that connecting logic and computing helps to motivate both mathematics and computer science majors. The students are surprised to learn that ideas developed by mathematicians studying abstract questions of logic turned out to be fruitful in the design of computers.

## Teaching Mathematics and Computer Science Majors

The needs, goals and attitudes of mathematics and computer science students are somewhat different. I have tried in various ways to satisfy both camps in my Discrete Mathematics classes. My main goal for the mathematics majors is to develop their logical reasoning. To that end, I have my students write many short proofs for homework. Although I emphasize that logical reasoning is also important in programming and theoretical computer science, most of my computer science majors do not see theorem proving as a priority. To show computer science majors how discrete mathematics will be useful to them, my class includes examples of applications to computer science in areas such as digital logic circuits, the binary number system, logical programming languages, and analysis of algorithms. However, I do not want to do so much of this that the mathematics majors feel they are in a computer science course.

Another excellent way to motivate both mathematics majors and computer science majors is to use history to show them the interconnections between the two subjects. I weave some of the history into my lectures, but I have found that the most effective way to get the students interested is to have them research a topic and give a short presentation. I assign a 10–15 minute talk in class on an application or person related to the math in the course. Many of the students choose to give brief biographies. They have chosen to speak about such contributors to mathematical logic and computing as Cantor, Hilbert, Gödel, Turing, Kleene, von Neumann, Chomsky and Julia Robinson. My students in other classes such as Number Theory also enjoyed learning the history of the subject, but mathematical logic and computer science have the advantage that most of the important results have come in the last 200 years. The students can relate better to lives that are not so remote in time from our own. I recognize that giving up 10–15 minutes of class time per student is a significant cost. However, I believe that if I can get my students interested and engaged in the subject, it is a worthwhile use of the time. In larger classes, I have the students work in pairs to keep the amount of time devoted to talks from becoming overwhelming.

I do not claim to be a professional historian or that my students do original historical research. We are consumers of history researched by others. I also recognize that some of the sources my students use may have questionable validity, particularly if the source is the first web-site the student finds. (There are, on the other hand, some excellent sources on the web such as the *MacTutor History of Mathematics Archive* maintained by the University of St. Andrews [53].) To combat this problem, I require my students to use multiple sources, and I sometimes respond to a student presentation with an opposing point of view I have read. In the end, I think the benefits of giving students a general idea of the history of the subject outweigh the danger that they will learn some questionable or even false claims. History does not have the surety of deductive logic, but that is no reason for mathematicians and computer scientists to avoid history. History helps us understand the organization of our subjects and the interconnections between them.

## Why History Works

Much of the logic, set theory and proof techniques studied in Discrete Mathematics derive from the work of Boole, Cantor, Frege and other nineteenth century logicians. The twentieth century brought fascinating new insights in logic and the invention of the computer. Some of the less technical aspects of twentieth century logic can be used to enliven a Discrete Mathematics course. The topics in my Theory of Computation class follow many of the historical developments in the twentieth century, although not always in chronological order, including work of Hilbert, Gödel, Turing, Kleene, Chomsky, Backus, Cook and others. Taking time to discuss the historical contributions of these mathematicians and computer scientists helps students see the interconnections between the two disciplines.

As discussed above, the impact is even greater if the students research topics and report to the class. For instance, in Discrete Mathematics, I let my students choose from a list of applications. A mathematician or computer scientist is associated with each topic. The student (or pair of students) may chose whether to talk about the application or give a biographical talk about the associated person. Those who choose biographical talks must still give examples of applications to discrete mathematics from their subject's work. Some examples of applications/biographies I have used are: Logic Circuits (Boole), Prolog (Frege), Conditionals and Loops in Programming (Ada Lovelace), Formal Languages (Chomsky), The Halting Problem (Turing), Finite State Automata (Kleene), Cardinality (Cantor) and Computability (Julia Robinson). I schedule the talks throughout the semester so they will roughly coincide with related topics in the course. I recommend that the students prepare transparencies or Power Point slides to accompany the talk and, while I don't require a formal paper, they must turn in their notes and other materials to me after giving the talk, including a list of references. I encourage them to discuss the talk with me in advance during office hours. I evaluate the talks on the following equally weighted criteria: clarity of presentation, organization, good examples, and mechanics (presentation materials, reference list).

Most students seem to enjoy the assignment, judging from their enthusiasm and comments on my teaching evaluations such as "The application project was cool!" I feel this assignment works well for two reasons. The first is "general intellectual interest." This is a phrase I borrowed from noted logician Harvey Friedman, who used it in a foundations of mathematics e-mail discussion to describe a quality of the work of Gödel and Turing. They are the only two mathematicians on *Time* magazine's list of the greatest scientists of the twentieth century. According to Friedman, the reason they achieved this level of fame is the general intellectual interest in their work [18]. To me, general intellectual interest means one can understand and be fascinated by their accomplishments, without knowing all the technical details. The work of many of the logicians discussed below has this quality since they attempted to answer philosophical questions such as:

- What is proof?
- What is computation?
- What is truth?
- What is intelligence?
- What is possible?

The second reason I think students enjoy learning the history of logic and computing is human interest. Our classes tend to emphasize logic and strip away the personal. There is beauty in an elegant proof or algorithm, but our students are human beings who are naturally interested in the lives of the people who created the subject. Since much of the history of mathematical logic and computing is recent, it is particularly easy for students to relate to these lives, and there are many inspirational, humorous and moving anecdotes to tell. My students often include anecdotes in their presentations, and I like to include them occasionally in my lectures as well. Anecdotes keep students interested in the lecture. They can also help motivate students to do mathematics by showing them that mathematics was created by a diverse group of people, many of whom had to persevere against difficult challenges in their lives. A few brief examples are listed below.

- Hilbert supported Emmy Noether's appointment at Göttingen by saying "I do not see that the sex of the candidate is an argument against her admission as a Privatdozent. After all, the [University] Senate is not a bath house." [38, p. 143]
- Gödel applied his logical analysis to the United States Constitution and discovered a "paradox" that would allow the United States to become a dictatorship. [15, pp. 179–181].
- Despite Turing's invaluable contributions to the war effort, he was later persecuted for homosexuality, leading to his tragic suicide. Turing's story is movingly told in Andrew Hodges excellent biography [28], and dramatized in Hugh Whitemore's play *Breaking the Code* [59].
- Von Neumann had a phenomenal memory, ability to do mental calculations, and facility with languages. He once amazed his colleague Herman Goldstine by reciting several pages of *A Tale of Two Cities* from memory [21, p. 166]. He was also a fun-loving gregarious person who gave legendary parties at his home in Princeton [36, pp. 97–122].
- A reporter called the University of California at Berkeley for information about Julia Robinson and was told she was "Professor Robinson's wife." He responded, "Well, Professor Robinson's wife has just been elected to the National Academy of Sciences!" [37, p. 79] (Constance Reid's beautiful little book on her sister, Julia Robinson, has been particularly inspiring to some of my female students.)

## Nineteenth Century Logic and Computing

As a convenience to readers who wish to use the historical connections between logic and computing in their teaching, this section and the next provide a brief survey. More extensive overviews along these lines can be found in *The Universal Computer* by Martin Davis [14] and in the *Companion Encyclopedia of the History and Philosophy of the Mathematical Sciences* edited by I. Grattan-Guinness [22, pp. 595–707]. This section surveys some of the developments in nineteenth century logic and computing that laid the groundwork for the invention of the modern computer in the twentieth century.

### Logic

George Boole (1815–64) gave mathematical foundations to propositional logic in his *The Mathematical Analysis of Logic* (1847) [3] and *An Investigation of the Laws of Thought* (1854) [4]. Building on work of Augustus De Morgan (1806–71), Boole's system was a great advance on the *Organon*, Aristotle's classic treatment of syllogisms. Boole showed that a slight variant on ordinary arithmetic could be used to represent logical arguments. He represented classes, such as the class of *sheep* or the class of *horned animals*, by variables and showed how to represent intersections, unions and complements of classes by algebraic expressions. He used 1 to represent the universal class and 0 to represent the empty class. Thus he started the practice followed in modern computing of representing *true* by 1 and *false* by 0. The truth tables of propositional connectives can then be expressed in algebraic terms, with $xy$ expressing *x and y* and $1 - x$ expressing *not x*. Since the arithmetic fact $1 + 1 = 2$ doesn't make sense in the new context, Boole used $x + y$ only for disjoint unions. Other logicians such as Stanley Jevons (1835–1882) and Hugh MacColl (1837–1909) revised Boole's system, reinterpreting $+$ as the inclusive or, and adding the new rule $1 + 1 = 1$ to Boole's laws. The resulting Boolean algebra gives students a unified approach to logic, set operations, and logic circuits.

Charles Dodgson (1832–98) and John Venn (1834–1923) popularized the use of diagrams for solving logic problems. Gottfried Leibniz (1646–1716) and Leonhard Euler (1707–83) had previously used diagrams in similar ways. The primary legacy of Dodgson and Venn is in logic instruction, but the use of diagrams points toward the mechanization of logic. Dodgson is better known under his pen name, Lewis Carroll. Carroll's *The Game of Logic* [8] gives an algorithm for students to solve logic problems by moving counters on a rectangular board. The process is similar to doing arithmetic on an abacus. Carroll's board is rarely used today, but is equivalent to Venn's circle diagrams. Carroll's famous *Alice* books contain many amusing references to logic, and are often quoted in logic text books. His own logic books include hundreds of problems that students still enjoy solving because of their quirky humor. Contrary to the views of other logicians, Carroll argued that the universal quantifier should imply existence. For instance, he thought the statement *All unicorns have a horn* should imply that at least one unicorn exists. Carroll was on the losing side of this argument: the mathematical properties of the quantifiers are simpler the other way. However, it is worth discussing the issue with students. Many of them agree with Carroll, and have to learn to accept the standard usage.

Georg Cantor (1845–1918), the founder of set theory, published his initial investigations of infinite cardinalities in 1874 [7]. He introduced the idea that two infinite sets have the same cardinality if there is a one-to-one correspondence between them. He showed the set of algebraic numbers is countable, that is, it has the same cardinality as the set of whole numbers. In devising one-to-one correspondences between the whole numbers and other sets, Cantor showed that many kinds of information can be coded by whole numbers, an important idea for computer science. Cantor developed the diagonal argument to show that the set of real numbers is not countable. Diagonal arguments have been used repeatedly in 20th century logic and computer science, notably in Turing's proof of the unsolvability of the Halting Problem. Cantor used the diagonal argument to show that the cardinality of a set is always less than the cardinality of its power set. He realized that this implied there could be no set of all sets, an observation that preceeded Russell's famous paradox. (See below.) For further reading on Cantor see Dauben's biography [12].

Symbolic logic was brought to fruition by Gottlob Frege (1848–1925), who developed the rules of predicate logic in his *Begriffsschrift* (*Conceptual Notation*, 1879) [16]. A predicate is a propositional function such as "$x$ is larger than $y$." The predicate takes on a truth value when appropriate words are substituted for the variables. Frege's system, unlike previous ones, included predicates with any number of variables, and he worked out laws for the universal quantifier, *for all x*. Thereby, Frege made it possible to express any mathematical proof formally. (The existential quantifier, *for some x*, can be defined from the universal quantifier and negation.) Frege used a two-dimensional notation that was rather cryptic and cumbersome. Most of the logical notation we use today comes from others, such as Charles Sanders Peirce (1839–1914), who independently studied predicates and quantifiers, and Giussepe Peano (1858–1932), who developed axioms for arithmetic in 1889.

In his *Foundations of Arithmetic* [17], Frege sought to go further than Peano and *define* the integers within logic. However, Frege was bitterly disappointed by a 1902 letter from Bertrand Russell (1872–1970) that arrived when the second volume of his work was already in press. Russell pointed out that within Frege's system one could define a predicate $w(v)$ that is true if and only if $v$ is a predicate and $v(v)$ is false. But if $w$ itself is substituted for $v$ this leads to a contradiction: $w(w)$ is true if and only if $w(w)$ is false. Frege was forced to add an appendix to his book with an attempt to work around the inconsistency, an attempt that was later shown to be unsuccessful. Russell's paradox struck a blow to Frege's program to define mathematics within logic. In the appendix, Frege wrote, "A scientist can hardly meet with anything more undesirable than to have the foundation give way just as the work is finished. In this position I was put by a letter from Mr. Bertrand Russell as the work was nearly through the press." On top of this disappointment came his wife's death in 1904. Frege was thrown into a depression and did little work for several years. While this strikingly melancholy anecdote is often told, it is important that students know that Russell's Paradox did not undermine Frege's success in developing predicate logic. Frege's achievement stands as one of the greatest advances in the history of logic. For further reading on Frege, see Harold Noonan's *Frege: A Critical Introduction* [34].

Russell noted that his paradox also arises if we consider the set of all sets that are not members of themselves. This, and other similar paradoxes, required a revision of set theory and stimulated much work in logic around the turn of the century. Russell and Alfred North Whitehead (1861–1947) developed type theory to work around Russell's paradox and continued Frege's work on the logical foundations of mathematics in their monumental *Principia Mathematica* (1911–13) [58]. However, type theory is cumbersome and Russell's paradox was resolved in a simpler way by restricting the set existence axioms in Ernst Zermelo's (1871–1953) axiomatization of set theory (1908) [62]. This effectively makes a distinction between sets and classes, where classes are collections that are "too big" to be sets, such as the class of all sets. Paradoxes were later used by twentieth century logicians such as Alfred Tarski (1902–83) and Kurt Gödel (1906–78) to prove limitations on formal systems. In 1933, Tarski used the ancient paradox of the liar ("This sentence is false.") to show that truth cannot be defined within a formal system [47]. Gödel used a variant of the liar paradox in the proof of his Incompleteness Theorems. Raymond Smullyan's puzzle books such as *What is the Name of this Book?* [43] and *Forever undecided: a puzzle guide to Gödel* [44] have lots of entertaining logic problems that can be used to introduce students to the kind of reasoning used by Tarski and Gödel. For a detailed outline of the developments in mathematical logic from Boole to Gödel, see *The Search for Mathematical Roots 1870–1940* by I. Grattan-Guinness [23].

## Computing

Desire to ease the burden of laborious calculation led many mathematicians and scientists to work on computing machines. Although analog devices such as slide rules, planimeters and eventually harmonic and differential analyzers played an important role in scientific computation, this survey will focus on digital computing. Mechanical calculating devices had been around since the 1600's when Blaise Pascal (1623–62) built a machine that could add and subtract, and Gottfried Leibniz (1646–1716) built one that could multiply and divide as well. Leibniz dreamed of a calculating machine similar to a modern computer. He thought such a machine would settle all logical arguments. In 1805 Joseph-Marie Jacquard (1752–1834) invented a programmable loom. The Jacquard Loom was controlled by a series of punched cards that were fed into the machine in sequence. Later, others would use Jacquard's innovation to advance toward Leibniz's dream.

George Babbage (1792–1871), an English mathematician and inventor, spent much of his life building computing machines. His many interests also included codes and mathematical notation. He used mathematical notation to describe switching mechanisms in his paper *On expressing by signs the action of machinery* [1]. In 1822, he began work on the Difference Engine, a machine designed to generate mathematical tables by solving difference equations. He never completed the project, despite substantial government funding, but a working Difference Engine based on Babbage's design was built in 1852. Meanwhile, Babbage had begun work on a more ambitious machine, the Analytical Engine. This was to be a calculating machine that could be programmed using Jacquard's punched cards. Lady Ada Lovelace (1815–52), who had studied logic and mathematics with DeMorgan, became fascinated with Babbage's machine and wrote notes explaining it, including example programs [33]. She used a nice analogy to describe the machine: "We may say most aptly that the Analytical Engine weaves *algebraical patterns* just as the Jacquard-loom weaves flowers and leaves." [33, Note A] Babbage received little support for the difficult task of assembling a device of such complexity, and he died in 1871 with only scattered pieces of the machine completed. Alan Turing later gave credit to Babbage for conceiving the universal computer, but the design of the Analytical Engine differed from modern computers in that it did not use stored programs. Still, Babbage anticipated many of the ideas later used in computing. He was ahead of his time, and was hindered by the limitations of the technology available to him. For more information on Babbage and Lady Lovelace, including historical documents, see the *Analytical Engine* web site [56].

Herman Hollerith (1860–1929), also inspired by the Jacquard loom, developed a tabulating machine that was used in the 1890 United States Census [21, pp. 65–71]. The machine used punched cards to enter the data. Hollerith formed the Tabulating Machine Company in 1896, which eventually became International Business Machines.

## The Twentieth Century

Although the earlier developments in logic and computing discussed above show some of the connections between the two subjects, it is in the twentieth century that the connections become most significant. We can trace the genesis of "computer code" from Hilbert to Gödel to Turing and von Neumann. Students in upper-level courses in Mathematical Logic or Theory of Computation encounter concepts such as *provability*, *consistency*, *decidability* and *computability*. In those classes, the history of twentieth century logic is directly relevant to the course material. The concepts listed above are ordinarily beyond the scope of Discrete Mathematics courses. Still, Discrete Mathematics students who research the lives of twentieth century logicians and computer scientists have sufficient background to appreciate the "general intellectual interest" in their work. Mathematics and computer science majors also discover a common heritage by learning the history of twentieth century logic and computing.

### David Hilbert

In 1900, David Hilbert (1862–1943) gave his famous speech outlining the most important mathematical problems for the next century. Hilbert was remarkably prescient in listing problems that turned out to be important avenues for mathematical investigation. Hilbert's second problem was one that might be considered rather esoteric: to investigate the consistency of the arithmetical axioms. Surprisingly, we can follow a trail from this question to the most dramatic technological change of our time, the invention of the computer. Hilbert had shown that the consistency of Euclidean geometry could be reduced to the consistency of arithmetic, so he now asked for a proof that arithmetic is itself consistent. His interest in the consistency of mathematical systems was strengthened in the years that followed by disputes with the intuitionists such as L.E.J. Brouwer (1881–1966), who challenged the validity of nonconstructive methods of mathematical reasoning such as proof by contradiction. Although Brouwer's objections were based on philosophical grounds, there is a more direct concern about proof by contradiction: what if mathematics itself is inconsistent? If so, any statement, true or false, could be proved by contradiction. As inconsistencies were discovered in Frege's attempt to formalize arithmetic and in Cantor's set theory, Hilbert wanted to put mathematics on a solid footing.

Hilbert's program was to use finitistic methods, acceptable even to intuitionists, to show that arithmetic, and eventually all of mathematics, is consistent. He began the new field of proof theory in which proofs are formalized and treated as mathematical objects. To keep the intended consistency proof from being circular, Hilbert realized that he needed two sorts of mathematics: the ordinary mathematics being studied and a finitary *metamathematics* in which to prove facts about mathematical proofs. By formalizing proofs symbolically, Hilbert started an intensive investigation of the use of strings of symbols to code statements. Furthermore, he called attention to decision problems. For instance, is there an algorithm to determine which statements of a mathematical system have proofs? This question led Alan Turing and others to formulate the definition of *computability*, a concept that preceded and inspired the construction of physical computers. It is certainly reasonable to say that, although Hilbert never worked on computing directly, Hilbert's program led others to important contributions in the development of computers and computer programming.

For further reading, see Constance Reid's *Hilbert* [38]. Hilbert's Problems and their influence have been discussed in *The Honors Class: Hilbert's Problems and Their Solvers* by Ben H. Yandell [61] and in recent articles by Ivor Grattan-Guinness [24] and Rüdiger Thiele [48].

## Kurt Gödel

In 1931, Kurt Gödel brought an end to Hilbert's Program by publishing his famous Incompleteness Theorems [19]. Gödel showed that any potential axiomatization of mathematics, or even of arithmetic, would necessarily be incomplete in the sense that there would be sentences expressible in the system that could neither be proved nor disproved in the system. In addition, Gödel showed that no consistent mathematical theory strong enough to prove the basic axioms of arithmetic could prove its own consistency. Thus metamathematics couldn't prove its own consistency, much less the consistency of arithmetic. Hilbert's Program was doomed to failure.

Gödel's Theorems are often described only as negative results showing the limitations of mathematics. However, an examination of Gödel's proof reveals novel ideas that were absorbed by some of the most influential visionaries of computing. Gödel took Hilbert's symbolic formalization of proofs a step further by assigning a number to each symbol and using those numbers to form code numbers for sentences. These code numbers could be combined to obtain a number coding an entire proof. Gödel numbering was a revolutionary new way to relate text to numbers. The longest part of Gödel's proof was to show that logical relationships between sentences could be mirrored by arithmetic relationships between their code numbers. Ultimately he showed that proof theoretic statements such as "The number $p$ codes a proof of the sentence with code $s$" could be expressed arithmetically. He was then able to show that in any potential axiomatization of arithmetic, one could form a sentence with code number $c$ that expressed "the sentence with code number $c$ is not provable." In other words, through coding, Gödel's sentence said "This sentence is not provable." Let's assume the system is *correct* in the sense that it doesn't prove any false sentences. Then the system cannot prove Gödel's sentence, since if it did the sentence would be false. So, Gödel's sentence cannot be proved, and is therefore a true sentence expressible in the language of arithmetic that can neither be proved nor disproved. (The discussion here has been simplified by the assumption that the system is correct. Gödel used a weaker, somewhat technical hypothesis. J. Barkley Rosser (1907–89) suggested a slight variation on Gödel's sentence that allowed the proof to go through assuming only that the system is consistent.) Although Gödel did not directly consider computing machines, his numerical coding of text and use of the computations of arithmetic to express logical relationships foreshadow the later development of computer code and the parsing of programming languages. In fact, Gödel influenced two of the founders of computer science, Turing and von Neumann.

Gödel also contributed to theoretical computer science in a more direct way. Not all axiomatizations of mathematics are useful. For instance, if one declares that every true sentence is an axiom, the result is a complete system, but a useless one, since there is no way to tell which sentences are axioms. Gödel's theorem does not apply to such useless axiomatizations, but only to ones where there is an effective method to recognize axioms. To clarify the hypotheses for his theorems, Gödel defined the *recursive functions* in 1934 [20], basing his definition on a suggestion of Jacques Herbrand (1908–31). The recursive functions are ones that can be computed by a sequence of substitutions according to a fixed set of rules. Recursive substitutions such as $f(x) = f(x-1) + f(x-2)$ are allowed. Gödel suspected that all functions that can be computed by following an algorithm are recursive, but he was not sure. A few years later, Turing and Kleene showed that Gödel's recursive functions coincide with the computable functions. Thus, in a sense, Gödel was the first to define computability, but he did not attempt to design a computing mechanism.

For further reading see Dawson's carefully researched biography [15]. Other books on Gödel's life, philosophy and work include a slim and accessible guide by Hintikka [27] and the dense, idiosyncratic musings of Gödel's friend Hao Wang [57]. Douglas Hofstadter's Pulitzer Prize-winning best seller, *Gödel, Escher, Bach* [30] provides an entertaining, nontechnical introduction to Gödel's theorems and their connections with computer science.

## Alan Turing

Alan Turing (1912–1954) studied Hilbert's Program and Gödel's Theorems in a course at Cambridge University in 1935. He became fascinated by the Entsheidungsproblem, a question of Hilbert's that had not been resolved by Gödel. (See Davis [14, pp. 146–7].) The Entsheidungsproblem was to find a procedure to determine whether or not a given statement could be proved from a given set of axioms using predicate logic. Turing suspected that there was no such decision procedure, but in order to prove this he needed to precisely define what a decision procedure is. Clearly a decision procedure should be an algorithmic process that does not involve creative thought, chance or magic. But how could one define an algorithm?

In 1936, Turing published *On Computable Numbers with an Application to the Entsheidungsproblem* [49], in which he defined a type of abstract machine and argued that any given algorithm could be carried out by one of his machines. Turing's article is so clearly written that the opening pages may be given to students as an introduction to Turing machines. A Turing machine consists of a control unit with a finite number of possible states, a paper tape divided into squares and a tape head that can read and write symbols. At each step of a computation the tape head scans one square of the paper tape. Depending on the symbol scanned and the current state of the control unit, the Turing machine may write a new symbol on the current square, move the tape head one square to the left or right and the control unit may enter a new state. The new state may cause the machine to halt or to continue as described above. Turing argued convincingly that any algorithmic process could be simulated using a series of these basic steps. He added weight to this argument by showing that Turing machines could perform the computations of arithmetic. Turing then gave an example of a problem with no algorithmic solution. The Halting Problem is to determine whether or not a particular Turing machine will ever halt. Turing used a variation on Cantor's diagonal argument to show that none of his machines could solve the Halting Problem. Turing's mathematical definition of Turing machines shows that each instance of the Halting Problem can be expressed as a mathematical statement. Furthermore, if the machine halts, a mathematical simulation of the machine proves that it halts. So, as Turing noted, the algorithmic unsolvability of the Halting Problem implies there is no decision procedure for the provable statements of mathematics.

Shortly after completing his paper, Turing learned that Alonzo Church (1903–95), a logician at Princeton, had independently settled the Entsheidungsproblem [9] [10]. Church had characterized algorithmic processes using his *lambda calculus*, and his student Stephen Kleene (1909–94) showed the functions definable in the lambda calculus were the same as Gödel's recursive functions [31]. On learning of Church's work, Turing soon showed that Turing machines could compute any function definable in the lambda calculus and vice versa [50]. Another independent characterization of algorithms came from Emil Post (1897–1954). In a 1936 paper [35], he gave a sketchy description of a type of machine similar to Turing's. Post machines were also shown to compute the same functions as Turing machines. Thus the approaches of Gödel, Turing, Church and Post all provided equivalent ways to define an algorithm. Turing's penetrating analysis of computation convinced Gödel, who had been unsure whether his definition captured all possible forms of computation, that the correct definition of computability had been found.

For the history of computing, the most significant result in Turing's 1936 paper was his construction of a Universal Turing machine. Turing showed that, rather than building a different Turing machine for each task, a single Universal machine could be used. To simulate the operation of a Turing machine, a code for the given machine is placed on the tape of the Universal machine along with the input data. Thus Turing gave the first description of a universal stored-program computer. Although Turing's paper was a theoretical one, and provided none of the engineering details required to build such a machine, the concept was clearly described and later served as an outline for Turing and others in designing physical computers.

Turing spent two years working with Church at Princeton and then returned to England. During World War II he worked with a team of code breakers at Bletchley Park. In the 1930's, Polish mathematicians had developed a method for cracking the German Enigma code [41]. However, the Enigma device could be set multiple ways and was frequently modified, so the code had to be cracked repeatedly. Turing made vital contributions to the design of an electro-mechanical computing device called a Bombe to speed up the process. Turing applied mathematical logic to achieve a much faster and more powerful method of cracking the Enigma code than had been possible before [29]. A large electronic code breaker called the Colossus was also built at Bletchley park. Turing was not directly involved in the construction of this device, but he surely saw the potential for using vacuum tubes to build a realization of his universal Turing machine. After the war Turing turned his attention to doing just that. In 1945 he drafted a report to the National Physics Laboratory giving detailed plans for a universal stored-program computer called the Automatic Computing Engine [52]. Because of difficulty obtaining funding, it was not until 1950 that a pilot version of the ACE was constructed, so Turing's ACE report was not as influential as it might have been. However, Turing clearly deserves recognition as a founder of computer science.

For further reading see Andrew Hodges excellent biography [28]. Hodges also maintains the Alan Turing home page on the World Wide Web [29]. This is a great source on Turing, including new information that has become available since the biography was published. *The Universal Turing Machine: A Half Century Survey*, edited by Rolf Herken [26] contains historical articles on Turing's work as well as articles on related research in mathematical logic and computer science. For further reading on the definition of computability see Robert Soare's article *Computability and Recursion* [45].

## John von Neumann

John von Neumann (1903–57) was a great mathematician with a wide range of interests, including logic and computing. As an undergraduate student he worked on clarifying Cantor's ordinal numbers, and in 1925 he published his own axioms for set theory [54]. He then worked with Hilbert on both quantum mechanics and proof theory. He achieved some partial results toward proving the consistency of mathematics, but when he heard Gödel speak about his Incompleteness Theorems at a symposium in 1930, he realized that Hilbert's Program was hopeless, and he abandoned his work in logic. In 1931, von Neumann moved to the United States to teach at Princeton University and he joined the Institute for Advanced Study in 1933. As mentioned above, Alan Turing worked with Church in Princeton from 1936 to 1938. During this period, von Neumann did not formally work with Turing, but it is apparent from discussions he had with others that he became familiar with Turing's work. In fact, von Neumann asked Turing to stay on as his assistant in 1938, but Turing returned to England instead.

During World War II, von Neumann became involved in several military projects, including the development of the atomic bomb at Los Alamos. This work involved many lengthy calculations, and von Neumann became interested in computing machinery. In 1944, von Neumann had a chance encounter with mathematician Herman Goldstine (1913–2004) who was supervising a computing project for the Army Ordnance Department [21, p. 182]. Goldstine invited von Neumann to see the Electronic Numerical Integrator And Computer, under development at the Moore School of Electrical Engineering at the University of Pennsylvania. The mammoth electronic computing device used 18,000 vacuum tubes, and was designed by a team led by engineer J. Presper Eckert (1919–95) and physicist John Mauchly (1907–80). The other members of the team were engineers, except for Arthur Burks (1915– ), a mathematician and logician. The ENIAC was not a stored program computer. It might best be described as an electronic version of a desk calculator that could be programmed by setting external switches and connecting patch cords. Numbers were entered from an IBM punched card reader and were represented internally in base 10.

At the time of von Neumann's visit, plans were already being made to design a more sophisticated computer, the Electronic Discrete Variable Computer. The Army began funding development of the EDVAC in January 1945. Von Neumann became heavily involved in the logical design of the machine, and in June 1945 he wrote the *First Draft of a Report on the EDVAC* [55]. Von Neumann never wrote another draft of the report, but the first draft was widely circulated and highly influential. The EDVAC would be a general purpose, binary, stored-program computer with what is now called the von Neumann architecture. It had five main parts: *Central Arithmetic, Central Control, Memory, Input* and *Output*. The operation of such a computer consists of a cycle in which the *Central Control* fetches an instruction from *Memory* at the location indicated by a program counter, increments the program counter, decodes the instruction, and performs the indicated operation. The operation could involve sending a command to *Central Arithmetic*, moving data in or out of *Memory*, changing the program counter (to cause a branch in program execution), or sending a command to the *Input* or *Output* devices. Note that the cycle is essentially the same as in a Turing machine, except that the machine von Neumann described is more efficient since there is random access to the *Memory* and numerical calculations can be rapidly performed by the *Central Arithmetic*. Von Neumann described the EDVAC using a neurological analogy he adapted from a paper by Pitts and MacCulloch, that in turn had been inspired by Turing's 1936 paper [14, p. 183]. The designs of many other computers around the world were based on the EDVAC and a more advanced computer built by von Neumann, Goldstein, Burks and others at the Institute for Advanced Study between 1946 and 1952.

For further reading on von Neumann see Norman Macrae's biography [32]. Herman Goldstine's *The Computer from Pascal to von Neumann* [21] contains a first-hand account of the development of the ENIAC, EDVAC and Institute for Advanced Study computers. Students may also enjoy Ed Regis's popular history of the Institute for Advanced Study, *Who Got Einstein's Office* [36]. It includes chapters on Gödel and von Neumann.

## Invention of the Modern Computer

It is not clear how much of the EDVAC Report represents original contributions by von Neumann since he was summarizing the work of the team. Eckert and Mauchly later attempted to patent both the ENIAC and EDVAC. Goldstine gave von Neumann primary credit for designing the first stored-program computer, but Eckert and Mauchly claimed they had conceived of stored programs in 1943 [46, pp. 74–75]. They said they chose not to use stored programs in the ENIAC because they wanted to build it quickly for war-time use. For patent purposes, the issue was moot, since government lawyers determined that the dissemination of the EDVAC Report had put the design in public domain [46, p. 97]. It is interesting to note that in 1948 von Neumann, with the help of Adele Goldstine and Richard Clippinger, managed to program the ENIAC to function as a limited stored program computer [21, pp. 233–235]. This is similar to Turing's use of the stored program concept in his construction of a universal Turing machine. The ENIAC could be programmed much faster as a stored-program machine, with it's switches and patch cords left in a particular configuration, and it continued to run this way until the end of its operating life in 1955.

Eckert and Mauchly did obtain a patent on the ENIAC. They later assigned the patent to Sperry Rand, but it was invalidated in a 1971–1973 court case between Honeywell and Sperry Rand [46, p. 34]. The judge's ruling was partly based on the fact that a physicist at Iowa State University, John Atanasoff (1903–95), had developed a special purpose electronic digital computer in 1940 [25]. Atanasoff's machine was not a general purpose computer but a device to solve systems of linear equations. Mauchly had visited Atanasoff in 1941 to see the computer and learn how it worked. Although Mauchly testified that he had begun his own studies of digital electronic circuits in the 1930's and didn't use any of Atanasoff's ideas in the ENIAC, the judge ruled otherwise. The judge also noted that although Mauchly and Eckert were the inventors of the ENIAC, "the work on the ENIAC was a group or team effort and that inventive contributions were made by Sharpless, Burks, Shaw and others." [46, p. 34]

It is impossible to assign credit for the modern (digital, electronic, stored-program) computer to one person. The teams that designed the ENIAC and EDVAC drew on numerous sources, including earlier mechanical calculating machines, Atanasoff's electronic computer, and the contemporaneous electromechanical relay-based machines such as Howard Aiken's (1900–73) Harvard-IBM Mark I, built in 1944 [11]. (German engineer Konrad Zuse (1910–95) had built a relay-based computer in 1941 [39], but it was unknown in the United States.) Clearly a good deal of credit is due to the brilliant engineers, such as Eckert, who developed the required technology. Still, over the course of time, the technology has changed, leaving only the fundamental logical structure. Turing seems to have been the first to clearly state the concept of a universal stored program computer and von Neumann seems to have been the first to disseminate a practical design for such a machine. Their backgrounds in mathematical logic, work on Hilbert's Program and understanding of Gödel numbering put them in a position to be leaders in the development of computing.

For further reading, Nancy Stern's *From ENIAC to UNIVAC* [46] attempts to assess the contributions of Eckert, Mauchly, von Neumann and others using original documents, interviews and information from the Honeywell-Sperry Rand trial. It includes a complete reprinting of von Neumann's EDVAC report. *The First Computers—History and Architectures*, edited by Raúl Rojas and Ulf Hashagen [40], provides technical information and history for the ENIAC and EDVAC as well as other machines designed by Atanasoff, Aiken, von Neumann, Zuse, and others.

## More Twentieth Century Logic and Computer Science

In addition to Hilbert, Gödel, Turing and von Neumann, a few of the many others who contributed to logic and computer science in the 20th century are listed below. Some of them have been discussed above. Any of them would make an excellent choice for a student's biographical report.

- Bertrand Russell: Russell's paradox, *Principia Mathematica*
- Alonzo Church: The lambda calculus, the undecidability of predicate logic
- Stephen Kleene: Recursion theory, finite automata
- Emil Post: Post machines, formal grammars
- Noam Chomsky (1928– ): formal grammars
- John Backus (1924– ): Backus-Naur (context-free) grammars

- Julia Robinson (1919–85), Martin Davis (1928– ), Hilary Putnam (1926– ), Yuri Matijasevic (1947– ): Undecidability of Hilbert's 10th problem (to determine which Diophantine equations have solutions)
- John McCarthy (1927– ): Lisp (based on the lambda calculus)
- Stephen Cook (1939– ): Complexity Theory, P=NP? (inspired by analogy to work of Church and Turing)

Students can find short biographies and lists of additional references for Russell, Church, Kleene, Post and Robinson in the *MacTutor History of Mathematics Archive* [53]. Another good source for Julia Robinson is Constance Reid's *Julia: A Life in Mathematics* [37]. *Noam Chomsky: A Life of Dissent* by Robert Barsky [2] is one of many books on Chomsky. *Out of Their Minds: The Lives and Discoveries of 15 Great Computer Scientists* by Dennis Shasha and Cathy Lazere [42] includes interviews with Backus, McCarthy and Cook.

## Conclusion

While computer science has established itself as a separate discipline from mathematics, the two subjects are still closely linked, with logic at the heart of both. Ironically, the links between mathematics and computer science have become less obvious since, as Turing foresaw, computers are now used for many purposes other than numerical computation. Today, students may think of computers primarily as machines for text-processing, communicating, playing music and playing games. While technological advances made possible the small, fast and inexpensive computers of today, the amazing range of computer applications is implicit in the theoretical concept of the universal stored-program computer. This concept, as we have seen, arose in work on Hilbert's program by Gödel, Turing and Von Neumann. Instructors can use this fascinating history in their classes to help mathematics and computer science students to appreciate the power of logic and the depth of the connections between the two disciplines.

## References

1. Charles Babbage, "On a Method of Expressing by Signs the Action of Machinery," *Philosophical Transactions of the Royal Society of London* **166** (1826) 250–265.

2. Robert Barsky, *Noam Chomsky: A Life of Dissent*, MIT Press, Cambridge, Mass., 1997.

3. George Boole, *The Mathematical Analysis of Logic*, Macmillan, Barclay, and Macmillan, Cambridge, UK, 1847.

4. ——, *An Investigation of the Laws of Thought*, Walton and Maberly, London, 1854; Dover Publications, New York, reprinted 1973.

5. T. Bynum, tr. and ed., *Conceptual Notation and Related Articles*, Oxford University Press, Oxford, 1972.

6. Martin Campbell-Kelly and William Aspray, *Computer: A History of the Information Machine*, Harper Collins, New York, 1996.

7. Georg Cantor, "Über eine Eigenschaft des Inbefriffes aller reellen algebraischen Zahlen," *Journal für die reine und angewandte Mathematik* **77** (1874) 258–62.

8. Lewis Carroll, *Symbolic Logic, The Game of Logic: Mathematical Recreations of Lewis Carroll: 2 Books Bound As 1*, Dover Publications, New York, reprinted 1958.

9. Alonzo Church, "An unsolvable problem of elementary number theory," *American Journal of Mathematics* **58** (1936) 345–63.

10. ——, "A note on the Entsheidungsproblem," *Journal of Symbolic Logic* **1** (1936) 40–41; correction 101–102.

11. I. Bernard Cohen, "Howard Aiken and the Dawn of the Computer Age," in [40] 107–120.

12. Joseph Warren Dauben, *Georg Cantor: his mathematics and philosophy of the infinite*, Harvard University Press, Cambridge, Mass., 1979.

13. Martin Davis, ed., *The Undecidable*, Raven Press, Hewlett, NY, 1965.

14. ——, *The Universal Computer*, W.W. Norton & Co., New York, 2000; reprinted as *Engines of Logic*, W.W. Norton & Co., New York, 2001.

15. John W. Dawson, Jr., *Logical Dilemmas: The Life and Work of Kurt Gödel*, A.K. Peters, Wellesley, Mass., 1997.

16. Gottlob Frege, "Conceptual Notation" in [5] 101-203. (Originally published in German in 1879.)

17. ——, *Grundgestze der Arithmetik* (vol. 1 of 1893 and vol. 2 of 1903, combined in one vol.), Olms, Hildesheim, reprinted 1962.

18. Harvey Friedman, *FOM: NYC logic conference/Wide Perspective* (web site), http://www.cs.nyu.edu/pipermail/fom/1999-December/003577.html

19. Kurt Gödel, "Über formal unentsheidbare Sätze der Principa Mathematica und verwandter Systeme I," *Monatshefte für Mathematik und Physik,* **38** (1931) 173–198; translated in [13].

20. ——, "On Undecidable Propositions of Formal Mathematical Systems," Lecture notes taken by Kleene and Rosser at the Institute for Advanced Study; reprinted in [13].

21. Herman H. Goldstine, *The Computer from Pascal to von Neumann*, Princeton University Press, 1972.

22. I. Grattan-Guinness, ed., *Companion Encyclopedia of the History and Philosophy of the Mathematical Sciences*, Routledge, New York, 1994.

23. ——, *The Search for Mathematical Roots 1870-1940: Logics, Set Theories and the Foundations of Mathematics from Cantor through Russell to Gödel*, Princeton University Press, Princeton NJ, 2000.

24. ——, "A sideways look at Hilbert's twenty-three problems of 1900," *Notices of the American Mathematical Society* **47** (2000) 752–757; corrections by G. H. Moore and response of Grattan-Guinness, *Notices* **48** (2001) 167.

25. John Gustafson, "Reconstruction of the Atanasoff-Berry Computer," in [40] 91–106.

26. Rolf Herken, ed., *The Universal Turing Machine: A Half-Century Survey*, Springer-Verlag, Wien-New York, 1995.

27. Jaakko Hintikka, *On Gödel*, Wadsworth/Thompson Learning, Inc., Belmont CA, 2000.

28. Andrew Hodges, *Alan Turing: The Enigma*, Simon & Schuster, New York, 1983.

29. ——, *The Alan Turing Home Page* (web site), http://www.turing.org.uk/turing/index.html

30. Douglas Hofstadter, *Gödel, Escher, Bach*, Basic Books, New York, 1979.

31. Stephen C. Kleene, "$\lambda$-definability and recursiveness," *Duke mathematics journal* **2** (1936) 340–353.

32. Norman Macrae, *John von Neumann*, Pantheon Books, New York, 1992.

33. L. F. Menabrea (Ada Augusta, Countess of Lovelace, translation with additional notes), *Sketch of The Analytical Engine Invented by Charles Babbage*, Bibliothèque Universelle de Genève, October, 1842, No. 82; available at [56].

34. Harold W. Noonan, *Frege: a Critical Introduction*, Polity Press, Cambridge, UK, 2001.

35. Emil L. Post, "Finite combinatory processes–formulation I," *Journal of Symbolic Logic* **1** (1936) 103–105.

36. Ed Regis, *Who Got Einstein's Office?*, Addison-Wesley Pub. Co., Reading, Mass., 1987.

37. Constance Reid, *Julia: A Life in Mathematics*, Mathematical Association of America, Washington, D.C., 1997.

38. ——, *Hilbert*, Springer-Verlag, New York, 1970.

39. Raúl Rojas, "The Architecture of Konrad Zuse's Early Computing Machines," in [40] 237–261.

40. Raúl Rojas and Ulf Hashagen, editors, *The First Computers – History and Architectures*, MIT Press, Cambridge, Mass., 2000.

41. Tony Sale, *Codes and Ciphers in the Second World War* (web site), http://www.codesandciphers.org.uk/index.htm

42. Dennis Shasha and Cathy Lazere, *Out of Their Minds: The Lives and Discoveries of 15 Great Computer Scientists*, Springer-Verlag, New York, 1998.

43. Raymond Smullyan, *What is the Name of This Book?*, Prentice-Hall, Englewood Cliffs, NJ, 1978.

44. ——, *Forever undecided: a puzzle guide to Gödel*, Knopf, New York, 1987.

45. Robert I. Soare, "Computability and Recursion," *Bulletin of Symbolic Logic* **2** (1996) 284–321.

46. Nancy Stern, *From ENIAC to UNIVAC*, Digital Press, Bedford, Mass., 1981.

47. Alfred Tarski, "The Concept of Truth in Formalized Languages," in [60] 152–278. (Originally published in Polish in 1933.)

48. Rüdiger Thiele, "Hilbert's twenty-fourth problem," *American Mathematical Monthly,* **110** (2003) 1–24.

49. Alan Turing, "On Computable Numbers with an Application to the Entsheidungsproblem," *Proceedings of the London Mathematical Society*, ser. 2 **42** (1936) 230–67. Correction: **43** (1937) 544–46; reprinted in [13].

50. ——, "Computability and λ-definability," *Journal of Symbolic Logic* **2** (1937) 153–163.

51. ——, "Proposal for Development in the Mathematics Division of an Automatic Computing Engine (ACE)" (Report to the Executive Committee of the National Physics Laboratory), 1946; reprinted in [52] 1–86.

52. —— (D.C. Ince, ed.), *Collected Works of A.M. Turing: Mechanical Intelligence*, North-Holland, Amsterdam, 1992.

53. University of St. Andrews, School of Mathematics and Statistics, *MacTutor History of Mathematics Archive* (web site), http://www-groups.dcs.st-andrews.ac.uk/ history/index.html

54. John von Neumann, "Eine axiomatisierung der mengenlehre," *Journal für die reine und angewandte Mathematik* **154** (1925) 219–240.

55. ——, *First Draft of a Report on the EDVAC*, unpublished manuscript, 1945; reprinted in [46].

56. John Walker, *The Analytical Engine* (web site), http://www.fourmilab.ch/babbage/contents.html

57. Hao Wang, *Reflections on Kurt Gödel*, MIT Press, Cambridge, MA, 1987.

58. Alfred North Whitehead and Bertrand Russell, *Principia Mathematica*, Cambridge University Press, Cambridge, UK, reprinted 1989.

59. Hugh Whitemore, *Breaking the Code*, Samuel French Ltd., London, 1988.

60. J. H. Woodger, ed., *Logic Semantics, Metamathematics: Papers from 1923 to 1928*, Clarendon Press, Oxford, 1956.

61. Ben H. Yandell, *The Honors Class: Hilbert's Problems and Their Solvers*, A K Peters, Natick, MA, 2002.

62. Ernst Zermelo, "Untersuchugen über die Grundlagen der Mengenlehre," *Mathematische Annalen* **65** (1908) 261–281.

# 13

# From the Tree Method in Modern Logic to the Beginning of Automated Theorem Proving*

**Francine F. Abeles**
*Kean University*

## Introduction[1]

In teaching an upper division elective course in mathematical logic for mathematics and computer science students, I have found that the class usually is divided evenly between these two groups of students, both of which suffer from insufficient experience in proving theorems. To remedy this insufficiency, I have chosen as the engine for the first part of the course a proof technique known as the tree method, an intuitively appealing and relatively simple approach for establishing the validity of arguments that works for a large subset of first order logic whose roots go back to the early part of the twentieth century. This method provides students with the opportunity to develop several important technical skills such as deriving conclusions from premises, exposing the inconsistency of a set of statements, and determining if an argument is sound. That the method became a crucial step in the development of automated theorem proving beginning in the 1950s serves to heighten the interest of the computer science students in the class, the group that needs to be even more engaged in theorem proving activities to provide the tools for constructing complex computer programs. What most distinguishes this course is that I present the method *and* its historical context so that the topics and their emergence are tightly coupled.

To facilitate this approach, I will present both the development of increasingly more sophisticated trees together with the relevant historical topics so that an instructor who adopts this approach using logic to teach the skills to prove mathematical theorems can choose just how much history to incorporate. Little of this historical content appears in logic texts or in standard histories of mathematics. At recent professional meetings, where there have been presentations or discussions about the role of logic in writing proofs in collegiate mathematics courses, participants have expressed a strong interest in knowing more about truth trees and their history. Although the presentation here is certainly not complete, it is sufficient for presenting this important logical tool historically, and it supplements the textbook for this first half of the course, the third edition (1991) of Richard Jeffrey's book, *Formal Logic: Its Scope and Limits*. [18]

To illustrate some of the many differences between the scope of logic before and after the mid nineteenth century, when modern logic emerges, the course begins with some quite complex syllogism problems

---

*This paper is dedicated to the memory of Richard Jeffrey (1927–2002) whose 1967 publication of *Formal Logic: Its Scope and Limits* was the first book to popularize the tree method of testing unsatisfiability.

[1]For assistance with biographical and bibliographical information throughout this article, I am indebted to Shirley Horbatt, librarian in the Kean University Library.

(soriteses) given by Charles L. Dodgson (Lewis Carroll) who also applied a tree method to solve them. Syllogistic reasoning still was considered in his time to be the archetype of correct reasoning. In July 1894, Carroll devised his Method of Trees to test the validity of these highly complicated multiliteral statements. His method was the earliest modern use of trees as a way of reasoning efficiently from syllogisms. It was, in effect, a mechanical test of validity through a reductio ad absurdum argument for a large part of propositional logic. [5, pp. 279–319; 1, pp. 25–35]

One of Carroll's problems, suitably altered for a modern audience, to obtain the strongest possible conclusion from the premises, is given in Example 1 for the universe of men. [5, pp. 135–37] Carroll considers these statements to be equivalent: some $x$ exist and none of them are $y$; all $x$ are not $y$.

## Example 1

1. All the policemen on this beat eat with our aunt.

2. No man with long hair can fail to be a poet.

3. Andrea Jones has never been in prison.

4. Our aunt's cousins all like cold meat.

5. None but policemen on this beat are poets.

6. None but her cousins ever eat with our aunt.

7. Men with short hair have all been in prison.

By applying the appropriate syllogistic rules to statements 1 and 5, we can conclude that all men not eating with our aunt are not poets. Next we use this conclusion and statement 2 to conclude that all men not eating with our aunt are not long-haired. This conclusion coupled with statement 6 allows us to say that all long-haired men are cousins of our aunt. This conclusion when used with statement 4 allows us to reason that all long-haired men like cold meat. Taking this statement together with statement 7 permits us to say that all men who don't like cold meat have been in prison. Finally, this conclusion when coupled with statement 3 leads to the ultimate conclusion that Andrea Jones likes cold meat.

In the next two sections of this paper I provide a concise history of the tree method so that the instructor who adopts this approach will not only understand its evolution and can include some of the historical context in a logic course, but also can choose to enrich the course with additional material such as Beth's semantic tableau method (which quickly became useful in the development of automatic theorem proving), or the role of *modus ponens* in the structure of a proof.

## Natural Deduction and the Roots of the Tree Method[2]

The principal developers of natural deduction for proving theorems were Gerhard Gentzen (1909–1945) and Stanisław Jaśkowski (1906–1965) who worked independently, yet published their results in the same year, 1934. [10; 17] Gentzen established formulations of predicate logic different from those developed by Gottlob Frege, David Hilbert, and Bertrand Russell in the sense that Gentzen considered his proof procedures to be more "natural" than those based on axiomatic systems. One of these formulations is known as Gentzen's calculus of natural deduction sequents. A sequent $A_1, A_2, \ldots, A_n \to B_1, B_2, \ldots, B_n$ is a one-dimensional array of two finite sequences of formulas connected by an arrow $\to$ whose meaning

---

[2]In this section and the next one, I have relied on the following pieces to supplement the original sources: [3, pp. 22–4, pp. 36–9]; [4, pp. 113–52]; [19, ch. 6].

is: the sequence on the right-hand side (RHS) is a logical consequence of the sequence on the left-hand side (LHS). The sequence on the left is the set of antecedents; the sequence on the right is the set of succedents. To arrive at a formal deduction, each line in the proof is either an axiom or is a derivation from a previous line according to a set of rules.

In their systems, Gentzen and Jaśkowski treated the existential quantifier differently. Gentzen included both the universal and the existential quantifiers explicitly; Jaśkowski included universal quantification explicitly, but existential quantification only indirectly by giving the existential quantifier only as the negation of the universal quantifier. Probably for this reason, logicians of his day preferred Gentzen's approach which became the one that led to further development.

The modern tree method as a decision procedure for classical propositional logic and for first order logic originates in Gentzen's pioneering work on natural deduction in 1934, particularly his formulation of the sequent calculus known as LK. But the route is not a direct one; the chief contributors being Evert W. Beth (1908–1964), and the contemporary logicians, Jaakko Hintikka (1929– ), Raymond Smullyan (1918– ), and Richard Jeffrey (1927–2002).

In 1955, Beth presented a tableau method he had devised consisting of two trees that would enable a systematic search for a refutation of a given (true) sequent. The tableau has the LHS of the sequent as the root of one tree, and the RHS of the sequent as the root of the other tree. Using the tableau method, one can obtain a cut-free proof (one that does not use the cut elimination rule which is a generalization of *modus ponens*) by systematically building a counterexample for the sequent, and finding that at some stage the process becomes blocked, i.e. there is no counterexample, confirming that the sequent is true. But we have to be sure that no counterexample at all exists, so the construction must be *formal*, i.e. based on a *syntactical* definition of logical derivation, one composed of a finite list of logical inferences, each inference resulting from an application of a formal rule. The problem of determining whether or not a given formula $V$ is a logical result of the (given) formulas, $U_1, U_2, \ldots, U_n$ can be determined by constructing an appropriate counterexample showing that $V$ does *not* logically result from $U_1, U_2, \ldots, U_n$. But if the construction is unsuccessful and a counterexample cannot be found, then $V$ *is* a logical inference from $U_1, U_2, \ldots, U_n$. A semantic tableau is a systematic method for constructing a counterexample, if one exists. It should be emphasized that if a semantic tableau shows that no counterexample exists, then its pieces can be rearranged to obtain a *semantic* derivation (in some natural deduction system), in the sense of Alfred Tarski's (1902–1983) semantic definition of the concept of logical consequence, and conversely. So, obtaining a derivation of V from $U_1, U_2, \ldots, U_n$ by applying the rules of some system of natural deduction, or showing that such a derivation is unobtainable using these rules, is a method equivalent to the method of semantic tableau. [26, ch. II] An example of Beth's semantic tableau method (Example 1A) is in the Appendix. [6, pp. 186–7, pp. 196–7]

In his 1934 paper, Gentzen proved that the use of the cut rule can be eliminated from any logical proof. The cut rule, or *modus ponens*, is a valid derived rule in natural deduction systems, but it is *not obviously* a valid derived rule in the tree method, i.e., if there are closed trees for both $A$ and $A \rightarrow B$, then it is not obvious that there is one for $B$ as well. (A tree is closed if it contains a matching pair of contradictory sub-formulas.) Indeed, Gentzen's LK without the cut rule *is* the tree method. In any natural deduction system, not using the cut rule can increase the length of derivations considerably. The efficiency of the tree method is preserved by adding a version of the cut rule, in the form of the law of the excluded middle, which permits any open branch of a tree to split into two, then allowing any sentence to be appended to the bottom of one of the new branches, and the negation of that sentence to be appended to the bottom of the other.

Some years before Beth published his work on the semantic tableau method, Jacques Herbrand (1908–1931) had made the equivalent observation in his work on quantification theory that *modus ponens* can be eliminated from the proof of any provable formula. In his proof of what has become known as Herbrand's fundamental theorem, he showed that if a formula $F$ is provable, then there is another proof of it that uses only subformulas of $F$. Indeed, Gentzen's sequent calculus is based on Herbrand's results, but Gentzen's elimination of his cut rule is not as general as Herbrand's elimination of *modus ponens*. [14]

## The Development of the Tree Method in the Second Half of the 20th Century

A tree is a left-sided Beth tableau in which all the formulas are true. The rules for decomposing the tree, i.e., the inference rules, are equivalent to Gentzen's rules in his sequent calculus LK.[3] In the first edition (1967) of his book, *Formal Logic* [18, p. 227], Jeffrey states that the sources of his tree method are Beth's semantic tableaux [6;7], and Hintikka's method of model sets [15; 16]. Jeffrey also asserts (on p. ix) that he used Smullyan's idea of one-sidedness to reduce a tableau to a tree, and he suggests that both Beth's semantic tableaux and Hintikka's model sets may derive from Herbrand's work.

In Hintikka's method, a Hintikka set is a set of formulas interpretable as a partial description of a model in which all the formulas are true. As in Beth's tableau method (which is more complicated than Hintikka's method), when a counterexample of a formula cannot be constructed, that formula is proved. [15; 16] Hintikka sets constitute models in which all the members are true when suitably interpreted. [15, p. 26]

In the period 1959 to 1968, Smullyan connected Hintikka's model sets and Beth's tableaux and gave a complete and systematic development of his own method of analytic tableaux, which are left-sided Beth tableaux or truth trees, as he called them. But Smullyan acknowledged that "the whole idea derives from Gentzen." [24, p. 15] In both the analytic tableau method and the semantic tableau method the construction of a counterexample is the goal. They differ in that an analytic tableau is composed of just one column, and the rules of its construction, the same rules governing the tree method listed in the next section, are rules for eliminating logical constants only. To decide if formula $B$ can be inferred from the formula $A_1, \ldots, A_n$, we begin with $A_1, \ldots, A_n$ and the negation of $B$. Applying the rules appropriately to $A_1, \ldots, A_n$, $B$, we construct a tree of sub-formulas. If each branch of the tree is closed, then we have shown that $B$ can be inferred from $A_1, \ldots, A_n$. [24]

In the Appendix there is an analytic tableau (Example 2A) constructed to answer the same question raised in Example 1A, is formula (3) a consequence of formulas (1) and (2)? [19, pp. 204–5]

## The Tree Method[4]

Before using the tree method to test the validity of arguments, or equivalently the consistency of a set of sentences, we first list the basic inference rules. In rules 2, 4, and 8, the symbols below the line (the conclusion) are connected by "or," i.e., in rule 2 these symbols are $\sim S$, $T$; in rule 4 they are $S$, $T$; in rule 8 they are $S$, $\sim T$. In rules 3, 5, 7, 9 and 10 the symbols below the line are connected by "and," i.e., in rule 3 they are $S$, $T$; in rule 5 they are $S$, $T$, and $\sim S$, $\sim T$; in rule 7 they are $S$, $\sim T$; in rule 9 they are $\sim S$, $\sim T$; in rule 10 they are $\sim S$, $T$ and $S$, $\sim T$.

---

[3]Gentzen developed two versions of first order logic, each in two flavors. He named one of these versions, natural deduction, of which NK is classical, and NJ intuitionistic. He named the other version the logistic calculus, and its two flavors are the classical LK and the intuitionistic LJ.

[4]All the examples in this section are problems and examples taken from the third edition of [18].

| 1. Contradiction. | $S$ | Given two premises, one the denial of the other, we infer the conclusion $x$ that the premises contradict each other, i.e., no counterexamples are determined when these two premises appear as full lines of a path in a tree. |
| --- | --- | --- |
| | $\sim S$ | |
| | $x$ | |

| 2. Conditional. | $S \rightarrow T$ | Given the premise the sentence $S$ implies $T$, conclude the sentence $\sim S$, or the sentence $T$. |
| --- | --- | --- |
| | $\sim S \quad T$ | |

| 3. Conjunction. | $S \wedge T$ | Given the premise the sentence $S$ and $T$, conclude the separate sentences $S$ and $T$. |
| --- | --- | --- |
| | $S$ | |
| | $T$ | |

| 4. Disjunction. | $S \vee T$ | Given the premise the sentence $S$ or $T$, conclude the sentence $S$ or the sentence $T$. |
| --- | --- | --- |
| | $S \quad T$ | |

| 5. Bi-conditional. | $S \leftrightarrow T$ | Given the premise $S$ implies $T$ and $T$ implies $S$, conclude the sentence $S$ and the sentence $T$ or the sentence $\sim S$ and the sentence $\sim T$. |
| --- | --- | --- |
| | $S \quad \sim S$ | |
| | $T \quad \sim T$ | |

| 6. Double negation. | $\sim\sim S$ | Given the negation of the negation of premise $S$, conclude the sentence $S$. |
| --- | --- | --- |
| | $S$ | |

| 7. Negation of the conditional. | $\sim(S \rightarrow T)$ | Given the premise the sentence not($S$ implies $T$), conclude the sentence $S$ and the sentence $\sim T$. |
| --- | --- | --- |
| | $S$ | |
| | $\sim T$ | |

| 8. Negation of conjunction. | $\sim(S \wedge T)$ | Given the premise not($S$ and $T$), conclude the sentence $\sim S$ or the sentence $\sim T$. |
| --- | --- | --- |
| | $\sim S \quad \sim T$ | |

| 9. Negation of disjunction. | $\sim(S \vee T)$ | Given the premise not($S$ or $T$), conclude the sentence $\sim S$ and the sentence $\sim T$. |
| --- | --- | --- |
| | $\sim S$ | |
| | $\sim T$ | |

| 10. Negation of the bi-conditional. | $\sim(S \leftrightarrow T)$ | Given the premise not($S$ implies $T$ and $T$ implies $S$), conclude the sentence $\sim S$ and the sentence $T$, or the sentence $S$ and the sentence $\sim T$. |
| --- | --- | --- |
| | $\sim S \quad S$ | |
| | $T \quad \sim T$ | |

To work with quantified statements we need four additional rules.

| 11. Universal instantiation (UI). | $\forall x \ldots x \ldots$ | Given the premise that something is true of all $x$, conclude that the something is true of any member of the domain over which the variable $x$ ranges. |
| --- | --- | --- |
| | $\ldots$ any name $\ldots$ | |

| 12. Existential instantiation (EI). | $\exists x \ldots x \ldots$ | Given the premise that something is true for at least one $x$, conclude that the something is true of some member of the domain over which the variable $x$ ranges. |
| --- | --- | --- |
| | $\ldots$ new name $\ldots$ | |

| 13. Negation of UI. | $\sim \forall x \, S$ | Given (as the premise) the denial that everything in a non-empty domain has property $S$, conclude that there is at least one member of that domain that lacks property $S$. |
| --- | --- | --- |
| | $\exists x \sim S$ | |

| 14. Negation of EI. | $\sim \exists x \, S$ | Given (as the premise) the denial of the existence of any member in a non-empty domain having property $S$, conclude that every member of that domain lacks property $S$. |
| --- | --- | --- |
| | $\forall x \sim S$ | |

Here is an example I develop in class. Consider these two statements: "There is something that is popular if it's a song." "Then there is something that is a popular song." The upper case letters $S$, $P$ are predicates; the lower case letter $a$ is a value (name), the lower case letter $x$ is a variable name; $Sa$ and $Pa$ are sentences. The first of these sentences expresses "there is something that is a song; the second expresses "there is something that is popular." If the first statement is the premise and the second the conclusion, is the argument if $\exists x(Sx \to Px)$, then $\exists x(Sx \land Px)$ a valid one? Recall that an argument will always be valid except when the premise is a true statement and the conclusion is a false statement. We construct a tree using the premise and the *denial* of the conclusion. If the set of these two statements is inconsistent, all the paths through the finished tree will be closed, and the argument is valid. On the other hand, if the set of these two statements is consistent (not contradictory), there will be at least one open path (counterexample) through the finished tree, and the argument is invalid. An open path represents a counterexample because it has the argument's premises and the denial of its conclusion as true statements. Example 2 is the tree test for this argument. Each line in the tree is a true statement. Inference rules are applied only to sentences that appear as full lines of open paths. It's understood that successive lines are connected by "and."

## Example 2

| | | | | | | |
|---|---|---|---|---|---|---|
| 1. | | $\exists x(Sx \to Px)$ | | | | Premise |
| 2. | | $\sim \exists x(Sx \land Px)$ | | | | Denial of conclusion |
| 3. | | $\forall x \sim (Sx \land Px)$ | | | | Rule 14 from line 2 |
| 4. | | $Sa \to Pa$ | | | | Rule 12 from line 1 |
| 5. | | $\sim(Sa \land Pa)$ | | | | Rule 11 from line 3 |
| 6. | | $\sim Sa$ | | $Pa$ | | Rule 2 from line 4 (The tree branches here.) |
| 7. | $\sim Sa$ | $\sim Pa$ | $\sim Sa$ | | $\sim Pa$ | Rule 8 from line 5 (The tree branches again.) |
| | | | | | x | Rule 1 from lines 6, 7 |
| | A | B | C | | D | |

There are four paths A, B, C, D through this finished tree, one of which (D) is closed (rule 1 from lines 6 and 7). So an "x" is placed below this path. Paths A and B are equivalent because since there are no songs ($\sim Sa$ on line 6), it doesn't matter if they are popular or not. But path C provides the counterexample: something is popular and it's not a song. The tree test shows that the argument is invalid and the set $\{\exists x(Sx \to Px), \sim\exists x(Sx \land Px)\}$ is consistent.

Another type of problem is given in Example 3. Is the statement $\exists x(Px \to \exists y Py)$ a logical truth? It will be if the tree test shows the argument given by the statement is a valid one. The argument has no premises, just a single statement which we take to be the argument's conclusion, so we begin with its denial.

## Example 3

| | | |
|---|---|---|
| 1. | $\sim\exists x(Px \to \exists y Py)$ | Denial of the conclusion |
| 2. | $\forall x \sim (Px \to \exists y Py)$ | Rule 14 from line 1 |
| 3. | $\sim(Pa \to \exists y Py)$ | Rule 11 from line 2 |
| 4. | $Pa$ | Rule 7 from line 3 |
| 5. | $\sim\exists y Py$ | Ditto |
| 6. | $\forall y \sim Py$ | Rule 14 from line 5 |
| 7. | $\sim Pa$ | Rule 11 from line 6 |
| | x | Rule 1 from lines 4, 7 |

The finished tree is closed, (signified by an "x" placed at the end of its single path) so the argument is valid, hence the statement is a logical truth, and the (singleton) set of statements is inconsistent (contradictory). What might such a statement actually mean? One possibility is, there is a person who is a parent providing there is another person who also is a parent.

The tree test can also be used when predicates have more than one value. For example, we can ask if this sentence is consistent: There is someone who shaves only those people who do not shave themselves. In logical notation we write, $\exists x \, \forall y (Sxy \leftrightarrow \sim Syy)$. If we construct a tree test beginning with the denial of this statement, the finished tree will be closed, so the statement, the Barber Paradox, is contradictory as we expect it to be.

With the tree test we can decide whether or not an argument is a valid one providing the initial set of premises and conclusion is a finite set, and the quantifiers apply to a single variable. Then the tree will be finished after a finite number of steps. However, an infinite tree becomes possible when more than one quantified variable is introduced. For example, the argument

$$\frac{\forall x \, \exists y \, Pxy}{Paa}$$

generates an infinite tree. In this case the tree does not terminate and the tree test is not a decision procedure.

## Connections with the Beginning of Automated Theorem Proving[5]

The results that we have described thus far motivated further work in the mechanization of reasoning. In this final section, I provide a short history of *resolution*, the most important of the early proof methods in automated theorem proving. The advent of computing machines in the 1950s spurred further the idea of automating proofs of theorems. This work itself has a rich history, beginning with Martin Davis's program to prove theorems in an arithmetic with addition as the only operation. The first of these theorems, which he proved in 1954 but did not publish, was that the sum of two even numbers is an even number. There are several additional milestones that are linked to the earlier work of Herbrand, Gentzen, and Beth. We describe those that were developed in the period 1954 to 1965.

Arguably, the first working proof procedure for first order logic, constructed by Paul C. Gilmore, was based on Beth's semantic tableau method. [11; 12] (Deciding whether or not *any* given argument is valid is only partially solvable in first order logic.) In 1958, Hao Wang developed the first set of programs to prove theorems in both propositional and first order logic. These were based on the methods of Herbrand and Gentzen. [28; 29] In the period 1960 to 1962, Martin Davis and Hilary Putnam, using Herbrand's theorem, created a program to reduce the test of the validity of a formula $F$ in first order logic to a sequence of tests of the validity of the disjunction of a finite number of propositional formulas (substitution formulas containing no quantifiers) obtained from $F$. Their procedure used simplification by cancellation to split a formula into two formulas that could again be simplified. [9] In 1960, Dag Prawitz improved Davis's and Putnam's scheme of listing the substitutions by using unification, a term coined by J. Alan Robinson (1930– ) for the method originally developed by Herbrand (Robinson cited Herbrand's thesis, specifically his "Property A Method," as the first use of unification), and independently rediscovered by Prawitz. [21] Robinson defined unification in this way: If $A$ is any set of well-formed expressions, and $\theta$ is a substitution, then $\theta$ is said to unify $A$. With a unification algorithm developed by M. Douglas McIlroy in 1962, Davis wrote the first program using unification to improve Davis's and Putnam's procedure. [8]

One of the most important milestones in the history of automated theorem proving, resolution, a single inference method for first order logic was announced by Robinson in 1963, and described fully by him in

---

[5]In this section, in addition to the source material, I have relied on [19, ch. 7], and [2].

1965. [22] Resolution is a refutation method operating on clauses containing function symbols, universally quantified variables, and constants. Recall that reducing efficiently from statements in first order logic to statements in propositional logic (ground clauses) requires unification. Herbrand's fundamental theorem permits clauses (finite disjunctions of literals or the empty clause) with variables to stand for all their ground instances (clauses whose members are only literals having no variables). So resolving two of these clauses requires resolving all their ground instances.

But we can bypass this reduction to ground clauses, and work directly with the original clauses by first using a substitution (which can cause several literals to collapse to a single literal) followed by resolution. The essence of the resolution method is that it searches for evidence of unsatisfiability (self-contradiction) in the form of a pair of clauses, one containing a literal, and the other its complement (negation). Resolution depends on using Gentzen's cut rule which ensures that the complexity of each step in a proof is not more complex than its conclusion.

Example 4 is a problem that I develop in class. It gives Robinson's resolution method in its simplest form (without quantifiers). [19, p. 217] To show that $U$ can be derived from the set of clauses, $(P \rightarrow S) \wedge (S \rightarrow U) \wedge P \wedge \sim U$, we begin by using the equivalent conjunctive form for the first two clauses: $(\sim P \vee S)$ and $(\sim S \vee U)$. Then we proceed to show that the formula composed of these four clauses is not satisfiable (inconsistent) by employing resolution alone.

## Example 4

| | | |
|---|---|---|
| 1. $\sim P \vee S$ | | Clause 1 |
| 2. $\sim S \vee U$ | | Clause 2 |
| 3. $P$ | | Clause 3 |
| 4. $\sim U$ | | Clause 4 |
| 5. $\sim P \vee U$ | | Resolvent of clauses 1 and 2 ($S$ and $\sim S$ are eliminated) |
| 6. $U$ | | Resolvent of clauses 3 and 5 ($P$ and $\sim P$ are eliminated) |
| 7. $[\ ]$ | | Resolvent of clauses 4 and 6 ($U$ and $\sim U$ are eliminated) |

Step 7 shows that $U$ is a logical consequence of the set $\{P \rightarrow S, S \rightarrow U, P\}$ because a set of clauses is not satisfiable if and only if there is a derivation (by resolution) of the empty clause ($[\ ]$), i.e., a contradiction. From this example we see that Robinson is applying a tree method to clausal Gentzen sequents.

The effect of Robinson's work on unification-based resolution has been profound. The programming of resolution-based proof procedures led to the birth of logic programming in the 1970s. Unification based resolution is a machine-oriented reasoning method whose implications we continue to develop.

## Conclusion

The tree method is a direct extension of truth tables, and migrating to trees from the tables is easy to do. Using truth tables to verify inconsistency is straightforward, but very inefficient, as anyone who has worked with truth tables involving eight or more cases knows. The truth tree method examines sets of cases simultaneously, thereby making it efficient to test the validity of arguments involving a very large number of sentences by hand or with a computer. To test the validity of an argument consisting of two premises and a conclusion, equivalently determining whether the set of the two premise sentences and the denial of the conclusion sentence is inconsistent, by the method of truth tables involving say, three terms, requires calculating the truth values in eight cases to determine whether or not there is any case where the values of all three terms are true. But a finished closed tree establishes that validity by showing there are no cases in which the three sentences are true.

I believe I have demonstrated that the topics in the first half of this course should be presented as sets of concepts and techniques that have evolved logically and historically in a reasonably understandable way. In standard texts, the works of Herbrand, Gentzen, and Beth, arguably three of the greatest logicians of the twentieth century, have either been largely ignored or they are treated superficially. Further, I have included enough material so that both mathematics and computer science students can choose themes for term papers from these historically informed topics. The extensive bibliography will point them appropriately.

In the fourth section, the heart of this paper, I described explicitly the extremely elegant and efficient proof procedure, the tree method, and provided several examples for classroom use. To explore this method more thoroughly, the reader should consult the third edition of Jeffrey's book. Jeffrey describes this mechanical procedure as "thrillingly easy to understand and use. It is this simplicity that lets students get control of the nuts and bolts of formal logic in a couple of months." [18, p. xi] As a practitioner of this method, I strongly agree.

# Appendix

## Example 1A

Decide whether or not the formula $(\exists z)[P(z) \wedge \sim S(z)]$      (3)
is a logical consequence of the formulas $(\exists x)[P(x) \wedge \sim M(x)]$   (1)
and $(\exists y)[M(y) \wedge \sim S(y)]$.      (2)

To decide this, construct a tableau placing (1) and (2) on the LHS; and (3) on the RHS. On the RHS, we attempt to construct a counterexample, which, if we can do it will yield a negative decision. We are interested in seeing how "truth" propagates downwards on both sides of the tableau by breaking down the formulas and their descendants into smaller parts. To construct the tableau, we use the following set of rules A through H. Rules D, E, and F eliminate connectives, so eventually each side of the tableau will consist only of sentences, those that were given or those produced by rules G and H.

**A.** The tableau has two columns: the LHS, labelled "valid," and the RHS, labelled "invalid." Arbitrary formulas can be inserted in either column.

**B.** If the same formula occurs in both columns of the same (sub)tableau, then that (sub)tableau is closed; if the two sub-tableaux of a (sub)tableau are closed, then that (sub)tableau is also closed.

**C.** If $\sim A$ occurs in a left column, then $A$ is inserted in the *conjugate* right column, i.e., in the right column of the same (sub) tableau; if $\sim A$ occurs in a right column, then $A$ is inserted in the conjugate left column.

**D.** If $A \wedge B$ occurs in a left column, then both $A$, $B$ are inserted in the same column. If $A \wedge B$ occurs in a right column, then that (sub)tableau splits into two sub-tableaux in the right columns of which we insert $A$, $B$, respectively. (A sub-tableau created from splitting other (sub)tableaux is *subordinate* to those (sub)tableaux; the formulas in both columns of a (sub)tableau occur in the corresponding columns of every subordinate sub-tableau.)

**E.** If $A \vee B$ occurs in a left column, then the (sub) tableau splits into two sub-tableaux. We insert $A$ and $B$, respectively in their left columns. If $A \vee B$ occurs in a right column, then both $A$ and $B$ are inserted in that column.

**F.** If $A \rightarrow B$ occurs in a left column, then the (sub) tableau splits. In the right column of one sub-tableau we insert $A$, and in the left column of the other, we insert $B$. If $A \rightarrow B$ occurs in a right column, then we insert $B$ in the same column, and $A$, in the conjugate left column.

**G.** If $(\forall x)A(x)$ occurs in a left column, then we insert $A(p)$ in the same column for each instantiation **p** of $x$. If $(\forall x)A(x)$ occurs in a right column, we introduce a new instantiation **p** and insert $A(p)$ in the same column.

**H.** If $(\exists x)A(x)$ appears in a left column, we introduce a new instantiation **p** and insert $A(p)$ in the same column. If $(\exists x)A(x)$ appears in a right column, then we insert $A(p)$ in the same column for each instantiation **p**.

|  | Valid |  |  | Invalid |  |
|---|---|---|---|---|---|
| 1. $(\exists x)[P(x) \wedge \sim M(x)]$ | Given | | 3. $(\exists z)[P(z) \wedge \sim S(z)]$ | | Given |
| 2. $(\exists y)[M(y) \wedge \sim S(y)]$ | Given | | 7. $M(a)$ | | Refer to step 6 |
| 4. $P(a) \wedge \sim M(a)$ | H from step 1 | | 11. $S(b)$ | | Refer to step 10 |
| 5. $P(a)$ | D from step 4 | | 12. $P(a) \wedge \sim S(a)$ | | H from step 3 |
| 6. $\sim M(a)$ | D from step 4 | | **branch(i)** | **branch (ii)** | |
| 8. $M(b) \sim S(b)$ | H from step 2 | | 13. $P(a)$ | $\sim S(a)$ | D from step 12 |
| 9. $M(b)$ | D from step 8 | | | ‾‾‾‾‾ | |
| 10. $\sim S(b)$ | D from step 8 | | 16. $P(b) \wedge \sim S(b)$ | | H from step 3 |

| **branch (i)** | **branch (ii)** | **branch (iii)** | **branch (iv)** | |
|---|---|---|---|---|
| ‾‾‾‾‾ | 15. $S(a)$ Refer to step 14 | 17. $P(b)$ | 18. $\sim S(b)$ | D from step 16 |
| | ‾‾‾‾‾ | | ‾‾‾‾‾ | |
| **branch (iii)** | **branch (iv)** | | | |
| | 19. $S(b)$ Refer to step 18 | | | |
| | ‾‾‾‾‾ | | | |

By rule C, $M(a)$ is inserted on the invalid side in step 7 because $\sim M(a)$ appears on the valid side in step 6. Observe that the tableau branches after step 12 into the sub-tableaux (i) and (ii), and using rule B, (i) is closed by the presence of $P(a)$ in step 5 on the LHS, and $P(a)$ in step 13 on the RHS. Sub-tableau (ii) branches after step 16 into the sub-tableaux (iii) and (iv), and again using rule B, sub-tableau (iv) is closed by the presence of $S(b)$ in step 11 on the LHS, and $S(b)$ in step 19 on the RHS. Sub-tableau (ii) is closed, again using rule B, because sub-tableau (iv) is closed. But sub-tableau (iii) is open. This open path gives the counterexample that the universe is composed of two items $a$ and $b$, with item $a$ having property $P$ but not property $M$, while item $b$ has property $M$ but not property $S$. So the answer to the initial question is no.

## Example 2A

1. $(\exists x)[P(x) \wedge \sim M(x)$          Premise
2. $(\exists y)[M(y) \wedge \sim S(y)]$          Premise
3. $\sim(\exists z)[P(z) \wedge \sim S(z)]$          Negation of conclusion
4. $P(a) \wedge \sim M(a)$          Instantiation of x in step 1
5. $P(a)$          Conjunction from step 4
6 $\sim M(a)$          Conjunction from step 4
7. $M(b) \wedge \sim S(b)$          Instantiation of $y$ in step 2
8. $M(b)$          Conjunction from step 7
9. $\sim S(b)$          Conjunction from step 7
10. $\sim[P(a) \wedge \sim S(a)]$          Instantiation of $z$ in step 3

**branch (i)    branch (ii)**

11. $\sim P(a)$    $\sim\sim S(a)$          Negation of conjunction from step 10
12. $\times$          $S(a)$          Double negation from step 11

The contradiction from steps 5 and 11 closes branch (i), but branch (ii) is open and complete (because no additional statements can be added to the branch). Again, the answer to the question is no because only if *every* branch of the tree is closed does each branch contain a pair of contradictory statements, from which we could then conclude that formula (3) is a consequence of formulas (1) and (2). Four of the eleven rules to decompose a tree are used in this example. These four are

existential instantiation in steps 4, 7, 10: $\dfrac{\exists x)A(x);}{A(a)}$          conjunction in steps 5, 6, 8, 9: $\dfrac{A \wedge B;}{\begin{array}{c}A\\B\end{array}}$

negation of conjunction in step 11: $\dfrac{\sim(A \wedge B);}{\sim A \quad \sim B}$          and double negation in step 12: $\dfrac{\sim\sim A.}{A}$

In each rule, the item(s) below the line can be derived from the item(s) above the line. Items below the line in the same row are connected by "or," while items below the line in the same column are connected by "and."

## References

1. F. Abeles, "Lewis Carroll's Method of Trees: Its Origins in *Studies in Logic*," *Modern Logic* 1, 1990, pp. 25–35.

2. ———, "Herbrand's Fundamental Theorem and the Beginning of Logic Programming," *Modern Logic* 4, 1994, pp. 63–73.

3. I. Anellis, "From Semantic Tableaux to Smullyan Trees: A History of the Development of the Falsifiability Tree Method," *Modern Logic* 1, 1990, pp. 36–69.

4. ———, "Forty Years of "Unnatural Deduction and Quantification: A History of First-order Systems of Natural Deduction, From Gentzen to Copi," *Modern Logic* 2, 1991, pp. 113–152.

5. W. W. Bartley III, ed., *Lewis Carroll's Symbolic Logic,* Clarkson N. Potter, New York, 1977.

6. E. W. Beth, *The Foundations of Mathematics*, 2nd revised edition, North-Holland, Amsterdam, 1965. (First edition 1959.)

7. ——, *Formal Methods*, D. Reidel, Dordrecht and Gordon and Breach, New York, 1962.

8. M. Davis, "Eliminating the Irrelevant from Mechanical Proofs," *Proceedings of Symposia in Applied Mathematics* 15, 1963, pp. 15–30.

9. M. Davis and H. Putnam, "A Computing Procedure For Quantification Theory," *J. of the Association for Computing Machinery* 7, 1960, pp. 201–215.

10. G. Gentzen, "Untersuchungen über das logische Schliessen," *Mathematische Zeitschrift* 39, 1934, pp. 176–210, pp. 405–431. (English translation in Szabo 1969, [25] pp. 68–131.)

11. P. C. Gilmore, "A Program for the Production of Proofs for Theorems Derivable within the First Order Predicate Calculus from Axioms," *Proceedings of the International Conference on Information Processing*, UNESCO, Paris, 1959.

12. ——, "A Proof Method for Quantification Theory: Its Justification and Realization," *IBM J. of Research and Development* 4, 1960, pp. 28–35.

13. W. Goldfarb, ed., *Jacques Herbrand: Logical Writings*, Harvard Univ. Press, Cambridge, 1971.

14. J. Herbrand, Recherches sur la théorie de la démonstration, Ph. D. thesis, Univ. of Paris, 1930. (English translation of ch. 5 containing Herbrand's fundamental theorem in van Heijenoort 1976, [27] pp. 525–581.)

15. K. J. J. Hintikka, "Form and Content in Quantification Theory," *Acta Philosophica Fennica* 8, 1955, pp. 7–55.

16. ——, "Notes on the Quantification Theory," *Societas Scientiarum Fennica Commentationes physico-mathematicae* 17, n. 12, 1955a, pp. 1–13.

17. S. Jaśkowski, "On the Rules of Supposition in Formal Logic," *Studia Logica* 1, 1934, pp. 5–32. (Reprinted in MacCall 1967, [20] pp. 232–258.)

18. R. Jeffrey, *Formal Logic: Its Scope and Limits*, McGraw-Hill, New York, 1991, 1981, 1967.

19. W. Marciszewski and R. Murawski, *Mechanization of Reasoning in a Historical Perspective*, Rodopi, Amsterdam and Atlanta, 1995.

20. S. MacCall, ed., *Polish Logic 1920–1939*, Clarendon Press, Oxford, 1967.

21. D. Prawitz, "An Improved Proof Procedure," *Theoria* 26, 1960, pp. 102–139.

22. J. A. Robinson, "A Machine-Oriented Logic Based on the Resolution Principle," *J. of the Association for Computing Machinery* 12, 1965, pp. 23–41.

23. ——, "Logic and Logic Programming," *Communications of the Association for Computing Machinery* 35, 1992, pp. 40–65.

24. R. M. Smullyan, *First-Order Logic*, Springer-Verlag, New York, 1968.

25. M. E. Szabo, ed., *The Collected Papers of Gerhard Gentzen*, North Holland, Amsterdam and London, 1970, 1969.

26. A. Tarski, *Introduction to Logic and to the Methodology of Deductive Sciences*, Oxford University Press, New York, 1965.

27. J. Van Heijenoort, *From Frege to Gödel. A Source Book in Mathematical Logic, 1879–1931*, Harvard Univ. Press, Cambridge MA and London, 1976, 1971, 1967.

28. H. Wang, "Proving Theorems by Pattern Recognition. Part I," *Communications of the Association for Computing Machinery* 3, 1960, pp. 220–234.

29. ——, "Proving Theorems by Pattern Recognition. Part II," *Bell System Technical J.* 40, 1961, pp.1–41.

# 14

# Numerical Methods History Projects

**Dick Jardine**
*Keene State College*

## Introduction

Hermite, Runge, Birkhoff, and Shoenberg are just four of the many mathematicians of the past 200 years who contributed to the field of numerical analysis. In the numerical methods course described in this article, students met Charles Hermite (1822–1901) while studying the interpolating polynomials bearing his name. Students became acquainted with Carl Runge (1856–1927) as they did the historical research to learn what applications motivated the development of his popular method for approximating the solution to differential equations. Isaac Shoenberg (1903–1990) and Garrett Birkhoff (1911–1996) had different purposes in developing and applying splines, just one of the mathematical methods created to support the American wartime effort of the 1940s. Students study splines in this and other numerical analysis and mathematical modeling courses ([1] and [5] are typical texts), but in this one they learn the context for the historical development of the mathematics.

Projects and other history-based course activities provide an opportunity to bring to life the topics of a numerical analysis course. A brief overview of history-based projects included in a numerical methods course is given here. All the projects are described in the paper; one of the course projects is contained in an appendix. The historical components of the projects are the focus, however other historical activities included in the course are also presented, hopefully providing the interested reader with resources and ideas for implementation in other courses.

## The projects

The first project was on the subject of root-finding methods: bisection, fixed-point iteration, secant method, and Newton's method. Although some of the methods are by no means modern, classroom discussion provided insight into the modern-day implementation of variations of those methods in their graphing calculators and the Maple software used throughout the course. Since the 2002 Winter Olympics occurred during the semester this course was implemented, the first project was an adaptation of an exercise from the text [1] modified to include an Olympic scenario. The project description and requirements are contained in the Appendix. Students applied and compared numerical root-solving methods in finding the time of flight for a ski jumper. In addition to the numerical problem, student teams investigated the history of root-finding methods applied by Newton, Raphson, Vieta, Al Kashi, and Paramesvara. The course instructor shared research resources with students, many of which are contained in the references [3], [4], [6], [7], [8].

In completing this and subsequent projects, student teams presented both written and oral reports of their solution. For their historical research, students were instructed to include at least one web source and, because of the unreliability of some web resources, at least one print reference. Good web sources include the MacTutor History of Mathematics Archive [8] or numerical analysis history sites [10], found using any of the better search engines, such as Google. Any good history of mathematics text, e.g., [7] and the *Dictionary of Scientific Biography* [4] are good print sources to begin historical research.

An additional historical activity on the subject of root-finding, which students found both educational and entertaining, was William Dunham's Pohle lecture [2] entitled Newton's (Original) Method. I chose to view the streaming video in class because of concerns about out-of-class accessibility, as not all KSC students have ready access to computers and the internet. (It would be appropriate to require students to view the video outside of class if the WWW is universally accessible, since the video is available at all hours on the web.) In viewing the video, students learn that Thomas Simpson (1710–1761) had a role in the modern version of Newton's method. This is the same Simpson of the numerical integration technique bearing that name. Later in the course, students learned that the Runge-Kutta method for integrating differential equations is a generalization of Simpson's rule. This historical connection between three major subject areas of the course is made apparent to the students through the use of web-streaming video, course projects, and instructor presentations. What is ordinarily seen as a sequence of disconnected topics can be readily linked with the history thread, providing a coherence that some students may not see any other way.

The second project continued the ski-jumping theme, for continuity but not out of necessity. Students were given a picture of an old ski jump, a two-dimensional graph of the curve modeling a proposed new ski-jump, and were challenged to construct an interpolating polynomial to match the given curve. The interpolant would subsequently be used to compute the arc length, using a numerical integration method for the purpose of estimating the cost of the ski jump's construction. The project combined the course topics of interpolation and numerical integration.

In that second project, each student group attacked the problem with just one method, assigned by the instructor. The four methods used were Hermite polynomials, Lagrange polynomials, clamped cubic splines, and free cubic splines. Each project team was also assigned one of the mathematicians—Hermite, Lagrange, Schoenberg, and Birkhoff—for the history component of the activity. With regard to the history portion, students were challenged to learn and report on the kind of problem the mathematician was attempting to solve using interpolation.

The third project was a course review activity with a history component, supporting all the numerical analysis topics covered over the entire semester. Each student team was assigned a numerical method and a corresponding mathematician. During the last two lessons of the course, students presented a brief introduction to their mathematician and completed an example problem employing their assigned numerical method to assist classmates' review for the final examination. The final examination was modified to incorporate variations of some of the student exercises. The students were assessed on the quality of their presentation, to include the historical piece, and the correctness and thoroughness of the example problem they provided for their peers. This project served to wrap the course together for the students with the common theme of the history of numerical analysis.

## Conclusion

A history thread helped tie this numerical methods course together, which made the experience more meaningful and appropriate for students at a liberal arts college. There is always a trade-off in the loss of specific mathematical course content during class time when the instructor adds activities such as a historical video, relates historical anecdotes, and requires history presentations of students. Some might argue the time would be better spent on mathematical problems that could be solved in that class period. Some students question the merit of doing even modest historical research and writing in a mathematics course, and some doubt the merits of the additional homework and project requirements. It is not hard to justify, nor was it difficult to implement, at a college in which most of the mathematics majors graduate to become teachers. The pre-service teachers appreciated the interdisciplinary connection of writing, history, and science afforded by the historical activities in this mathematics course since they will be expected to do similar work in their classrooms.

It is my opinion that the historical component of the projects contributed to increasing student motivation to learn the mathematics they practiced during this numerical methods course. Course evaluations and reflections revealed that most students felt that the projects were very significant in their learning. One student wrote, "I also enjoyed how with each project you added in a history section so we were not just doing a math part but also finding out about the history… This was my favorite part to all the projects." Adding the history part affected different students in different ways, and I am confident that it made most realize that this branch of mathematics had real people associated with its development, people solving real problems of this and previous centuries.

Special thanks go to V. Frederick Rickey for his inspiration, motivation, and recommendations for the improvement of this paper.

## References

1. Richard Burden and J. D. Faires, *Numerical Analysis,* 7th ed., Brooks/Cole, Pacific Grove, CA, 2001, 65.

2. William Dunham, "Newton's (Original) Method, or Though this be Method, yet there be madness in't," available at http://www.pohlecolloquium.org/ accessed February 14, 2003.

3. J. Dutka, "Richardson Extrapolation and Romberg Integration," *Historia Mathematica* 11, 1984, 3–21.

4. Charles Couston Gillespie, editor-in-chief. *Dictionary of Scientific Biography,* 14 vols., supplement I, and index. Charles Scribner's Sons, New York, 1970–80.

5. Frank Giordano, M. Weir, and W. Fox, *A First Course in Mathematical Modeling,* 3rd ed., Brooks/Cole, Pacific Grove, CA, 2003.

6. Herman Goldstine, *A History of Numerical Analysis From the 16th Through the 19th Century,* Springer, New York, 1977.

7. Victor Katz, *A History of Mathematics: An Introduction,* 2nd ed., Addison-Wesley, Reading, MA, 1998.

8. MacTutor History of Mathematics Archive (http://www-history.mcs.st-and.ac.uk/history/) accessed February 14, 2003.

9. V. Frederick Rickey, Historical Notes for the Calculus Classroom, prepared for the Institute in the History of Mathematics and Its Use in Teaching, The American University, 3-21 June, 1996, 57–8.

10. Kees Vuik, History of Mathematicians, (http://dutita0.twi.tudelft.nl/users/vuik/wi211/hist.html), accessed February 14, 2003.

# Appendix

## Project Requirements

### MATH260 PROJECT 1:    Time to Fall

With the approach of the Winter Olympics, interest in ski jumping has increased. From your physics courses, you know that there are equations of motion that are readily derived for the motion of a falling object. But those models that you encountered in the introductory courses often assumed away viscous forces that definitely affect the descent of a falling object (a ski-jumper leaving a ramp, for example).

Assume that an object with mass $m$ is dropped from a height $s_0$ and that the height of the object after $t$ seconds is:

$$s(t) = s_0 - \frac{mg}{k}t + \frac{m^2 g}{k^2}\left(1 - e^{-kt/m}\right)$$

where gravity $g = 32.17$ ft/s$^2$, the coefficient of air resistance $k = 0.25$ lb-s/ft, the mass is 200 lb, and the initial position $s_0 = 300$ feet. Find the time it takes for the jumper to hit the ground to within .01 seconds.

1. Find the time using the four methods used in the course to date: bisection, fixed-point iteration, Newton's method, and the secant method, if possible. Create a table showing the performance of each method used in reaching the solution

2. Compare the performance of the methods, and select the best method for solving this problem. The criteria for selection include the complexity of the operations (operations count), the speed of convergence, and the assurance that the method will in fact converge.

3. Root-finding methods have an interesting history. Part of the project is to do historical research on root finding methods, so consider one of the topics listed below. Actual assignments of the specific topics will be done in class. The historical portion is to be integrated into both your written report and oral presentation.

   (a) Write a brief biography of Isaac Newton and discuss the problems he solved with what we now call Newton's method.

   (b) Newton's method is often called the Newton-Raphson method. Why? Who was Raphson and what was his connection to this numerical method?

   (c) The root-finding methods are *algorithms*. What is the origin of the word algorithm? What is the connection between the origin and its present day use?

   (d) Iterative methods were found on clay tablets of the ancient Mesopotamian culture. What problems were inscribed in clay for posterity? Is there a connection between them and Newton's method?

   (e) The method of false position (or *regula falsi*) harkens back to ancient Egypt. Trace the appearance of the method through the history of mathematics.

   (f) Francois Vieta (16th century) also had an iterative method for solving equations. What was it and how did it compare with Newton's (or other) methods?

   (g) Al Kashi (ca. 14th–15th century) was credited with using fixed-point iteration. Who was Al Kashi and what kind of problems was he interested in solving with this method?

   (h) Paramesvara (ca. 14th–15th century) was credited with using the secant method. Who was Paramesvara and what kind of problems was he interested in solving with this method?

# 15

# Foundations of Statistics in American Textbooks: Probability and Pedagogy in Historical Context

## Introduction

The last two decades have seen increasing interest within the mathematics community in reforming the undergraduate curriculum. Efforts at reform have embraced a wide range of issues, including teaching methods, content, assessment, and administration. Among the several mathematical disciplines receiving close scrutiny under the lens of reform, statistics—especially at the introductory level—has received much attention, focused particularly on pedagogy, technology, and the content of introductory courses. Recommended changes in content include an increased emphasis on data production and analysis, with less time given to "recipes and derivations" [5]. Some statisticians also suggest treating data ethics and introducing students to the nature of official statistics and the organization of national statistics offices. This shift in emphasis marks a significant departure from traditional approaches to the subject, even at the introductory level. As one spokesman for reform efforts in statistics has put it, "What we want beginners to learn about statistics has changed dramatically in the past generation" [23, p. 126].

Statisticians have reached a fairly high level of agreement about the appropriate content and emphases in introductory statistics courses [23]. Others who teach the subject, however (mathematicians, sociologists, economists, for example) aren't always familiar with the recommendations of the statistics community, or in agreement with them. One set of recommendations focuses on the place of probability in elementary courses. Recognizing that additions to the list of topics in a course nearly always require some deletions, many reformers argue for a diminished role for probability theory [5, 23], pointing out that "only an informal grasp of probability is needed to follow the reasoning of standard statistical inference" [23, p. 128]. Mathematicians, on the other hand, may be inclined to preserve the more traditional emphasis on probability.

The conversation about the role of probability and how to teach it in statistics courses is not a recent phenomenon, but has been going on since the introduction of statistical inference in the early decades of the twentieth century. The treatment of probability in four statistics monographs and textbooks from the 1920s through the 1940s reflects some important aspects of that conversation and its place in the context of the formation of the statistics community in the United States in the first half of the twentieth century. These texts do not offer models that today's introductory statistics textbooks should follow; rather, all are written at an advanced level and only the later ones fully take into account the measure-theoretic framework of probability formulated by Andrei Kolmogorov in 1933 [21]. They do not provide a blueprint for today's curriculum. Indeed, the voices of the past seldom answer today's questions directly. Instead, an understanding of the past provides a broader perspective from which to evaluate the needs and choices of the present. In this case, recent history of mathematics can inform teaching by shedding light on some of the factors that influence pedagogical choices. Such light may enable today's teachers of statistics to see more clearly the range of forces impinging on their own choices, equipping them to respond to those perspectives that most accurately reflect their own values and priorities.

## The Books

The four books under consideration here reflect to some extent the values and priorities of what was becoming in the 1930s a professional community of mathematical statisticians. Three were written by early members of that community, Henry Rietz and Samuel Wilks, and the timing of their appearance closely paralleled the progress the community made in the 1930s and 1940s toward establishing the institutional structures of their discipline. Rietz published his MAA Carus monograph, *Mathematical Statistics*, in 1927 [26], two years before the initial publication of the *Annals of Mathematical Statistics*, the first American journal devoted to the subject. His student, Wilks, published two books based on the notes from his Princeton statistics courses [37, 34] in the decade following the founding of the Institute of Mathematical Statistics (IMS), the main professional organization for the discipline. Both the *Annals*, first published in 1929, and the Institute, founded in 1935, played important roles in the process of the formation of the mathematical statistics community in the United States [17]. Rietz and Wilks wrote their books in the context of that process. They were members of a group of researchers seeking to define their interests in the theoretical aspects of statistics and to set those interests apart from those of the broader academic and scientific communities to which the group belonged [18]. That group had made significant progress toward establishing its legitimacy by the time the fourth book was published by Princeton in 1946. Harald Cramér's *Mathematical Methods of Statistics* [6] was a revised and translated edition of the author's 1945 Swedish book. It was the first textbook in English treating probability and statistical inference in terms of measure theory. The book highlighted the close connection between statistics and mathematics, a connection that the American mathematical statistics community had worked to strengthen since its earliest days.

Before the formation of the American mathematical statistics community, methods of data analysis were most often used in the United States by researchers outside of mathematics—in agriculture, economics, and psychology, for example. The main professional organization promoting the application of statistics to these fields was the American Statistical Association (ASA). In part, the establishment of the mathematical statistics community in the United States resulted from its members' desire for more institutional support for its interests—interests that were perceived by some of the leadership of the ASA as too technical and theoretical for the mainstream of the community it represented. The mathematical statisticians, as they called themselves, wanted to set their discipline apart from these fields of application and align it more closely with mathematics [17, 18]. They accomplished this goal, to some extent, by emphasizing the importance of the mathematical foundations of statistics, expecially in the training of statisticians. Increasingly, the foundations of statistics were presented using mathematical standards of reasoning and exposition, and this approach depended in part on a careful formulation of the definition of probability.

The books treated here drew from three versions of the definition of probability: the so-called classical definition, the frequentist definition, and the measure-theoretic definition. The classical approach, as these writers thought of it, came from the ideas of the French mathematician Pierre Simon de Laplace. According to the characterization of probability in his 1814 *Essai philosophique sur les probabilit és*, "[t]he theory of chance consists in reducing all the events of the same kind to a certain number of cases equally possible."[1] Games of chance such as rolling dice, drawing cards from a deck, or tossing coins provide the simplest context in which to explain this classical definition of probability.[2] Rietz incorporated the classical definition into his informal discussion of probability, but pointed out some limitations of its use, preferring to combine it with a frequentist definition.

The frequentist approach had received its most elaborate initial formulation in the works of the Austrian mathematician, Richard von Mises. He first wrote on this subject in 1919, when he published two papers discussing the lack of mathematical rigor and precision in then current probabilistic thinking [31, 32]. He proposed to define probability in terms of infinite sequences of elements representing observable events. Certain postulates served to define these sequences. In particular, for each event in question, the limit of its relative frequency in the sequence had to exist. Furthermore, in any randomly selected subsequence, the limit of the relative frequency of any event had to be the same as in the original sequence. This limit is then the probability of the event.

Samuel Wilks based his 1937 work on a thorough articulation of von Mises's axiomatic system, attempting to deduce results with careful reference to his definitions and postulates. He had, however, abandoned the frequentist point of view by the time his 1943 text appeared, although he maintained his commitment to a rigorous deductive framework for probability. The framework in the second book had some elements in common with the measure-theoretic approach that would firmly take hold by the end of the decade.

This definition of probability emerged from generalizations of the ideas of measure theory and integration first formulated by Émile Borel and Henri Lebesgue at the turn of the twentieth century. Although Kolmogorov was not the first to consider the ideas of probability in the context of measure theory, his *Grundbegriffe* of 1933 [21] provided a definitive statement of the measure-theoretic approach, in which a $\sigma$-field $\mathcal{F}$ of subsets of some abstract sample space $\Omega$ forms the set of events, and the probability of an event is given by a countably additive set function $P : \mathcal{F} \to [0, 1]$.

Many of the tools of statistical inference had appeared before mathematicians became concerned with building an axiomatic structure for probability. The basic ideas had been formulated to model observed experience and were well established by the time the foundations of mathematical probability began to receive attention. For this reason, the choice of axioms usually had only minor implications for the methods of statistics. For example, the books of Wilks and Cramér began with different sets of axioms, yet their explanations of hypothesis testing scarcely differed beyond their notation. The significance of the choice of axioms rested, rather, in what it revealed about the efforts of mathematical statisticians to define the scope of their discipline. Their increasing emphasis on a logical, deductive structure for probability reflected their desire to set their subject apart from applied statistics. Consequently, this examination of the textbooks of Rietz, Wilks, and Cramér will focus not only on the definitions and axioms they gave, but on the changing importance the authors placed on a clear, complete articulation of those assumptions as mathematical statistics became a more well-defined, specialized discipline.

## Rietz's Text: Informal Empirical Foundations

When Henry Rietz published his Carus Monograph, the statisticians who would eventually establish the American mathematical statistics community were just beginning to call attention to the special nature of their subject. Rietz was a key player among those mathematical statisticians pointing out the unique

---

[1]From [22, p. 6], quoted in [7, p. 29].

[2]This notion made up only part of probability for Laplace and his contemporaries, however, for they combined it with ideas about intensity of belief and strength of arguments. Their definition of probability was thus broader than what early twentieth-century mathematicians usually attributed to them. Along with an objective measurement of certainty, it included subjective elements that most American mathematical statisticians rejected [7].

educational requirements of their discipline and the importance of meeting the professional needs of the growing number of people whose research centered on this area [18]. Nevertheless, his book did not make explicit use of probability theory as a means for distinguishing *mathematical* from applied statistics. Instead, he noted that the classical definition could address some questions of probability, while others required an informal frequentist definition based on the limit of the relative frequency of an event. Rietz acknowledged that the existence of such a limit was "an empirical assumption whose validity [could] not be proved," but that seemed reasonable and useful in light of "experience with data in many fields" [26, p. 8].

Rietz's definitions did not take the form of explicit axioms or assumptions, so in that sense, the foundations of his discussion had little formal structure. However, he seemed to make a clear distinction in his text between theoretical and empirical probability. He regarded the assumption of the existence of a limit of the relative frequency as an idealization of experience that, "for purposes of definition, [was] in some respects analogous to the idealization of the chalk mark into the straight line of geometry" [26, p. 9]. In geometry, one can discuss a line without reference to any physical representation, and Rietz seemed, to some extent, to view probability in this more abstract way. For example, in outlining the basic ideas of his subject, he "consider[ed] these concepts in pairs as follows: relative frequency and probability; observed and theoretical frequency distributions, arithmetic mean and mathematical expectation; mode and most probable value; moments and mathematical expectations of a power of a variable"[26, p. 5]. In each pair, the first element is a number or function computed from a set of data, while the second is a mathematical idealization of that computation.

Later in the text, Rietz gave a proof of the *theorem of Bernoulli*, a version of the law of large numbers.

In a set of $s$ trials in which the chance of a success in each trial is a constant $p$, the probability $P$ of the relative discrepancy $\frac{m}{s} - p$ being numerically as large as any assigned positive number $\epsilon$ will approach zero as a limit as the number of trials $s$ increases indefinitely, and the probability, $Q = 1 - P$, of this relative discrepancy being less than $\epsilon$ approaches 1 or certainty [26, p. 28].[3]

He first remarked that "this theorem is an immediate consequence of the definition" of probability as the limit of the relative frequency of success. Then he pointed out that "[w]ith a definition of probability other than the limit definition, the theorem may not follow so readily," and went on to give a proof using Chebyshev's inequality.[4] The proof assumed, as Rietz put it, "that we have any definition of the probability $p$ of success" [26, p. 28]. This assumption reflected Rietz's awareness that one can treat the probability of an event as an abstract concept, and that properties such as the law of large numbers do not depend on particular empirical interpretations for their proofs. Thus, while his exposition did not rely entirely on precise deductive reasoning from a well-developed axiomatic system, it accomplished Rietz's purpose of "shifting the emphasis and point of view in the study of statistics in the direction of the consideration of the underlying theory" [26, p. vi].

Rietz did not manage to shift probability entirely into the realm of abstract mathematics, and indeed, did not claim to have such a goal. From his point of view, probability and statistics had distinctive, mathematical qualities, but also served as tools in other branches of research. Careful, deductive reasoning had a role to play in the development of those tools—but as a means, not as an end in itself. Deductive reasoning and the formulation of axioms had become important aspects of mathematical research by the time Rietz wrote his book in 1927. In the late nineteenth and early twentieth centuries, groups and fields had received axiomatic treatment, while ideas about the abstract notion of a vector space, first formulated in the 1880s, became the focus of much research in the 1920s. David Hilbert, in his 1899 *Grundlagen der Geometrie* [13], chose a set of axioms from which the theorems of Euclidean geometry would follow. His

---

[3]Here $\frac{m}{s}$ is the ratio of successes to number of trials.

[4]Rietz's derivation of Chebyshev's inequality, as well as his use of it in the proof of the law of large numbers, resembles those found in more modern texts such as [15].

interest in foundations surfaced again the following year, when he spoke to the International Congress of Mathematicians, outlining twenty-three problems "from the discussion of which an advancement of science may be expected"[14, p. 445]. Suggesting several criteria by which to judge the quality of a problem's solution, Hilbert recommended that "it shall be possible to establish the correctness of the solution by means of a finite number of steps based upon a finite number of hypotheses which are implied in the statement of the problem ... . [This] is simply the requirement of rigor in reasoning" [14, p. 441].

By the 1920s, the axioms of other branches of mathematics had received intense scrutiny as well. In a paper of 1908, Ernst Zermelo had introduced axioms for set theory [39]. Maurice Fréchet's study of function theory in 1906 presented axioms for abstract metric spaces [11], while the postulates for what are now called Banach spaces appeared in the first years of the 1920s [1]. In setting forth those assumptions, Stefan Banach explained his motivation for the axiomatic approach. Rather than demonstrating that certain theorems held for particular examples, he "consider[ed] in a general fashion the set of elements on which [he] postulate[d] certain properties, [he] deduce[d] theorems on it," and then showed that the postulates held for specific cases.[5] This general method of reasoning became the standard model for mathematics, but Banach's description suggests that it was not yet commonplace in the early 1920s. Certainly probability theory had not reached the level of abstraction and rigor attained by other mathematical subjects. Rietz's approach, however, seems to reflect an emerging interest in establishing probability, and thus statistics, on an axiomatic foundation.

It stood in contrast to the typical presentation of statistical methods made by the users of statistics. For example, in 1923, psychologist Truman L. Kelley began his introductory work, *Statistical Method*, by noting that two schools of statisticians had recently arisen. "The first school is that represented by mathematicians who start with certain elementary principles and deduce therefrom facts of distribution, frequency and relationship. . . . The second school is best represented by those biometricians and economists who start with observed data and endeavor so to group them and treat them that the constant features of the data are made apparent" [20, p. 1]. Kelley, a professor of education at Stanford University, had studied mathematics with Henry Rietz as an undergraduate at the University of Illinois. After some graduate work there in psychology, he went on to get his Ph.D. at Columbia under Edward Thorndike. Kelley belonged to the second school of statisticians, and his book developed statistical methods by starting with what he called "concrete problems." It consisted primarily of explanations of the methods for calculating such quantities as the mean, standard deviation, and correlation coefficient of sets of data, and the standard errors of these quantities.

Rietz's work reflected the general view of mathematical statistics emerging in the 1920s from the work of members of Kelley's "first school." A few mathematicians and statisticians had begun to point out the special nature of a subject that combined theoretical reasoning of a mathematical nature with the methods of analyzing data that social scientists had been applying since the turn of the century [17]. These mathematical researchers had not yet drawn a clear line demarcating their interests from other disciplines, but Rietz's monograph had a more theoretical quality to its reasoning than methodological books such as Kelley's. It highlighted certain mathematical aspects of statistical ideas, but maintained an observational, empirical point of view that preserved the ties between the subject and its application.

## The Textbooks of Samuel Wilks: Foundations in Transition

In his 1937 textbook, [37], Samuel Wilks went beyond Rietz in formulating abstract mathematical axioms for the theory of probability. Like Rietz's work, his took a frequentist approach to defining probability, but outlined more completely the principles first given by von Mises. Wilks motivated his definitions

---

[5]From [1, p. 134], quoted in [3, p. 72]

and axioms with empirical examples such as those illustrated by rolling dice and tossing coins, but his assumptions had a more precise, mathematical quality that Rietz's lacked, and the reasoning that followed them was more firmly grounded in the definitions than in experience.

Kolomogorov had published his measure-theoretic approach to probability four years before the appearance of Wilks's book, and while Wilks did not cite this work, he probably knew of it, and may have been influenced by the model it provided of deductive, mathematical probability.[6] By the late 1930s, the abstract axiomatic approach to mathematics that Banach had felt compelled to explain in 1922 had become more common in research and exposition. For example, in the mid 1930s, the group of French mathematicians who eventually wrote under the pseudonym Nicolas Bourbaki began to discuss a joint project that would "define for 25 years the syllabus for the certificate in differential and integral calculus by writing, collectively, a treatise on analysis."[7] The first volume of Bourbaki's on-going work appeared in 1939 [4], and aimed at developing rigorously the fundamental theorems of mathematics from a set of clearly formulated axioms. Wilks's *Statistical Inference*, with its attention to propositions and their logical implications, while somewhat unusual for a statistics textbook, would not have startled readers becoming increasingly accustomed to the style of mathematics that Bourbaki's work would soon come to epitomize.

Wilks based his theory of probability on three assumptions concerning objects he called *populations*. A population $K$ "consist[s] of a sequence of elements $e_1, e_2, e_3, \ldots$ each element being characterized by $k$ numbers. Thus each element in $K$ corresponds to a set of $k$ numbers, and can be represented by a point in $k$-dimensional *variate space* $S_k$, which will be regarded as Euclidean" [37, p. 3].[8] The elements represent, for example, the outcomes of repeated tosses of a coin ($k = 1$), the height and weight of each child in a population of school children ($k = 2$), or the results of throws of three dice ($k = 3$).

Probability does not treat all conceivable populations, and Wilks gave two assumptions that specified the relevant sequences. For a given $K$ and a subset $A$ of $S_k$, let $n_A$ equal the number of the first $n$ elements of $K$ that lie in $A$. The first restriction on $K$, Wilks's *Assumption I*, postulated the existence of $\lim_{n \to \infty} \frac{n_A}{n}$. Wilks then defined this limit, denoted $W_A$, to be the probability "that an element drawn at random from $K$ is represented by a point in $A$" [37, p. 3]. The second assumption required that the probabilities associated with a population stay the same if a new population is formed by a *random selection* of elements from an existing population, where a subsequence is selected randomly from a population $K$ if the inclusion and exclusion of elements does not depend on the value of the variate associated with the element. Following the statements of his first two assumptions, Wilks proved what he called the *additive property* (or *law*) of probability, appealing to the definition of probability: if $A$ and $B$ are mutually disjoint subsets of the variate space, since $n_{A+B} = n_A + n_B$, we have $W_{A+B} = W_A + W_B$.

*Assumption III* postulated the existence of an almost-everywhere continuous probability density function for a population associated with a continuous variate space. In the one-dimensional case, for a point $x$ in the variate space $S$, the probability density $w(x)$ is defined as

$$w(x) = \lim_{(x_2 - x_1) \to 0} \frac{W_{[x_1, x_2]}}{x_2 - x_1},$$

where $[x_1, x_2]$ is an interval containing $x$, and $W_{[x_1, x_2]}$ is the probability that an element from the population $K$ belongs to $[x_1, x_2]$. In other words, one assumes that for every $x$ in $S$, this limit exists and defines a function $w(x)$ that is continuous almost everywhere.

---

[6]American mathematicians could have learned of Kolmogorov's work through Henry Rietz's review in the *Bulletin of the American Mathematical Society* [27], or through a research paper of Joseph Doob [8] that Wilks included in the bibliography of his textbook. Wilks based his textbook on his lectures for an introductory graduate course in statistics. He may not have had time to incorporate Kolmogorov's ideas into those courses by 1937, and his students would not necessarily have had sufficient mathematical background in measure theory to understand them.

[7]From the records of the group's first meeting in [2, p. 28].

[8]In the vocabulary of von Mises, the sequences are called *collectives*, and the numbers associated with the elements are *labels*.

For both the continuous and the discrete case, Wilks defined the distribution function $W(x)$ of the real variable $x$ associated with a one-dimensional variate. For a discrete distribution, $W(x) = \sum_{i=1}^{r(x)} p_i$, where $p_i$ is the probability that the variate takes on the value $x_i$, for $i = 1, 2, \ldots, r(x)$, and $r(x)$ is the largest integer for which $x_{r(x)} \leq x$. In the case of a continuous distribution, $W(x) = \int_{-\infty}^{x} w(x)\,dx$ [37, p. 6].

In addition to these three assumptions about populations, Wilks's axiomatic system consisted of four fundamental operations—*selection*, *mixture*, *partitioning*, and *conjunction*—that expressed the rules for combining probabilities. These rules allowed one to solve "problems of deriving probabilities of compound events and contingencies from sets of simple probabilities." Such problems, in Wilks's view, made up the central part of the calculus of probabilities. For example, "[t]he problem of finding the probability $P_6$ of throwing a 6 with a die," according to Wilks, "is not one in the calculus of probabilities, but, given $P_6$, to find the probability of getting a 12 in two throws of a die is such a problem" [37, p. 13]. The fundamental operations on populations provided the means of solving these problems.

The operations each form a new population $K'$ from one or more given populations. In addition to specifying the elements of $K'$ in terms of the elements of the original populations, they specify the variate associated with $K'$, and therefore the probability distribution of that variate. Thus each operation consists of a description of the new set of elements, the new variate, and the means of determining the probabilities associated with the new population. *Assumption II*, described above, explained *selection*. A population formed by *mixture* has the same elements as $K$, but its variates are single-valued functions of the variates of $K$. This operation provides a means of computing probabilities of compound events. Conditional probabilities can be calculated using the operation of *partitioning*.

Given populations $K' : (e'_1, e'_2, \ldots)$ and $K'' : (e''_1, e''_2, \ldots)$, with $K'$ characterized by the $k'$ variates $x'_1, \ldots, x'_{k'}$ and $K''$ characterized by the $k''$ variates $x''_1, \ldots, x''_{k''}$, the new population $K : (e_1, e_2, \ldots)$ is formed by *conjunction* from $K'$ and $K''$ if $e_i$ consists of $e'_i$ and $e''_i$ jointly, that is $e_i = (e'_i, e''_i)$. The population $K$ is characterized by the $k = k' + k''$ variates $x'_1, \ldots, x'_{k'}, x''_1, \ldots, x''_{k''}$. If the variates of $K'$ are distributed independently of those in $K''$, the probability density of $K$ is the product of the densities of $K'$ and $K''$. Otherwise, the density is given by the product of the marginal and conditional densities according to the rules associated with the operation of partition.

Wilks's construction of a precise definition of a random sample illustrates the role these operations played. In the framework of *Statistical Inference*, one thinks of a random sample of observations from a population $K$ as an element in a new population $K^*$ formed from $K$ by the operations of selection and conjunction. First, consider $n$ populations $K_a : (e_{a1}, e_{a2}, \ldots)$, $a = 1, 2, \ldots, n$, formed by selection from $K : (e_1, e_2, \ldots)$ with $e_{ai} = e_{(i-1)n+a}$, where $e_{(i-1)n+a}$ is the $[(i-1)n + a]$th element of $K$. For example

$$K_1 : (e_1, e_{n+1}, e_{2n+1}, \ldots)$$
$$K_2 : (e_2, e_{n+2}, e_{2n+2}, \ldots)$$
$$\vdots$$
$$K_n : (e_n, e_{2n}, e_{3n}, \ldots).$$

Now form $K^*$ by conjunction from $K_1, K_2, \ldots, K_n$, so that $K^* : (e^*_1, e^*_2, \ldots)$ has elements $e^*_i$ that consist of $e_{(i-1)n+1}, e_{(i-1)n+2}, \ldots, e_{(i-1)n+n}$, jointly. For example,

$$e^*_1 = (e_1, e_2, \ldots, e_n)$$
$$e^*_2 = (e_{n+1}, e_{n+2}, \ldots, e_{2n})$$
$$e^*_3 = (e_{2n+1}, e_{2n+2}, \ldots, e_{3n}).$$

Wilks called the elements $e^*_i$ *random samples of $n$ observations from $K$* [37, p. 21]. If $K$ has a one-dimensional variate space, the variate associated with $K^*$ has values in $n$-dimensional Euclidean space, and a set of values for a particular element is called a *sample $O_n$*. Assuming $K$ has a distribution, then $K^*$ has a distribution $f(x_1, x_2, \ldots, x_n)$. If each variate is independent of all the others, then

$$f(x_1, x_2, \ldots, x_n) = f'(x_1) f'(x_2) \cdots f'(x_n),$$

where $f'(x)$ is the distribution of $x$ in $K$.

After explaining the concepts associated with random samples, Wilks applied them to the derivation of the binomial distribution. As the context for his derivation, Wilks used the example of tossing a coin, pointing out, however, that the derivation had an abstract meaning not associated with any particular application. He began by supposing "a coin is tossed $n$ times and this process is repeated indefinitely. We get a population, each element being a sample, $O_n : (x_1, x_2, \ldots, x_n)$, where each $x$ can take on only the values 1 or 0; 1 for a head, 0 for a tail" [37, p. 22]. In terms of his earlier discussion of random samples, the population here is formed by selection and conjunction from the population associated with tossing a coin.[9] The distribution of the population of samples is given by the product $f(x_1) f(x_2) \cdots f(x_n)$, where $f$, the distribution associated with the original binomial population, is given by $f(1) = p$ and $f(0) = 1 - p$.

In order to find the distribution of the number of 1's in the samples, "[b]y the mixture principle, we let $\alpha$ be the number of 1's (a new variate) associated with each of these points" [37, p. 22]. Thus $\alpha$ is the variate associated with a population formed from the sample population by the operation of mixture. This new population has the same elements as the sample population, but the value $\alpha$ of the variate associated with an element is the sum of the values associated with the corresponding element in the sample population. The additive law then determines the distribution of $\alpha$:

$$f_n(\alpha) = \frac{n!}{\alpha!(n-\alpha)!} p^\alpha q^{n-\alpha}, \text{ for } \alpha = 0, 1, 2, \ldots, n,$$

and the derivation is complete. In this discussion, Wilks clearly relied on the axiomatic system that he had established earlier in the text. He referred to the "population of samples" and its distribution, and both the operation of mixture and the additive law entered into the computation of the probability function.

In these examples, Wilks made a more comprehensive effort to establish the foundations of probability, and therefore the foundations of statistics, than Rietz had ten years earlier. Nevertheless, statistical methods of inference remained the focus of his book. The probabilistic foundations occupied only one-fifth of the work, while the rest concerned the estimation of distribution parameters and hypothesis testing. His book appeared soon after the mathematical statistics community had formally established itself and as it was beginning to define its discipline's place among other scientific fields. Just as the institutional structures of the IMS and the *Annals* had revealed the community's desire to set itself apart from the statisticians represented by the ASA, Wilks's textbook, with its technical, axiomatic reasoning, helped to emphasize the mathematical nature of a discipline that, until recently, had been promoted primarily by social scientists.

In his introduction of the formal, theoretical approach to the foundations of probability that Wilks took in this exposition, he indicated that the theory served to model observed phenomena. He would explain this connection in more detail in his textbook of 1943, *Mathematical Statistics* [34]. There, however, his theoretical framework would start with a different set of concepts and assumptions. In the meantime, some of his other writings revealed that he was taking an interest in the role of mathematical theories as models of physical phenomena. These writings suggest some reasons for the shift in his theoretical point of view reflected in the 1943 textbook.

In 1936, Wilks wrote a review of the first volume of *Statistical Research Memoirs*, a periodical edited by Jerzy Neyman and Egon Pearson that published papers prepared by members of the department of

---

[9]Strictly speaking, each $O_n$ is actually the value of the variate associated with an element, while the elements themselves are $n$-tuples of elements from the binomial population.

statistics at University College, London.[10] Wilks praised the editors for providing an outlet for research on statistical theory, and observed that "[t]here is a definite need, particularly for academic purposes, of systematically building the theory of statistical inference on *a few clearly expressed principles*; otherwise problems which can be handled methodically by commonplace methods will continue to appear to require fresh intuition for their solutions." [36, p. 762, emphasis added].

Two years later, in 1938, Wilks reviewed another work of Neyman that made progress toward meeting that need. *Lectures and Conferences on Mathematical Statistics* presented the text of lectures Neyman had given the previous year at the graduate school of the United States Department of Agriculture. Some of the presentations had explained Kolmogorov's measure-theoretic framework for probability theory. "From a mathematical standpoint," Wilks remarked, "this is perhaps the most elegant and logically self-consistent theory of probability which has yet been devised" [35, p. 478–79]. The elements of the theory, as Wilks noted, had come "from the empirical field," but, as "between any other theory of probability and its application," there was a "gap which must be bridged by intuition, between this theory and its application to actual observations" [35, p. 479]. Neyman's lectures warned of "the danger of confusing the mathematical theory with actual empirical occurrences," and emphasized that "the mathematical theory of probability should be considered as a model to assist in making successful predictions about phenomena in which there is an element of 'randomness' present" [35, p. 479].

This distinction between theory and "actual empirical occurrences" quite likely came to the attention of Wilks again at the 1940 summer meeting of the IMS when Richard von Mises and Joseph Doob, another member of the American mathematical statistics community, debated the relative merits of the frequentist and the measure-theoretic approaches to probability.[11] In his address, Doob promised that "[t]he formal and empirical aspects of probability [would] be kept carefully separate" [9, p. 206]. To this end, he identified two sets of questions that he referred to as *Problem I* and *Problem II*. The first focused on "setting up a formal calculus to deal with (probability) numbers. Within this discipline, once set up, the only problems will be mathematical. The concepts involved will be ordinary mathematical ones, constantly used in other fields. The words 'probability,' 'independent,' etc. will be given mathematical meanings" [9, p. 206]. In contrast, the questions of *Problem II* involved "finding a translation of the results of the formal calculus which makes them relevant to empirical practice" [9, p. 206].

Reminding his audience that "[t]he classical probability investigators did not separate Problems I and II carefully," he noted that the measure approach eliminated the empirical elements, "thus removing all aspects of Problem II" [9, p. 206]. In the comments he made in the discussion that followed the two talks, Doob noted that "[t]he principal objection the measure advocates have to the frequency approach is that it is awkward mathematically" [33, p. 216]. He acknowledged, however, that von Mises's ideas had made important advances in understanding some of the connections between the physical phenomena and the probability model. As he put it, "the measure advocates consider the contribution of Professor von Mises' approach to be a contribution to a solution of Problem II, not to Problem I, the mathematical problem" [33, p. 217].

Doob's remarks may have persuaded Wilks that the frequentist approach did not adequately serve his pedagogical purposes, for when his second textbook, *Mathematical Statistics* [34], appeared in 1943, it contained none of the assumptions about populations or the fundamental operations that had occupied the foundations of his earlier work. He seemed to have a clearer sense in 1943 of the role of an axiomatic

---

[10]Neyman, a Polish mathematician, had studied with Pearson's father, Karl Pearson, at the Galton Biometric Laboratory of University College while on leave in 1925 from his posts as lecturer at the University of Warsaw and the Central College of Agriculture. He met Egon Pearson there and in 1934 joined the Department of Applied Statistics at University College, a year after Egon Pearson became its head. The two statisticians struck up a fruitful collaboration in the theory and methods of statistical inference.

[11]According to [25, p. 473], Wilks attended this meeting, along with 41 other IMS members.

structure in modeling the behavior of physical objects. Perhaps he had developed this point of view as he reviewed works such as those of Neyman and Pearson, and as he encountered the ideas discussed by Doob and von Mises. In the first chapter of *Mathematical Statistics*, Wilks discussed the same phenomena that his 1937 work had modeled, describing "empirical sequences which arise from 'randomizing' processes," such as tossing coins or drawing chips from a bowl, and seeking to isolate the features that could "be abstracted into a mathematical theory" [34, p. 2].

The features he pointed out were just those that his populations and the assumptions about them had idealized in *Statistical Inference*, and the discussion was reminiscent of that earlier work. For example, repeated rolling of a die produces a sequence of numbers, each one having the value $1, 2, 3, 4, 5,$ or $6$. Representing that sequence of numbers by $X_1, X_2, \ldots, X_n, \ldots$, Wilks defined the *empirical cumulative distribution function* $F_n(x)$ to be the fraction of the first $n$ numbers in the sequence less than or equal to $x$. He then noted that "[i]t is a matter of experience that as $n$ becomes larger and larger $F_n(x)$ becomes more and more stable, *appearing* to approach some limit, say $F_\infty(x)$ for each value of $x$" [34, p. 2]. In the 1937 text, the assumption that in a *population*, $\lim_{n \to \infty} n_A/n$ exists served to mathematize this phenomenon. Here, Wilks did not postulate that $\lim_{n \to \infty} F_n(x)$ exists. Instead he simply observed that in actual random processes, $F_n(x)$ *appeared* to stabilize. His next comment echoed the ideas of the 1937 text even more forcefully: "[i]f any subsequence of the original sequence is chosen 'at random' (i.e., according to any rule which does not depend on the values of the $X$'s) then a corresponding $F_n(x)$ can be defined for the subsequence, and again we know from experience that as $n$ increases, $F_n(x)$ for the subsequence *appears* to approach the same limit for each value of $x$ as in the original sequence" [34, p. 3]. This observation reiterated *Assumption II* of the 1937 work.

In *Statistical Inference*, populations and the three assumptions about the limits of relative frequencies comprised the mathematical model idealizing experience. In *Mathematical Statistics*, Wilks merely pointed out what seems to happen in experience. As he noted, "[t]he matter of $F_n(x)$ appearing to approach some function $F_\infty(x)$ as $n$ increases is purely an empirical phenomenon, and not a mathematical one" [34, p. 3]. However, the empirical phenomenon suggested a way of modeling random processes. Wilks did not base the model on the sequences themselves as he had in the 1937 text, but defined an object that idealized the empirical cumulative distribution function instead. A non-decreasing, right-continuous function $F(x)$ with $\lim_{x \to -\infty} F(x) = 0$ and $\lim_{x \to +\infty} F(x) = 1$ is called a *cumulative distribution function (c.d.f.)*. For any value of $x$, $F(x)$ gives the probability that its associated random variable $X$ is less than or equal to $x$: $F(x) = Pr(X \le x)$.[12] The function $F(x)$, then, is a purely mathematical object with well-defined properties, that, while it models the empirical $F_n(x)$, does not depend on experience for its meaning. Wilks made another assumption about the function $Pr$, called the *law of complete additivity*: "[l]et $E_1, E_2, \ldots$ be a finite or enumerable number of *disjoint* point sets on the $x$-axis $\ldots$ [then] $Pr(X \in E_1 + E_2 + \cdots) = Pr(X \in E_1) + Pr(X \in E_2) + \cdots$" [34, p. 6]. In this framework, what had appeared as a theorem in Wilks's first text is thus formulated as an axiom.

After giving these preliminary definitions, Wilks next distinguished between two classes of distribution functions, *discrete* and *continuous*, calling $F(x)$ a *continuous c.d.f.* if there exists a non-negative function $f$ with

$$F(x) = \int_{-\infty}^{x} f(\xi) \, d\xi.$$

The function $f$ is the *probability density function* of the random variable associated with $F$. A discussion of the theory of the Stieltjes integral followed these definitions. This integral, as Wilks pointed out, provided a common notation for both discrete and continuous distributions: $Pr(X \in E) = \int_E dF(x)$ [34, p. 17–23]. He had noted this fact in the 1937 work, but gave no discussion there of properties or results

---

[12]So the function $Pr$ is defined in terms of $F$. Wilks never defined *random variable*.

associated with the Stieltjes integral. Its extended treatment in the second book emphasized the dependence of statistics on mathematical theory.

Jerzy Neyman, whose use of measure theory had attracted Wilks's attention in 1938, noted in a review of *Mathematical Statistics* that the book helped to fill a gap between the two categories in which most statistics and probability texts fell. Many books describing statistical methods, according to Neyman, "frequently ignor[ed] the basic ideas and the mathematics behind these methods, occasionally misinterpreting them," while others "deal[t] with mathematical theory of probability with only occasional glimpses on some particular questions pertaining to statistical theory" [24, p. 41]. Wilks presented the fundamental ideas of probability as a mathematical subject in the context of its relationship to statistical methods, and his textbook provides a glimpse of the ideas that were becoming a central part of the training of American mathematical statisticians. He himself taught the material of the book in an introductory graduate course (like his earlier text, this one was based on his lecture notes for the course). Jacob Wolfowitz, who taught mathematical statistics at Columbia University, also used *Mathematical Statistics* in his course [38, p. 177].

Wilks's two textbooks illustrate the transitional nature of the period between the founding of the Institute of Mathematical Statistics in 1935 and the end of World War II. The mathematical statistics community had formally established itself, but was just beginning to define its place amidst other scientific endeavors. In the coming decade, the curriculum of statistics programs would become more standardized as members of the mathematical statistics community discussed their concerns about undergraduate and graduate education and their role and status relative to other research communities.

The changes Wilks made in his foundational systems between his first and second books demonstrated that questions about the most appropriate approach were still unsettled. Indeed, some writers eschewed such formal structures as those found in the works of Wilks. James V. Uspensky published a probability textbook in 1937 based on his lectures at Stanford University [30]. He included many of the more recent ideas of the theory but did not find it necessary to incorporate them into a formal, deductive system. In his view, "modern attempts to build up the theory of probability as an axiomatic science [were] interesting in themselves as mental exercises," but were not of fundamental importance to the subject [30, p. 8]. By the end of the next decade, however, probability as an axiomatic science seemed to have gained a permanent place in the mathematical literature. Harald Cramér's 1946 textbook [6], which integrated that theory into statistics, demonstrated that both subjects could be treated from the theoretical, axiomatic point of view that had become standard in mathematical exposition.

## Harald Cramér and Measure-Theoretic Probability

Cramér's *Mathematical Methods of Statistics*, combined in a single volume an exposition of the theoretical mathematics of measure theory, its use in establishing the foundations of probability, and the applications of probability to statistical methods of inference. One reviewer of the book, William Feller, hailed it as an "unconventional" treatment that combined three topics not usually found in the same exposition. He noted that "[t]he emergence of statistical theory and methodology as an exact science, firmly grounded in mathematical probability, [was] only of recent date," and that "unified guiding principles and methods . . . ha[d] not yet found expression in the textbook literature." Cramér's work thus "close[d] a serious gap in the literature and [would] greatly facilitate both teaching and research" [10, p. 136].

In his preface, Cramér indicated that he intended the book to be "an exposition of the mathematical theory of modern statistical methods, in so far as these are based on the concept of probability" [6, p. vii]. He devoted the first quarter of the book to some mathematical ideas, primarily the theories of measure and integration. The next part, approximately one third of the book, comprised a discussion of random variables and particular probability distributions. The remainder of the book dealt with statistical inference, including sampling distributions, significance tests, and estimation. Throughout the discussion, Cramér "presented with rigor," as one reviewer put it, "subjects usually developed in a mathematically questionable way" [29, p. 734].

In order to establish a rigorous axiomatic system for probability theory, Cramér described the concepts of Lebesgue measure and the Lebesgue integral. He then turned to the more general non-negative additive set functions on the Borel sets, and to the Lebesgue–Stieltjes integral. In particular, he considered bounded, non-negative, countably additive set functions on the Borel sets of $\mathbf{R}$. Any such set function $P$ corresponds to a unique (up to additive constant), non-decreasing point function $F(x)$ with the property that for any interval $(a, b]$,

$$F(b) - F(a) = P[(a, b]].$$

The function $F$ is right-continuous, and its discontinuity points form at most an enumerable set [6, pp. 50–51].

After giving these results, Cramér narrowed his focus to those set functions for which $P(\mathbf{R}) = 1$. In this case, he specified $F$ as $F(x) = P[(-\infty, x]]$, and noted that $0 \leq F(x) \leq 1$, $\lim_{x \to -\infty} F(x) = 0$, and $\lim_{x \to +\infty} F(x) = 1$. Either $P$ or $F$ could be used to define what Cramér called a *distribution*, $P$ referred to as the *probability function* of the distribution, while $F$ was called the *distribution function*. If $F$ has a derivative, $f$, it is called the *probability density* or the *frequency function* of the distribution [6, p. 57].

A chapter explaining the Lebesgue–Stieltjes integral and deriving some of its properties followed the discussion of distributions, after which Cramér extended all of these concepts to cases involving Euclidean space $\mathbf{R}^n$ for arbitrary $n$. Three chapters treating Fourier integrals, matrices, determinants, and quadratic forms, and "miscellaneous complements" (including Stirling's formula, orthogonal polynomials, and the gamma and beta functions) concluded the first section of the text. The next section used these mathematical ideas to formulate an axiomatic system for probability theory.

In an approach that resembled the introductory chapter of Wilks's 1943 text, Cramér began this second section with a discussion of what he called random experiments—experiments for which one's knowledge lacks the precision necessary to make exact predictions about the results, but that exhibit a quantifiable regularity over many repetitions. He then devoted several pages to explaining how to formulate a mathematical theory that modeled "some group of observable phenomena," such as random experiments. This process, as he described it, involved "choos[ing] as [a] starting point some of the most essential and most elementary features of the regularity observed in the data." Axioms of the theory give these features a simplified and idealized form. "From the axioms, various propositions are then obtained by purely logical deduction, without any further appeal to experience," and the system of propositions constitutes the mathematical theory [6, p. 145]. In other words, one first singles out the essential aspects of some physical phenomenon, formulating definitions and axioms that model those features. Then, independently of the empirical observations that motivated the axioms, one deduces propositions according to logical, mathematical reasoning.

Cramér chose the notions of *random experiment, random variable*, and *probability* as his essential, elementary features for the theory of probability. He observed that each repetition of a random experiment produced a result that could be described by a certain number of real quantities. Modeling this observation, he defined a *random variable* as a vector in $\mathbf{R}^k$ having associated with it a non-negative, countably additive set function $P$ with $P(\mathbf{R}^k) = 1$. For a Borel set $S$ in $\mathbf{R}^k$, "$P(S)$ represents the probability of the event (or relation) $\xi \in S$" [6, p. 152].

These ideas, together with a brief discussion of functions of random variables, constituted the chapter Cramér called "Fundamental Definitions and Axioms." Summarizing the contents of the chapter, he noted that he had "laid the foundations for a purely mathematical theory of random variables and probability distributions" [6, p. 164]. The following chapters worked out the details of this theory, and then used the results for the purposes of statistical inference. Cramér thus placed the ideas of statistics within an axiomatic, mathematical framework. He first explained how probability theory could be treated as a branch of mathematics, and then made it clear, as mathematician Mark Kac put it in a review of the book, "that the mathematical methods and techniques of statistics are but a part of the mathematical apparatus of probability theory." As a result, his book highlighted the close connection between statistics and mathematics. On the other hand, "he emphasize[d] with equal clarity that the problems and motivations of statistics are peculiarly its own" [19, p. 39]. Thus he pointed out the distinctiveness of the discipline and reinforced the boundaries that mathematical statisticians had drawn around it.

A professor using Cramér's textbook in a university course on mathematical statistics would have had to be familiar both with the mathematics of measure theory and integration and with the theory and practice of statistical inference. Such a professor would most likely have been found among the members of the American mathematical statistics community. This community had come together in the 1920s and 1930s, drawing most of its originial members from the mathematical research community. They argued that the teaching and practice of their discipline required sophisticated mathematical training that went beyond any basic courses given to most social scientists, the primary users of statistics (see, for example, [28]). Mathematical statisticians in the 1940s made this argument even more forcefully [16]. Cramér's book certainly supported the claim. As he presented it, however, mathematical statistics was not simply a subfield of probability—the discipline's distinctiveness went beyond its ties to mathematics. The decade in which Cramér's text appeared found the American mathematical statistics community working to establish separate departments of statistics in universities. Just as the *Annals* and the IMS had given the community some independence from the ASA and the social scientists, these departments would help establish its disciplinary autonomy from mathematics, and further establish statistics as an independent discipline.

## Addressing Today's Questions: Voices from the Past

Cramér, like Wilks and Rietz before him, made his choices about the foundations of probability in part because of the broader concerns of a disciplinary community. He and the others wrote their books in the context of the emergence of the mathematical statistics community in the United States.[13] That community of researchers worked very explicitly from the late 1920s through the 1940s to set their discipline apart from those it bordered, particularly from the social sciences where the ideas of statistics had first been used in the United States.

The increasing emphasis on abstraction and axiomatic reasoning in the texts analyzed here reflected the growth of the mathematical statistics community and its efforts to establish a boundary between its discipline and the social sciences. That is not to say that any direct link of cause and effect necessarily existed between the changes in these textbooks and the emergence of the community. Complicated factors combined to influence both. For example, on the one hand, the push toward axiomatization had shaped developments of various mathematical subjects since the late 1800s. As part of the mathematical culture of the early twentieth century, high standards of rigor and the striving for simple, complete systems of axioms surely had an influence on probability theory that transcended the emergence of the mathematical statistics community. On the other hand, while the issues surrounding the education of mathematical statisticians shaped the growth of that community in important ways, those issues emerged in conjunction with ideas about the role of statistics in government and industry and with the creation of professional structures of support for mathematical statistics such as the *Annals* and the IMS. Thus the community emerged and grew not only as the knowledge and pedagogy of its technical aspects advanced and because of changes in

---

[13]Recall that Cramér, though Swedish, produced his 1946 textbook for an American publisher.

social connections and institutions, but also as a result of the interaction between the institutional structures of the discipline and its technical content.

Today's teachers of introductory statistics face choices about the content of their courses: what concepts to emphasize, how much formal probability theory to cover, what use to make of technology, for example. Their choices are shaped in part by their understanding of what is central to the subject and their perceptions of which explanations and pedagogical tools make the material most accessible to their students. If the factors shaping teachers' choices were merely matters of centrality and effectiveness, teachers would agree about the best choices. Indeed, statisticians in recent years seem to have achieved widespread concensus about "the guiding principles" for making those choices [23, p. 162]. However, many teachers of introductory statistics are not statisticians, but psychologists, economists, sociologists, and mathematicians, among others. Guided by their perceptions of the place of statistics in their own disciplines, they may choose differently.[14] Mathematicians conceiving of statistics as a subdiscipline of mathematics may emphasize probability theory, derivations of formulas, or proofs about distributions. A psychologist using statistical methods to analyze data may focus on particular tests and how to apply them.

These different perspectives and choices overlap; they are not mutually exclusive. Still, they are to some degree in conflict with one another (why else would various departments at colleges and universities want to expend their resources to offer their own introductory courses, tailored to their particular disciplines?). The voices of the past heard here do not speak directly to that conflict. Rietz, Wilks, and Cramér faced different options, taught different students, had different goals. Certainly, as today's teachers do, they structured the probabilistic foundations of their texts with attention to what seemed to them of central importance to the subject. Their choices were surely shaped by a desire to teach and explain with clarity. Yet the emergence and development of the community to which they belonged suggest that concerns about the place of that community in the larger academic and scientific world played some role in the changing approaches Rietz, Wilks, and Cramér took to presenting the ideas of probability.

An understanding of the past can illuminate the present. The professional, institutional, and social milieu of Rietz, Wilks, and Cramér may differ significantly from that in which today's teachers of statistics work. Still, their experiences point out the importance of context. They speak to the relevance of underlying concerns and assumptions about the place of statistics in the broader intellectual endeavor. They call today's teachers to consider those assumptions and to make decisions about teaching that are informed by an awareness and understanding of their own presuppositions and priorities.

## References

1. S. Banach, "Sur les opérations dans les ensembles abstraites et leur application aux équations intégrales," *Fundamenta Mathematicae* **3** (1922), 133–81.

2. L. Beaulieu, "A Parisian Café and Ten Proto-Bourbaki Meetings (1934–1935)," *The Mathematical Intelligencer* **15** (1), 27–35.

3. M. Bernkopf, "The Development of Function Spaces with Particular Reference to their Origins in Integral Equation Theory," *Archive for History of Exact Sciences* **3** (1966), 1–96.

4. N. Bourbaki, *Éléments de mathématique: Les structures fondamentales de l'analyse*, Hermann, Paris, 1939.

5. G. Cobb, Teaching Statistics, in *Heeding the Call for Change*, ed. L. Steen, Mathematical Association of America, Washington DC, 1992, 3–34.

6. H. Cramér, *Mathematical Methods of Statistics*, Princeton University Press, Princeton, 1946.

7. L. Daston, *Classical Probability in the Enlightenment*, Princeton University Press, Princeton, 1988.

8. J. L. Doob, "Probability and Statistics," *Transactions of the American Mathematical Society* **36** (1934), 759–75.

---

[14]For results and analysis of a survey of responses to reform efforts among various teachers of introductory statistics, see [12].

9. ——, "Probability as Measure," *Annals of Mathematical Statistics* **12** (1941), 206–14.

10. W. Feller, "Review of *Mathematical Methods of Statistics*, by H. Cramér," *Annals of Mathematical Statistics* **18** (1947), 136–39.

11. M. Fréchet, "Sur quelques points du calcul fonctionnel," *Rendiconti del Circolo Matematico di Palermo* **22** (1906), 1–74.

12. J. Garfield, R. Hogg, C. Schau, D. Wittinghill, "First Courses in Statistical Science: The Status of Educational Reform Efforts," *Journal of Statistics Education* **10** (2002), www.amstat.org/publications/jse/v10n2/garfield.html.

13. D. Hilbert, *Grundlagen der Geometrie*, B. G. Teubner, Leipzig, 1899.

14. ——, "Mathematical Problems: Lecture Delivered before the International Congress of Mathematicians at Paris in 1900," trans. Mary F. Winston, *Bulletin of the American Mathematical Society* **8** (1902), 437–479.

15. R. V. Hogg and A. T. Craig, *Introduction to Mathematical Statistics*, 2d ed., The Macmillan Company, New York, 1965.

16. H. Hotelling, "The Teaching of Statistics," *Annals of Mathematical Statistics* **11** (1940), 457–72.

17. P. W. Hunter, "Drawing the Boundaries: Mathematical Statistics in Twentieth-Century America," *Historia Mathematica* **23** (1996), 7–30.

18. ——, "An Unofficial Community: American Mathematical Statisticians before 1935," *Annals of Science* **56** (1999), 47–68.

19. M. Kac, "Review of *Mathematical Methods of Statistics*, by H. Cramér," *Mathematical Reviews* **8** (1947), 39–40.

20. T. L. Kelley, *Statistical Method*, The Macmilllan Co., New York, 1923.

21. A. N. Kolmogorov, *Grundbegriffe der Wahrscheinlichkeitsrechnung*, Springer-Verlag, Berlin, 1933.

22. P. S. de Laplace, *A Philosophical Essay on Probabilities*, trans. F. W. Truscott and F. L. Emory, Dover, New York, 1951.

23. D. S. Moore, "New Pedagogy and New Content: The Case of Statistics," *International Statistical Review* **65** (1997), 123–165.

24. J. Neyman, "Review of *Mathematical Statistics* by S. S. Wilks," *Mathematical Reviews*, **5** (1944), 444–45.

25. Report of the Hanover Meeting of the Institute, *Annals of Mathematical Statistics* **11** (1940), 473–75.

26. H. L. Rietz, *Mathematical Statistics*, Mathematical Association of America, Chicago, 1927.

27. ——, "Reveiw of *Grundbegriffe der Wahrscheinlichkeitsrechnung*, by A. N. Kolmogorov," *Bulletin of the American Mathematical Society* **40** (1934), 522–23.

28. H. L. Rietz and A. Crathorne, "Mathematical Background for the Study of Statistics," *Journal of the American Statistical Association* **21** (1926), 435–40.

29. H. Scheffé, "Review of *Mathematical Methods of Statistics*, by H. Cramér," *Bulletin of the American Mathematical Society* **53** (1947), 733–35.

30. J. V. Uspensky, *Introduction to Mathematical Probability*, McGraw-Hill, New York, 1937.

31. R. von Mises, "Fundamentalsätze der Wahrscheinlichkeitsrechnung," *Mathematische Zeitschrift* **4** (1919), 1–97.

32. ——, "Grundlagen der Wahrscheinlichkeitsrechnung," *Mathematische Zeitschrift* **5** (1919), 52–99.

33. R. von Mises and J. L. Doob, "Discussion of Papers on Probability Theory," *Annals of Mathematical Statistics* **12** (1941), 215–17.

34. S. S. Wilks, *Mathematical Statistics*, Princeton University Press, Princeton, 1943.

35. ——, "Review of *Lectures and Conferences on Mathematical Statistics*, by J. Neyman," *Journal of the American Statistical Association* **33** (1938), 478–80.

36. ——, "Review of *Statistical Research Memoirs*, ed. J. Neyman and E. S. Pearson," *Journal of the American Statistical Association* **31** (1936), 760–62.

37. ——, *The Theory of Statistical Inference*, Edwards Brothers, Ann Arbor, MI, 1937.

38. J. Wolfowitz, "Review of *Mathematical Methods of Statistics* by H. Cramér," *Journal of the American Statistical Association* **42** (1947), 176–79.

39. E. Zermelo, "Untersuchungen über die Grundlagen der Mengenlehre I," *Mathematische Annalen* **65** (1908), 261–81.

# IV

## History of Mathematics and Pedagogy

*Life is good for only two things, discovering mathematics and teaching mathematics.*

—Simeon Poisson

# 16

# Incorporating the Mathematical Achievements of Women and Minority Mathematicians into Classrooms

**Sarah J. Greenwald**
*Appalachian State University*

## Introduction

There are many references for activities that incorporate general multiculturalism into the classroom (e.g., [4, 21, 29]). There are also numerous "women in mathematics" courses that focus on history and equity issues. Yet, except for a few sources such as [19], [23], [24], [26], and [27], sources that discuss women and minorities in mathematics do not include related activities for the classroom that contain significant mathematical content, and those that do are mainly aimed at the middle grades or high school level. This is unfortunate since students benefit from the inclusion of the achievements of women and minorities in mathematics classes, as "the result is that students will see mathematics as a discipline that transcends culture, time, and gender, and as a discipline for everyone, everywhere." [19, page xi]

Since there were only a handful of known women and minority mathematicians before the last 200 years, an effective study of them must focus on recent history. Projects that include such recent history are harder to create because of scarce resources for the classroom, but they are beneficial because students more readily identify with these mathematicians for the reasons stated in the next section. After examining the importance of incorporating the achievements of women and minority mathematicians into classrooms, we will discuss the methodology of historical projects about mathematicians and their mathematics. Then we will explore the implementation of these projects and the inclusion of some women and minority mathematicians as we give examples related to three living mathematicians: Andrew Wiles, Carolyn Gordon, and David Blackwell.

## Importance of Incorporating Women's and Minorities' Achievements into Classrooms

Many students never learn about women and minority mathematicians although research shows that this is beneficial to all students, including white males [5, page 248]. Since interviewers often ask women and minority mathematicians specific questions related to the common themes which emerge from their lives and work, and these themes are particularly engaging for students, the availability of this information and the openness of these mathematicians about their personal struggles allows students to readily identify with them.

## Common Themes

1) While subtle gender or racial discrimination can be as devastating to the career of a mathematician as overt discrimination, the support of family, teachers or mentors (male or female) at any stage of the educational process can help overcome these difficulties.

2) The number of women and minority mathematicians has greatly increased in recent years, probably at least in part due to increased educational opportunities.

3) Most early women mathematicians did not marry. More recently, many successful women mathematicians have balanced a career and family.

4) Successful mathematicians have diverse styles, and many of them struggle with mathematics in much the same way students do.

When we included the achievements of women and minority mathematicians into an introductory class for non-majors, one student sent an unsolicited e-mail message stating: "I think it is wonderful that you are spending time in class discussing the discrepancies between men and women in the world of math and sciences. I think it is very informative and very important so thanks!"

We will now discuss the importance of including information related to the above themes in the classroom, and the reasons why women and minority mathematicians are particularly well suited for the introduction of these themes.

## Importance of Including Mentoring Issues

Students relate personally to mentoring issues, and including them encourages the students to reflect on their own mathematical experiences, which is especially useful for math-phobic students and future teachers. As one student commented in evaluations: "The stereotypes that some women encountered as a woman practicing mathematical sciences are quite real to me. My father never encouraged me to succeed in math; it was always okay if I did not do my best in math. He felt that math was a subject that men naturally excelled in, leaving women to literature and sentence structure."

While white male mathematicians have also benefited from mentoring experiences, they are less likely to discuss this fact in interviews, because interviewers do not usually ask them questions related to mentoring. However, interviewers do generally ask women and minorities questions related to the support of teachers, family and mentors, and so including women and minorities in the classroom is a natural way to introduce the importance of mentoring in mathematics.

## Importance of Including Living Role Models

Including some living mathematicians and women and minority mathematicians in historical projects allows students to examine issues related to the historical progression of mathematicians, such as the number of women and minority mathematicians and the ability of women mathematicians to balance a successful career and family life. In addition, our students identify more easily with living mathematicians and, in the case of women students, women mathematicians, so it is important to include them. Over the course of two years, five introductory classes were asked which mathematician they most identified with. While more than half of the mathematicians we studied were no longer living, a higher percentage of students in every class identified with a living mathematician. In addition, even though we studied an equal number of white female and white male mathematicians, along with several minority male mathematicians, in each class a disproportionately higher number of women students identified with women mathematicians. These facts are not surprising since people often identify with those who share common experiences and characteristics.

## Importance of Including Mathematicians with Diverse Styles

Exploring the diverse styles of mathematicians in the classroom is especially rewarding for a number of reasons. One student reflected, "Learning about mathematicians and their styles helps you appreciate all the hard work they did to make your life easier." Not only do students appreciate and identify with these mathematicians, but including mathematicians who have different mathematical styles encourages students to reflect on how they do mathematics and helps to improve their self-confidence. As one student commented, "I have also never studied mathematicians before. Studying them made me realize that even they struggle with math. This made me feel better about my own struggles." In addition, numerous students have told us that studying the diverse ways mathematicians conduct research helped them to begin thinking of themselves as mathematicians.

While there are white male mathematicians who have discussed their mathematical styles and struggles in interviews, in-depth information related to the way they do mathematics is more often found in interviews about women and minority mathematicians.

## Importance of Including Mathematics

Students relate to the above themes which arise naturally from the lives and work of women and minority mathematicians, and so including some of these mathematicians into classrooms allows students to identify with them, which in turn helps them in their own studies. However it is a disservice if we reduce these mathematicians only to their gender or race. As Julia Bowman Robinson once said "What I really am is a mathematician. Rather than being remembered as the first woman this or that, I would prefer to be remembered as a mathematician should, simply for the theorems I have proved and the problems I have solved." [7, page 271] Women and minority mathematicians should be discussed in the context of their mathematics, as advocated by Robinson. Not only does this help students view these women and minorities as real mathematicians, but it also helps students relate to the mathematics. As one student noted,

> Seeing the individuals in the mathematician section gave me a glimpse of the people in math. This was the most interesting to me. It showed the personal side of math. I think people see math as the most impersonal and generic of the subjects. This is far from the case. By seeing the work these people do I have a new respect for the math that makes much of life possible. If all people were to see these examples and applications, they would think the same.

## Historical Project Methodology

When we first incorporated historical projects about an individual's life story and mathematics into classrooms, students created presentations and in some classes they also wrote a paper. While students found their projects interesting and rewarding, they were passive listeners during other presentations and did not effectively learn the related mathematics, even though they were responsible for the material on quizzes and tests. We then incorporated a classroom worksheet component, and this aspect has been extremely successful at engaging the class. Students find it rewarding to create a classroom activity sheet that is designed to engage the class with material they have presented. This type of assignment benefits the entire class and is especially appropriate for courses aimed at future teachers.

A worksheet assignment is also appropriate for courses that have been designated as "writing intensive" by the university. Students end up writing a significant amount because part of the assignment is for students to include information about their mathematician as related to the segment themes. In addition, this gives students a chance to encounter issues related to teaching and grading in a creative format. Since they create the worksheets themselves, and then hand it out to the rest of the class, they develop an ownership of the material and a pride in their writing not usually seen in writing designated courses. In some courses, we have replaced the paper component of the historical project with the worksheet assignment in order to ensure that the workload of the students does not increase.

Since these are meant as long term projects, students need to begin the projects early in order to have time to master the material. If they do not allow for this, they may present material they do not understand. Some students who have worked hard in order to understand the mathematics may present the material too quickly. In order to ensure an effective learning environment, we repeatedly expose the students to the mathematics. The class completes pre-readings on the mathematics so that they will have been exposed to it before the presentations. We also review the material after each presentation in order to correct any historical or mathematical inaccuracies and reinforce the material. Students complete the classroom worksheets and are also responsible for the material on tests.

## Including Women and Minorities in Historical Projects

Historical projects that incorporate the achievements of women and minority mathematicians can be implemented in a variety of levels of college courses. It is easy to incorporate women and minorities into historical projects because of the abundance of interviews that have been published recently. Because our students easily relate to the information contained within these interviews, these projects work extremely well. It is sometimes more challenging to incorporate the related mathematics because of a lack of classroom resources designed to do so. Here we will discuss how to incorporate these projects into several different kinds of classes.

### Projects in an *Introduction to Mathematics* Course

In an introductory class for non-majors, we created a three-week segment called *What is a Mathematician?* that is designed to expose students to a survey of topics in mathematics, the mathematicians who worked on these topics and the historical context. We assigned each group of two students a mathematician to be studied closely. If there were an odd number of people in the class, we asked for a volunteer to work alone. We chose the mathematicians in order to expose the students to specific mathematical content and a variety of research styles. We selected a range of mathematicians from the 1700s until today. We included women and minority mathematicians for the reasons outlined in the first section; it was easy to find women and minority mathematicians whose mathematics related to the course goals.

Because this assignment required a significant amount of time outside of class, we gave the students the resources [15] they would need in order to complete their projects so that they would not also have to search out useful references. In addition to reducing the workload, this ensured that the students used quality references.

Since this was an introductory class, we did not expect the students would have the mathematical maturity needed to determine the depth of mathematics appropriate for the class. Hence, we created a specific list of mathematics questions that related to each mathematician [13]. We asked the students to answer these questions and also asked the groups to include information related to issues of diversity, influences, support, barriers, and the mathematical style of their mathematician.

Students prepared Microsoft PowerPoint presentations and classroom worksheets in order to engage the rest of the class with mathematics related to their mathematicians. Students learned the material themselves in order to teach it to the rest of the class.

We gave the students presentation and worksheet checklists [14, 17] so that they had guidelines to follow. Since this was a teaching assignment, students graded the worksheets that they had created after the students completed them. We gave them suggestions for improvement which included suggestions from other students. Because this course is designated "writing intensive," the students were expected to revise and improve the worksheet they designed.

As the class learned about the mathematics and the mathematicians from the student presentations, we highlighted the validity and success of diverse styles and the changing role of women and minorities over time.

## Projects in a *Modern Algebra* Course

Each student in our *Modern Algebra* course was assigned a mathematician and mathematics to present to the rest of the class [9]. The course had many women students. Since women often relate more easily to women mathematicians, it was worthwhile to include women in the list of mathematicians. In addition, it was easy to do so, because there are numerous women who have done important work in this area. We chose topics that complemented course material and exposed the class to more current topics than are often presented in standard *Modern Algebra* courses, such as Sophie Germain and her work on Fermat's Last Theorem, Marjorie Lee Browne and her work on the matrix groups O(n) and U(n), and Emmy Noether and the ascending chain condition on sets and rings. While there is not normally enough time to cover these types of topics in this course, historical projects allow students to explore them without taking too much class time. Students presented biographies and mathematics after the related prerequisite course material was covered in class. For example, Sophie Germain's work on Fermat's Last Theorem was presented after modular arithmetic was covered in class.

## Projects in a *Women and Minorities in Mathematics* Course

Presentations, worksheets and papers based on student projects were used to run a senior level seminar class on women and minorities in mathematics [8, 10]. The three assignments were divided up according to when the mathematician was born: (1) the 18th and 19th centuries, (2) between 1900 and 1925, and (3) after 1925. We gave the students references and directed questions for the first project, but as the students became more independent, with our guidance they chose their own mathematicians, found their own references and decided what mathematics they wished to focus on. Many future high school teachers took this class. This was an especially useful activity for them because as teachers they will routinely decide the level of mathematics that is appropriate for their classrooms. Most of the students were very motivated to study the related mathematics, perhaps because the methodology engaged them and they enjoyed the interdisciplinary nature of the subject area. The course was successful, and it ran a second time by student request.

## Modeling the Worksheet Process in the *Introduction to Mathematics* Class

As students in the introductory mathematics course worked on their projects outside of class, in class we began the *What is a Mathematician?* segment by contrasting the diverse mathematical styles of Andrew Wiles and Carolyn Gordon.

We chose mathematicians whose research could be incorporated into a classroom without needing much background, and we specifically chose living mathematicians so that students would be forced to confront the commonly preconceived notion that mathematics is a dead field, with issues resolved long ago.

We chose Andrew Wiles, a white male mathematician, in order to take advantage of NOVA's *The Proof* video [25] on his solution of Fermat's Last Theorem. The video is extremely well done and addresses his influences, support, barriers, and mathematical style. In addition, there are many resources available to help incorporate his mathematics into classrooms [30]. Students viewed this video and then answered questions about Andrew Wiles related to the segment themes [16]. The next day, they compared their ideas with a worksheet that modeled a write-up of these themes and engaged them with related mathematics [11].

Since the video mentioned the fact that Sophie Germain was the only woman to work on Fermat's Last Theorem, and there were virtually no women seen in the video, we then examined detailed statistics on women and underrepresented minorities in mathematics [28, 32].

We wanted to contrast Andrew Wiles with a woman mathematician whose mathematics could be incorporated into the classroom. While we could have chosen Sophie Germain, we wanted to choose a living mathematician, so Sophie Germain was assigned to students instead. In addition, we wanted to choose as a model a woman who worked with many people, in order to contrast her with Andrew Wiles' solitary style, and we also wanted someone who was interested in gender issues, as we wanted to introduce the class to these issues. There are many possible choices of women mathematicians who satisfy these criteria, who also can be assigned as student projects.

We chose Carolyn Gordon, a well-known and respected geometer. While she is best known for her groundbreaking work on hearing the shape of a drum, she continues to do research as a leader in the related field of spectral geometry. Since there are only a few resources available for Carolyn Gordon and creativity is needed in order to engage students with ideas related to her mathematics, it is better to use her mathematics as a model for the class rather than as a student project topic. Carolyn Gordon satisfied our other criteria because she has many co-authors and is involved in AWM (Association for Women in Mathematics), and because she agreed to be interviewed for the worksheet so we could obtain information related to the segment themes. The contrasting styles of Carolyn Gordon and Andrew Wiles make them excellent choices for the beginning of the *What is a Mathematician?* segment.

## Selections from the Andrew Wiles and Carolyn Gordon Worksheets

An informal style of writing is used in our worksheets on Andrew Wiles and Carolyn Gordon in order to encourage students to identify with the mathematicians and not be intimidated by the mathematics. As part of this informal style, we do not use a formal reference system, but we do give students a list of references at the end of each worksheet and comment on how we used them (see Appendix A).

Here we present the mathematical style section of the Andrew Wiles worksheet that we give to our students (see [11] and [25]).

### Mathematical Style (from the Andrew Wiles Worksheet)

Andrew Wiles describes mathematical research as follows:

> Perhaps I could best describe my experience of doing mathematics in terms of entering a dark mansion. One goes into the first room, and it's dark, completely dark. One stumbles around bumping into the furniture, and gradually, you learn where each piece of furniture is, and finally, after six months or so, you find the light switch. You turn it on, and suddenly, it's all illuminated. You can see exactly where you were. Then you move into the next room and spend another six months in the dark. So each of these breakthroughs, while sometimes they're momentary, sometimes over a period of a day or two, they are the culmination of --- and couldn't exist without --- the many months of stumbling around in the dark that proceed them.

Andrew Wiles needs intense concentration in order to do mathematics. He reads books or articles to see how similar problems have been solved and he tries to modify other people's techniques to complete his own research. He uses by-hand calculations in order to look for patterns. When he is stuck on a problem, he tries to change it into a new version that is easier. Sometimes he must create brand new techniques to solve a problem and he is not sure where these come from. If he is stymied, he plays with his children or walks by the lake in order to relax and allow his subconscious to work.

While Andrew Wiles worked alone in secrecy for seven years, working with others has still been important to him. For example, before he publicized his proof of Fermat's Last Theorem, he first explained it to Nicholas Katz. After he could not fix the fundamental error later found in his proof, he called in Richard Taylor to help him. The two eventually fixed the problem. Hence, we see that while Wiles likes working alone, collaborative efforts have also been essential to his mathematics.

Andrew Wiles is a devoted and persistent mathematician. Even though he had no idea whether he could ever find a complete proof, especially because a proof of Fermat's Last Theorem might have required methods well beyond present day mathematics, he never gave up.

We wanted the students to engage the related mathematics. In the case of Andrew Wiles, it was easy to engage students with activities from NOVA's website [30]. For example, we used an activity where students demonstrate the Pythagorean Theorem by cutting out and matching up the relevant squares. Since the rest of the worksheet [11] was adapted from NOVA's website, we will not present it here.

Creating good classroom activities at the proper level can be a challenge if classroom resources on the mathematician and related mathematics are not available, as was the case with Carolyn Gordon's mathematics. Sometimes creativity is needed in order to engage the students. Here we present an in depth look at selections from the Carolyn Gordon worksheet and the activities we created to engage our students (see [12] and Appendix A).

## Gender Issues (from the Carolyn Gordon Worksheet)

During the entire time that she was an undergraduate, and during her first couple of years as a graduate student, Carolyn Gordon never met another woman mathematician. In addition, her mathematics courses consisted almost exclusively of men. Yet she was not aware that this was an issue that bothered her until she attended a conference and went to an AWM gathering during her third year in graduate school. When she walked into that room full of women mathematicians, she was shocked by the experience and recognized her previously unacknowledged sense of isolation.

Today she balances a successful research and teaching career at Dartmouth College with her family. She is married to a mathematician, David Webb, and they have a daughter. In addition, she is heavily involved in AWM. Carolyn Gordon sees how AWM has helped women who have encountered barriers and active discrimination, both firsthand and through the experiences of friends and colleagues. She has seen the importance of role models, and has become one herself, mentoring many women students and young faculty.

## Mathematical Style (from the Carolyn Gordon Worksheet)

Even though she is a geometer, she describes herself as being "terrible" at visualization and also as having a bad memory. While she is good with numbers, she says that this skill does not help in her research.

The following story also gives us insight into her mathematical style. The breakthrough that led to her research on hearing the shape of a drum occurred during her talk at a conference when a member of the audience asked her a related question. David Webb said that the question "was like a cold shower. It really made us sit up and think about this." Afterwards, the pair spent days making models and checking to see if they worked. Carolyn Gordon recalls, "We got these huge [paper] castles. They took up our living room."

According to Carolyn Gordon, sometimes she needs to step away from a problem and let her subconscious work. Then, while she is partially occupied with something else, new ideas will come to her. She compares the process of doing research to being in a maze. "Sometimes, when you are completely lost, you have taken a wrong turn, and you must back away and try a new direction. Other times, you will reach a door, find a way to open it, and discover that you have made progress and entered a deeper, more significant part of the maze."

## Mathematics (from the Carolyn Gordon Worksheet)

During the 19th century, Auguste Comte, a French philosopher, hypothesized that the chemical composition of stars would always be beyond the reach of scientists. He was incorrect, and soon afterwards scientists developed the related field of spectroscopy. In spectroscopy, the pitches at which a molecule rings are used to attempt to identify the structure. Given a set of vibration frequencies, an important research topic is to ask what can be inferred about the object's composition.

In 1966, mathematician Mark Kac asked a related question about whether one can always hear the shape of a drum. In other words, if you close your eyes and listen to differently shaped drums being played, Mark Kac wanted to know if you could distinguish the shape by the sound or vibration frequencies you hear. A mathematical drum is not a standard musical instrument; it is any shape in the plane that has an interior and a boundary, such as a circle, a square, or a triangle. The interior vibrates with each strike while the boundary frame remains rigid. Imagine that we had a machine that could tell us the exact frequency of the sound of the drum vibrations. Then we could check and see whether the machine could always distinguish the sounds of differently shaped drums.

In 1911, Hermann Weyl proved that one can always hear the area of a mathematical drum. It makes sense that we can hear the area since the bigger the area of the drum, the lower the tone. Later, in 1949, Subbaramiah Minakshisundaram and Ake Pleijel proved that one can always hear the length of the boundary, or the perimeter of the drum. It was then thought that the sound of a drum might determine its shape, and Mark Kac asked whether this was true.

The problem challenged researchers. Finally, in the spring of 1991, Carolyn Gordon, her husband David Webb, and Scott Wolpert, the audience member that we previously mentioned, came up with the answer: No! One can sometimes, but not always hear the shape of a drum. They found two mathematical drums that have different shapes, but still had the same vibration frequencies, therefore making exactly the same sound.

Figure 1. Carolyn Gordon and husband David Webb with their drums in 1991.
Printed with permission of Washington University.

Mathematicians would not have been satisfied with experiments to show that the drums sound the same because calculations and experiments cannot be exact and a very small frequency difference could escape experimental detection. Instead, Carolyn Gordon and her collaborators used mathematics to prove that her drums sounded the same without actually testing them in real-life. Later on, physicists created the drums and tested them in real life. They found that the drums sounded nearly the same with an error attributed to the experimental procedures.

## Classroom Activities (from the Carolyn Gordon Worksheet)

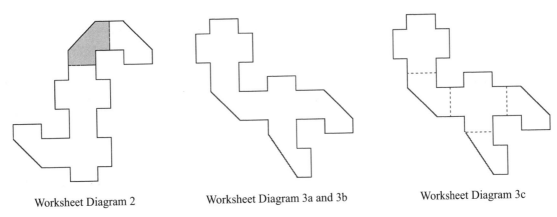

Worksheet Diagram 2                Worksheet Diagram 3a and 3b            Worksheet Diagram 3c

Figure 2. Drums for Classroom Activities. Reprinted with Permission of Ivars Peterson.

1) Cut along the boundaries of the sound-alike drums in Diagrams 3b and 3c (see Figure 2) which are duplicate copies of Diagram 3a. If the drums in Diagrams 2 and 3a have the same shape, then you will be able to place one on top of the other. Place Diagram 3b on top of Diagram 2 above, and try to move Diagram 3b (via rotating, translating or flipping it around) so that it matches Diagram 2. Question: Do these figures have the same shape?

2) Take Diagram 3c (the one with the dashed lines marked on it) and cut this drum along the dashes. Notice that you will have 5 pieces total. Try to fit these cut pieces onto Diagram 2. Notice that you won't be able to do so. In fact, no matter how you might cut Diagram 3c (don't try this now), you won't be able to fit it onto Diagram 2. This contradicts the fact that they must have the same area in order to sound the same. Let's try and resolve the apparent conflict by investigating the accuracy of the models represented. Identify which piece does not fit properly onto Diagram 2 above. Put a star on this piece on Diagram 3a and also put a star on this piece on Diagram 3b, which is the drum that you cut out along the boundary but left in one piece.

3) Take Diagram 3b and compare it to the drum Carolyn is holding (see Figure 1). Compare the piece that you starred with the same piece on Carolyn's drum. Notice that the vertical edge next to the part that Carolyn is holding is the same length as the vertical edge opposite it, but that this is not true of your starred piece and its opposite edge. The drum that Carolyn is holding is drawn correctly, but the starred piece on Diagrams 3a and 3b was not correctly drawn to scale and it is this error in scaling that causes the contradiction in 2). In Diagram 3a above, try to fix the problem piece and edge so that it is drawn to scale by adding to Diagram 3a to show how you would have drawn the piece. Take one of the other similar but correctly scaled pieces, place it on top of the problem piece and trace the correctly scaled piece in order to fix the problem piece.

The point of this exercise is to have you engage the models instead of just hearing about them (no pun intended). There are dangers in relying on models since it is difficult to create physical models representing abstract figures with precisely determined sides. These models were found on a webpage that discussed Carolyn Gordon's solution of the problem and the physicists' work that followed. The drums in the picture of Carolyn Gordon and David Webb are drawn correctly, and they have the same area and perimeter, but are shaped differently. If you close your eyes and listen to them being played, you cannot tell that they have different shapes, since they sound exactly the same.

Carolyn Gordon's research answers Mark Kac's question, but it raises many new issues. For example, now that we know that one cannot hear every property of a drum, what properties besides its area and perimeter really are audible? In addition, we know a great deal more about the vibrations of sound than about the vibrations of light. In the field of spectroscopy, we hope to recover the chemical composition of stars from their vibration fingerprints. Carolyn Gordon's research on hearing the shape of a drum is a small step in this direction. It also shows us that a mathematical proof does not need to be constructive and that there is not always just one conclusion that can be reached from a complete set of measurements.

### Student Reactions to the Andrew Wiles and Carolyn Gordon Worksheets

Students report that they enjoy the worksheets and the activities and that they find these extremely helpful as models for their own worksheet. Students respond well to the openness of Andrew Wiles and Carolyn Gordon about their mathematical styles and struggles, including Carolyn Gordon's feelings about gender issues in mathematics. The statistics on women and minorities then feel more personal to them. In order to cultivate this identification with gender issues, students read an article on the web called *Fifty-Five Cultural Reasons Why Too Few Women Win at Mathematics* [22].

At the beginning of the semester, students were asked to define mathematics. Since many of them reported that mathematics is the study of numbers or equations, these activities force them to re-examine their notion of what mathematics is. Student presentations begin during the week following the introduction of these worksheets. Once students have confronted their preconceived notions about mathematicians and mathematics, the rest of the mathematician segment is ordered so that we move forward in time to trace the historical progression of mathematics and the changing roles of women and minorities.

## Applying the Worksheet Process across a Variety of Courses: Using the Example of David Blackwell

We will discuss some ideas for incorporating David Blackwell and his mathematics into two different levels of college courses.

David Blackwell is considered to be one of the greatest African-American mathematicians [31], and yet, perhaps because he is still alive, resources on minorities in mathematics that are designed for use in the classroom do not include him. This is unfortunate since students relate to Blackwell's openness about racial issues and his mathematical style. In addition, Blackwell has done work in several fields such as game theory, Markov matrices, statistics and probability, and so ideas related to his mathematics can easily be incorporated into a variety of levels of classes [18].

Figure 3. David Blackwell. Printed with permission.

### David Blackwell Project in an *Introduction to Mathematics* Course

David Blackwell is the first living mathematician presented in this class as a student project. Students enjoy learning about him and his mathematics enough that they report this in their evaluations.

After class discussions on strategies for zero-sum games, where there is a clear winner and loser, we wanted to expose the students to the complexity of other types of games. In one interview [1, pages 25–26],

David Blackwell discusses the Prisoner's Dilemma and its relationship to the arms race with the Soviet Union, so we created a list of related mathematics questions for the students to answer in their project.

## Directed Mathematics for David Blackwell

What is game theory?

What is the Prisoner's Dilemma? How does this relate to what David Blackwell worked on? What matrix of payoffs represents the Prisoner's Dilemma?

If a person is deciding what to do, why does it make sense (when looking at the possible cases) for him to confess?

What are the applications of game theory to real life?

We provided the students with the resources that they would need to complete this project (see Appendix B). Students do struggle with the idea of a selfish strategy in the Prisoner's Dilemma, as opposed to a cooperative strategy. In one class, a student struggling with these ideas stated, "If I were a criminal and was going to rob a bank with someone, then I would make sure that he was honest and that I could trust him not to turn me in." When we pointed out the student had just used the words "criminal" and "honest" in the same sentence, the class laughed and then re-examined their beliefs. The arms race with the Soviet Union, which David Blackwell discussed in the interview mentioned above, is especially effective at helping students differentiate between the various strategies.

## Sample Student Worksheet for David Blackwell

The following classroom worksheet was completed by students Ross Bryant and Zachary Lesch-Huie from the Fall 2001 *Introduction to Mathematics* course. This worksheet is a nice response to the assignment. Since they used our Andrew Wiles and Carolyn Gordon worksheets as models, they turned in a list of references with comments (see Appendix C) instead of using a formal reference system. We have left the worksheet in the form turned in by the students in order to provide a true sample of student work as they handed it out to the rest of the class, including the minor errors contained within.

### Introduction (from the David Blackwell Student Worksheet)

David Blackwell is the first truly contemporary mathematician the class will study. Historical accounts of person's lives are doomed to appear hazy with age. Blackwell is interesting for this reason, because he is the only person thus far whose actions are not obscured by time. We have been able to look over Blackwell's own words through interviews to get a personal perspective on his incredible life and work. For one so accomplished, Blackwell certainly comes across as a man who is truly humble, genteel, staggeringly intelligent, and deeply in love with his work. At an early age Blackwell discovered an ability and love of math, and with the support of family and teachers he eventually rose to become the honored expert in statistics and professor at the University of Berkeley that he is today. His main interest has always been investigating and teaching set theory and probability theory. He has also made significant contributions to Bayesian statistics and game theory. Blackwell has been awarded the von Neumann Prize by the Operations Research Society of America, and in 1986, he received the most prestigious award in the field of statistics, the R.A. Fisher Award. He also has the distinction of being the first African-American mathematician elected to the National Academy of Sciences in 1965. In this report, we will explore these aspects of Blackwell's life: influences and support, issues of race, mathematical style, and game theory.

### Influences and Support (from the David Blackwell Student Worksheet)

Blackwell traces his mathematical abilities to his grandfather. While Blackwell never knew him, he was inspired by his legacy: it was in the library of books that his grandfather left to his family that Blackwell found his first algebra book. He also remembers an uncle who could add three columns of numbers in one step. His parents were quite supportive of his math and had hopes that he would become a grade school teacher. His ability however, would see him surpass these aspirations in ways his parents never dreamed of. In high school his mathematics club advisor would give the students problems from the School Science and Mathematics journal and submit their solutions. Blackwell was published. Blackwell entered college at the tender age of 16 and achieved a BA in a mere 3 years. It was during his graduate studies at Illinois University that Blackwell met his most important mathematical influence Joe Doob. It was Doob who directed Blackwell's research and steered him in the direction of probability theory. Later, in 1945 Blackwell claims he got started in statistics after listening to one lecture by Abe Girshick. At the time of the lecture Dr. Abraham Girshick was an already famous mathematician, but Blackwell thought that he had caught a mistake in his lecture. Blackwell wrote to him describing the error. Though the supposed error turned out to be unfounded, the newcomer impressed Girshick, and the two began a partnership. They wrote a book on statistical theory which was published in 1954.

## Racial Issues (from the David Blackwell Student Worksheet)

When one reads of Blackwell's childhood it becomes clear how hard his family tried to shelter him from the prejudice of early twentieth-century America. The family had had their share of difficulty; the reason his fast-adding uncle never went to school was because his grandfather feared he would be mistreated because of his race. Though the grade school Blackwell attended in Centralia, Illinois was integrated, there were two segregated schools in the same town. Blackwell makes clear he didn't feel discriminated against early on, but this was a result of his parent's protection. Still, racist culture continued as a presence in Blackwell's life. Even when he had proved himself as a brilliant mathematical student there was tension. When he was being considered for membership in the Institute of Advanced Study it was customary for members of the Institute to be appointed honorary members of the faculty at Princeton. Princeton objected to a black man being an honorary member of their faculty. The Director of the Institute objected on Blackwell's behalf, and Princeton backed down. Even this struggle was only partly known to Blackwell at the time. Says Blackwell: "Apparently there was quite a fuss over this, but I didn't hear a word about it." Even after Blackwell received his Ph.D. from Illinois (at age 21) his aspirations were limited by the culture of racism. "It never occurred to me to think about teaching in a major university since it wasn't in my [racial] horizon at all." Blackwell fired off 105 applications, one to every black college at the time. He accepted the first offer he got, from Southern University in Louisiana, and later moved to Clark College in Atlanta, and finally to Howard University in Washington, D.C., the top Black institution at the time. In 1950 Blackwell took a leave of absence to work at the RAND Corporation. In 1954, he was asked to join the faculty of the University of California at Berkeley. The position was similar to one he had been interviewed for some years before, when his race had been an issue.

## Mathematical Style (from the David Blackwell Student Worksheet)

Blackwell describes himself as a mathematical dabbler. "I've worked in so many areas — I am sort of a dilettante." To some this may imply that he lacks the ability to focus, but this is simply not the case. He is in fact well known for presenting his theorems in elegantly clear ways. Perhaps this is because of his slightly different way of appreciating math. For example, instead of describing the process of math technically, he uses more common, almost sentimental language to describe his own mathematical aesthetic. "There is beauty in mathematics at all levels" he explains. "The whole subject [is] just beautiful." Blackwell is also one of the voices of a radical type of statistics known as Bayesian statistics. Whereas "Classicist" statisticians see probabilities as entirely objective, Bayesians believe that there is always a suspected set of probabilities in the mind of the experimenter. When tackling a math problem, Blackwell finds that his main motivation is for understanding, a desire particularly meaningful to him. "I am not interested in doing research and I never have been. I'm interested in understanding, which is quite a different thing," he says. This rather broad desire often places Blackwell in a wide range of mathematics (which accounts for his wide range of contributions), filling in "holes" in certain theories so that his understanding can be "rounded out." By no means does this limit him to working alone. In fact, MathSciNet showed 90 published papers with Blackwell's name on them many of them were with one or more partners. Indeed, in his search for understanding, Blackwell finds both working alone and with others helpful.

## The Math (from the David Blackwell Student Worksheet)

It was while working at RAND and Girshick that Blackwell first began working with statistics and probability theory. These areas of interest led him into the field of game theory, the subject that he is perhaps best known for. Game theory is a branch of statistics in which game-like situations are "played" out with specific objectives. "The decision making options of the players are then statistically analyzed." Blackwell applied game theory most famously to old-style pistol duels. If two armed people are advancing on one another, when should they shoot? While Blackwell investigated the problem, a RAND economist consulted him about drawing up a budget proposal for the coming Cold War years. Was war imminent or not? The economist wanted to know. If so, they should begin budgeting for a short-term solution to the Soviet "problem." Blackwell's hypothetical "duelists" had just been placed in the context of World War 3. Blackwell, of course could not provide an answer, but found the problem none-the-less troubling. RAND saw the real-world application of game theory as a means to outwit their political opponent. However, game theory is for the most part much more benign, and can be used to understand and explore methods of decision-making and problem solving. Consider the classic game theory scenario called "The Prisoner's Dilemma."

## The Prisoner's Dilemma (from the David Blackwell Student Worksheet)

Imagine that two criminals, Zach and Ross, have raided a farmer's grain silo in a fiendish attempt to make off with the farmer's sweet, sweet sorghum. The two are apprehended by the police as they leave the scene of the crime, separated at the station house, and shaken down by the coppers. Each crook has to choose whether or not to confess and implicate the other. If neither man confesses, they both will serve one year on a charge of possessing illicit sorghum. If each confesses and implicates the other, both will go to prison for 10 years. However, if one crook confesses and implicates the other and the other crook does not confess, the one who has ratted out his accomplice with go free, while the other crook will rot in the pokey for 20 years on the maximum charge.

The strategies of the "players" or crooks are: confess or don't confess.

The payoffs (in this case, penalties) are the sentences served.

In game theory, data is expressed in a "payoff table" (or payoff matrix). Here is the payoff table for the Prisoner's Dilemma:

|  |  | Ross | |
|---|---|---|---|
|  |  | Confess | Don't Confess |
|  | Confess | 10, 10 | 20, 0 |
| Zach |  |  |  |
|  | Don't Confess | 0, 20 | 1, 1 |

Both crooks choose one of the two strategies. Or, Zach chooses a column and Ross chooses a row. The two numbers in each cell represent the years in jail for the prisoners when those two strategies are chosen. In each cell (a,b), "a" represents the amount of jail time for Zach, and "b" represents the amount of jail time for Ross.

Solution: One must go inside the mind of the prisoner. Zach might think: "Two things can happen, Ross can rat or clam up. If Ross confesses, I get 20 years if I don't confess, 10 years if I do, so in that case it's better to confess. Then again, if Ross stays quiet, and I do too I get a year; but in that case, if I do confess I can go free. Either way, it's best if I confess. So I'll confess." One can assume that Ross' line of thinking will go the same way. So if they both act "rationally" and confess they both go to prison for 10 years, putting up with horrible abuse and all for a sack of sorghum. Here's the thing, if they had both acted "IRRATIONALLY," and kept quiet, they could have gotten off with one year each, and in minimum security too. When both prisoners confess, they have fallen into what is called Dominant Strategy Equilibrium. Because each crook chooses the option with the best payoff, they choose the "dominant" strategy. But in this case these dominant strategies have made both the prisoners worse off. This is an interaction that can be applied to many aspects of modern life. Arms races, pollution, over hunting, etc., are all instances where (it seems) the individually rational action leads to horrible results for each person.

## Class Exercises (from the David Blackwell Student Worksheet)

1. Break the class in half. Then break each half into pairs. Each member of a pair is a "player" in The Prisoners' Dilemma. One half of the class' prisoners are "separated" and cannot communicate with one another. The other half of the class has a chance to talk together before they make a decision. Write down whether you can talk with your partner or not.

2. Given the circumstances of the payoff table, what will you choose to do, confess or hold out? Why?

3. Relate this information with what you learned about the nature of the Prisoners' Dilemma and Dominant Strategy Equilibriums at the end of the presentation.

We encourage the rest of the class to be skeptical learners as they are reading the worksheet and completing the activities, and we ask them for suggestions for improvement. This is especially important because even though we ask students to turn in preliminary responses to the mathematics questions and thematic issues in order to receive feedback before their presentations, errors may still appear in the presentations and worksheets. While this student worksheet contains only some minor errors, in some other cases students present or hand out work containing major errors. Hence, it is essential for the instructor to go over the mathematics and correct any historical or mathematical inaccuracies after each presentation.

## David Blackwell Project in a Women and Minorities in Mathematics Course

David Blackwell and his mathematics can also be incorporated into higher level classes. In the *Women and Minorities in Mathematics* course, mathematical questions and resources were given to students for the first paper assignment in the course that included mathematicians born before 1900. Since David Blackwell was born in 1919, he was a possible choice for the second paper assignment, which focused on mathematicians born between 1900 and 1925. We chose this time frame because both women and minority mathematicians were just beginning to gain acceptance but shared common issues in the context of society at the time. Since we wanted students to develop their own research skills, with our help, they chose their own mathematician, found their own references and decided what mathematics they wished to focus on.

One student chose David Blackwell. Since he has published over 90 papers and books in various areas of mathematics, there were many possible related mathematical topics to choose from. She chose to discuss the idea of a Markov matrix, which was the topic of Blackwell's doctoral thesis [2], and she also chose to explore a paper of Blackwell's called *The Big Match* [3], in which Blackwell looks at an infinite game, that is, a game with an infinite numbers of moves. Since Blackwell also discusses infinite games in [6, pages 46–47], this was a good choice for a topic. The student created a presentation, classroom activity sheet and a paper [20]. In the process she learned a lot of mathematics along with the context of the thematic issues.

## Student Responses to David Blackwell Projects

Students admire Blackwell's attitudes about racial issues and his mathematical style [1, 6] and they enjoy learning about him and his mathematics. In the *Introduction to Mathematics* class, where each student wrote about a mathematician that he or she related to, one female student commented:

I never thought that I could relate my mathematical style to a famous mathematician because first of all I didn't really think that I had a mathematical style. Doesn't having a mathematical style assume that one is greatly interested in mathematics? Well, not in my case. Math really has not been one of my strong points. Although I am interested in solving problems, I'm more concerned with understanding what I'm learning. I finally found a mathematician that I can identify with and that person is David Blackwell.

## Conclusion

Projects that incorporate the mathematical achievements of women and minorities into the classroom provide a rich mathematical environment for examining beliefs about what mathematics is and the diverse ways that people can be successful at doing mathematics. Because most women and minority mathematicians that we know about lived within the last 200 years, there are not as many resources that bring their mathematics into the college classroom in a meaningful way as there might be for earlier mathematicians. Yet the inclusion of the recent mathematical achievements of women and minorities is important, not only to ensure that these mathematicians are remembered for their mathematics instead of just their race or gender, but also because students relate to these mathematicians. With creative planning, their achievements can be incorporated into the classroom, and the result can be rewarding for all students.

One student from the *Women and Minorities in Mathematics* class, who planned to become a high school mathematics teacher, commented that this was one of the few classes in the department that he could walk away from and actually say, "I will use this in my classroom." Another student commented that this course should be taken by both math majors and future teachers, while a third student commented that she felt the projects helped her in advancing her ability to do research and speak about math, and that "it was very encouraging." Students from this class are now teaching in the middle grades and high school and have reported successful implementation of projects on women and minority mathematicians in their own classrooms.

One *Introduction to Mathematics* student commented, "Some students might say that learning about present and past mathematicians is useless, however, I learned that not all mathematicians sit at a desk doing equations. Some, such as Gordon, work on drums, while others, such as Blackwell, work with theories such as game theory." Another student reflected, "I thought that it was great that my mathematical style was so close to hers. It made me appreciate her way of thinking better, as well as my own way of thinking. I had never before really labeled my train of thought, but because of our studies I am aware of the way my mind works, and I find it very handy when I come across difficulties that I must deal with. Thanks!" The strength of this reaction is exactly the reason that we use these projects. This is the kind of response that we hope for in all of our students.

## Acknowledgements

Special thanks go to Ann Bies for her invaluable assistance with various aspects of this article. Thanks also to David Blackwell, Carolyn Gordon, Garalee Greenwald, and Amy Shell-Gellasch, to Math 1010 students at Appalachian State University for working through these projects and helping to refine them, and to Ross Bryant and Zachary Lesch-Huie who gave permission for their worksheet to appear here.

## References

1. D.J. Albers and G. Alexanderson, "David Blackwell," in *Mathematical People: Profiles and Interviews*, D.J. Albers and G. Alexanderson, eds., Birkhäuser, Boston, MA, 1985.

2. David Blackwell, *Some Properties of Markov Chains*, Ph.D. thesis, University of Illinois-Urbana Champaign, 1941.

3. D. Blackwell and T.S. Ferguson, "The Big Match," *Annals of Mathematical Statistics*, 39 (1968) 159–163.

4. M.A. Campbell and R. Campbell-Wright, "Towards a feminist algebra," in *Teaching the Majority: Breaking the Gender Barrier in Science, Mathematics and Engineering*, Sue Rosser, ed., Teachers College Press, Columbia University, New York, 1995.

5. C. Davis and S. Rosser, "Program and curricular interventions," in *The Equity Equation: Fostering the Advancement of Women in the Sciences, Mathematics, and Engineering*, Cinda-Sue Davis et al, ed., Jossey-Bass Publishers, San Francisco, CA, 1996.

6. Morris DeGroot, "A Conversation with David Blackwell," *Statistical Science*, 1 (1986) 40–53.

7. William Dunham, *The Mathematical Universe*, John Wiley & Sons, New York, 1994.

8.   Sarah J. Greenwald, *Women in Math Course* [online], 1999.
     Available: http://www.mathsci.appstate.edu/~sjg/womeninmath/

9.   ———, *Modern Algebra* Course [online], 2000.
     Available: http://www.mathsci.appstate.edu/~sjg/class/3110/

10.  ———, *Women and Minorities in Math Course* [online], 2001.
     Available: http://www.mathsci.appstate.edu/~sjg/wmm/

11.  ———, *Andrew Wiles and Fermat's Last Theorem Worksheet* [online], 2002.
     Available: http://www.mathsci.appstate.edu/~sjg/class/1010/mathematician/wiles2.doc

12.  ———, *Carolyn Gordon and Hearing the Shape of a Drum Worksheet* [online], 2002.
     Available: http://www.mathsci.appstate.edu/~sjg/class/1010/mathematician/carolyngordon.pdf

13.  ———, *Mathematician Segment* [online], 2002.
     Available: http://www.mathsci.appstate.edu/~sjg/class/1010/mathematician/

14.  ———, *Mathematician Segment Presentation Checklist* [online], 2002.
     Available: http://www.mathsci.appstate.edu/~sjg/class/1010/presentationchecklist.html

15.  ———, *Mathematician Segment References* [online], 2002.
     Available: http://www.mathsci.appstate.edu/~sjg/class/1010/mathematicianreferences.html

16.  ———, *Mathematician Segment Theme Questions on Andrew Wiles* [online], 2002.
     Available: http://www.mathsci.appstate.edu/~sjg/class/1010/andrewwiles2.html

17.  ———, *Mathematician Segment Worksheet Checklist* [online], 2002.
     Available: http://www.mathsci.appstate.edu/~sjg/class/1010/worksheetchecklist.html

18.  ———, *NCCTM Centroid Column and Activity Sheets on David Blackwell and Game Theory* [online], 2003.
     Available: http://www.mathsci.appstate.edu/~sjg/ncctm/activities/

19.  Art Johnson, *Famous Problems and their Mathematicians*, Teacher Ideas Press, Englewood, CO, 1999.

20.  Elizabeth Johnson, *David Blackwell* [online], 2001.
     Available: http://www.mathsci.appstate.edu/~sjg/wmm/final/blackwellfinal/main.htm.

21.  Karen Karp, *Feisty Females: Inspiring Girls to Think Mathematically*, Heinemann, Portsmouth, NH, 1998.

22.  Patricia C. Kenschaft, "Fifty-five cultural reasons why too few women win at mathematics," in *Winning Women into Mathematics*, P.C. Kenschaft and S. Keith, eds., Mathematical Association of America, Washington, DC, 1991.

23.  Beatrice Lumpkin, *Geometry Activities from Many Cultures*, J. Weston Walch, Portland, ME, 1997.

24.  B. Lumpkin and D. Strong, *Multicultural Science and Math Connections: Middle School Projects and Activities*, J. Weston Walch, Portland, ME, 1995.

25.  NOVA, *The Proof* [video], 1997.

26.  Marla Parker (Ed.), *She Does Math! Real-Life Problems from Women in the Job*, Mathematical Association of America, Washington, DC, 1995.

27.  Teri Perl, *Math Equals: Biographies of Women Mathematicians + Related Activities*, Addison-Wesley, Menlo Park, CA, 1978.

28.  J.E. Taylor and S.M. Wiegand, *AWM in the 1990s: A Recent History of the Association for Women in Mathematics* [online], 2003. Available: http://www.awm-math.org/articles/199812/awm1990s/

29.  Janet Trentacosta (Ed.), *Multicultural and Gender Equity in the Mathematics Classroom: The Gift of Diversity*, National Council of Teachers of Mathematics, Reston, VA, 1997.

30.  WGBH Science Unit, *NOVA Online—The Proof* [online], 1997. Available: http://www.pbs.org/wgbh/nova/proof/

31.  Scott Williams, *Mathematicians of the African Diaspora: Who are the Greatest Black Mathematicians?* [online], 2002. Available: http://www.math.buffalo.edu/mad/madgreatest.html

32. ———, *Mathematicians of the African Diaspora: Statistics on the Numbers of Blacks Receiving Mathematics Ph.D.s* [online], 2003. Available: http://www.math.buffalo.edu/mad/stats/index.html

## Appendix A: References and Comments on How We Used Them for Carolyn Gordon

We do not use a formal reference system for worksheets, but we do give students a list of references at the end of each worksheet and comment on how we used them in designing the worksheet. We gave the following references to the *Introduction to Mathematics* class at the end of the Carolyn Gordon worksheet.

Cipra, Barry (1992), "You can't always hear the shape of a drum," *Science*, March 27, 1992, Volume 255, No. 5052, p. 1642–1643.
This magazine article had an overview of the problem and a description of the reaction to Wolpert's question, their model building, and transatlantic work.

Cipra, Barry. (1997a). You can't always hear the shape of a drum [On-line].
Available: http://www.ams.org/new-in-math/hap-drum/hap-drum.html
This is a great website that I used to find an overview of the history and solution of the problem.

Gordon, Carolyn and David Webb, "You Can't Hear the Shape of a Drum," *American Scientist*, January-February, 1996, Volume 84, No. 1, p. 46–55.
This magazine article had a great summary of the solution of the problem, and some information about the authors. They recently received The Chauvenet Prize for writing this article, given for an outstanding expository article on a mathematical topic by a member of the Mathematical Association of America.

Mathscinet search on Carolyn Gordon. (2001) [On-line].
Available: http://www.ams.org/mathscinet
I used this site to find her published papers and collaborators.

Personal communication with Carolyn Gordon (2001).

Peterson, Ivars. (1997a). Ivars Peterson's MathLand: Drums that sound alike [On-line].
Available: http://www.maa.org/mathland/mathland_4_14.html
I used this site to find the pictures in Figure 3 and 4, and it also contained information about the physicists who made the drums and performed experiments to show that they sounded the same.

Weintraub, Steven. (1997). What's new in mathematics: June 1997 cover [On-line].
Available: http://www.ams.org/new-in-math/cover/199706.html
This website contains the pictures of Carolyn Gordon and David Webb holding their drums. It also contains links to an animated picture of the frequency and waves when the drums are struck.

I could not find information in books about Carolyn Gordon.

## Appendix B: References Provided to the Students for David Blackwell

We gave the following resources to the students assigned to David Blackwell in the *Introduction to Mathematics* class.

Albers, D. and Alexanderson, G. eds., *Mathematical People: Profiles and Interviews,* Birkhäuser, Boston,1985, p. 19–32.

DeGroot, Morris. "A Conversation with David Blackwell," *Statistical Science,* 1 (1986) 40–53.

O'Donnell, Michael. The Prisoner's Dilemma: A Fable [online] (1998).
Available: http://www.classes.cs.uchicago.edu/classes/archive/1998/fall/CS105/Project/node2.html

Williams, Scott. David Blackwell — Mathematicians of the African Diaspora [online] (2002).
Available: http://www.math.buffalo.edu/mad/PEEPS/blackwell_david.html

Young, Robyn. "David Blackwell" in *Notable Mathematicians : From Ancient Times to the Present* p. 62–64.

## Appendix C: Reference Section from the Student Worksheet

Students Ross Bryant and Zachary Lesch-Huie submitted the following references at the end of their David Blackwell worksheet.

Albers, D. and Alexanderson, G. eds., *Mathematical People: Profiles and Interviews,* Birkhäuser, Boston, 1985, p. 19–32.
This gave us background insight on Blackwell's life as a young man as well as Blackwell's feelings about working on the Prisoner's Dilemma for the RAND corporation and the relationship to the arms race with the Soviet Union.

DeGroot, Morris. "A Conversation with David Blackwell," *Statistical Science,* 1 (1986) 40–53.
This interview revealed Blackwell's personal experiences and dreams as a young man. He also discusses his mathematical techniques and inspirations.

O'Donnell, Michael. The Prisoner's Dilemma: A Fable [ online] (1998).
Available: http://www.classes.cs.uchicago.edu/classes/archive/1998/fall/CS105/Project/node2.html
This website focused on Tucker and the Prisoner's Dilemma and is where we got most of our information on this subject and its significance.

Williams, Scott. David Blackwell — Mathematicians of the African Diaspora [online] (2002).
Available: http://www.math.buffalo.edu/mad/PEEPS/blackwell_david.html
We learned all about Blackwell's research and game theory at this website.

Young, Robyn. "David Blackwell" in *Notable Mathematicians : From Ancient Times to the Present* p. 62–64.
This selection helped in giving us a general overview of Blackwell's life and some insight regarding his work with the theory of duels and game theory in general.

# 17

# Mathematical Topics in an Undergraduate History of Science Course

**David Lindsay Roberts**
*Laurel, Maryland*

## Introduction

An undergraduate survey course in the history of science presents numerous opportunities to discuss the role of mathematics in scientific developments, especially with regard to physics and astronomy. These courses are most commonly taught in a history department, or a history of science department. Although there may be some modest prerequisites in terms of science and mathematics, these are history courses, not mathematics courses, and it is usually inappropriate to treat mathematics topics in rigorous detail, and especially inappropriate to evaluate student solutions of mathematics problems as part of the course grade. But no such course can be considered complete without some reference to mathematics. Some description of the possibilities for treating mathematics within a history of science course may be beneficial for teachers of more mathematically oriented history courses, who may derive some ideas for incorporating history of science into their mathematics presentations.

Restricting attention to the last 200 years leaves no shortage of topics, but increases the difficulty of providing accurate but comprehensible explications of the mathematical issues. Historians of science may approach topics from a more philosophical standpoint, or indulge in more hand-waving, than some teachers of mathematics would find agreeable. Mathematicians may find that a history of science survey course is too much of a romp through the decades and centuries, flitting from one topic to the next without ever getting to the details. The following are examples of treatments of mathematical topics in the history of science classroom, assuming students have a basic understanding of algebra and geometry. An acquaintance with the major ideas of calculus is sometimes helpful as well. A few useful references are provided, without any attempt at completeness.

## Celestial Mechanics

This topic looms large in any history of science course that covers the work of Kepler and Newton in the seventeenth century, but it is also an important topic for the period after 1800. A survey of the period from Newton to Laplace is probably necessary to set the stage, culminating in Laplace's nebular hypothesis and the issue of the long-term stability of the solar system. This can lead into discussions of what it means to solve a differential equation with specified initial conditions, and the extreme difficulty of solving the 3-body problem. One can then race through the nineteenth century to get to the work of Henri Poincaré, possibly pausing to comment on the discovery of Neptune, or for a little hand-waving on perturbation theory. The payoff is that Poincaré's celestial mechanics provides an entry into the modern theory of chaos. Fortunately there are excellent popular accounts of this that are quite appropriate for such a course. The books by Ruelle and Ekeland [12, 3] are especially recommended. A mathematics instructor can readily supplement the approach sketched here with more technical details on topics such as differential equations and convergence of infinite series.

## Electromagnetism

Any discussion of the developing comprehension of electromagnetism in the nineteenth century will naturally emphasize the work of both Michael Faraday and James Clerk Maxwell. The usual account is that Maxwell translated Faraday's nonmathematical intuitions into mathematical notation, specifically into the language of partial differential equations. But Maxwell himself put the matter in a slightly different way: "As I proceeded with the study of Faraday, I perceived that his method of conceiving the phenomena was also a mathematical one, though not exhibited in the conventional form of mathematical symbols" [10, p. ix]. This can give rise to interesting class discussion on the nature of mathematics and how important symbolic notation is to the subject. To delve deeply into the particular case of Faraday and Maxwell requires considerable background in both electromagnetic phenomena and in vector analysis, but even general consideration of the issue is worthwhile.

## Quaternions

Mention of vector analysis in connection with Maxwell's theory of electromagnetism is an anachronism. If one looks into Maxwell's work one finds that he expresses his famous equations in terms of quaternions. The history of this topic is intriguing, and there is a good source [1] for the story of the rise and fall of quaternions in mathematical physics and the triumph of vector analysis. The technical details are formidable, but there are enough fascinating characters involved (Hamilton, Gibbs, and Heaviside, to name only three) to keep the interest of a general audience. Starting with a brief recall of complex numbers, quaternions are readily described for this audience, and the failure of the commutative law can lead to a fruitful exploration of general algebraic systems. The particular problem that plagued Hamilton, "why can't I multiply triples?" and the provocative answer that an associative multiplication of $n$-tuples is only possible for $n = 1, 2,$ or 4, is worth mentioning, even though the details (essentially the Frobenius theorem; see [7, pp. 326–329]) may be beyond the scope of the course. On the other hand, for the instructor of a course where such details are appropriate, a discussion of the history of quaternions can serve to humanize this abstract material and link it with developments outside mathematics.

## Statistical Mechanics

The nineteenth-century development of the kinetic theory of gases is an excellent vehicle for exploring how physicists learned to understand the behavior of large collections of particles even when knowledge of the behavior of individual particles is minimal. More generally, it is an illuminating example of the importance of model-building in physics, and some of the ways in which the mind-set of a physicist may differ from that of a mathematician.

In the nineteenth century physicists began to model gases as collections of tiny particles moving at high speed. Here is an appropriate example for a history of science class, using nothing more than simple algebra. (See [9, pp. 205–206], or other elementary physics books, for more details). Assume that we have $N$ identical molecules of mass $m$ in a cubical box of side $a$. Let us further suppose that the molecules all travel with the same velocity $v$ and that one third of the molecules bounce back and forth between top and bottom of the box, another one third bounce back and forth between front and back, and the last third bounce back and forth between the right and left sides. Finally, let us assume that no molecule ever hits another molecule. Using Newton's laws, and defining the pressure in terms of the force exerted by the molecules as they bump into the sides of the box, one can derive Boyle's Law: there is an inverse relation between pressure and volume. Given the drastic simplifying assumptions many mathematicians would be skeptical that this derivation means anything at all, but physicists generally find it encouraging. It should be emphasized that this is only the beginning, that nineteenth-century physicists proceeded to complicate the model by making more and more realistic assumptions. They partitioned the molecules of a gas into a richer array of classes and employed concepts such as the mean-free path (the average distance a molecule travels before hitting another molecule). Nevertheless the simple model above displays the core triumph of statistical mechanics: large scale behavior of matter as observed in experiments can be explained theoretically by the behavior of invisible particles.

The degree to which one might wish to explore this topic in more detail depends on the mathematical knowledge of the class. Some mention of Brownian motion and Einstein's role in explaining this is appropriate. Further commentary on conceptual issues in statistical mechanics can be found in the books by Ruelle and Ekeland earlier mentioned [12, 3]. Richard Feynman [4] is a good source for insight into how the thinking of physicists often differs from that of mathematicians.

## Special Relativity

This topic is usually a highlight of any history of science course that reaches the twentieth century. Einstein's recasting of the world picture has retained its power to surprise. Fortunately many of the basic arguments, such as those on simultaneity, are accessible with simple algebra, as shown by Einstein's own popular account [2]. Additional details can be worked up to suit the instructor and the class. Good source materials abound. Friedrichs [5] shows how to reach $E = mc^2$ using only simple vector analysis to treat the physics of momentum and energy. Taylor and Wheeler ([13], a superb book) have nice treatments of the Lorentz transformation, invariance of interval, and of Minkowski's spacetime approach. There is an opportunity to delve deeper into transformational geometry if desired. Minkowski's world-lines and light-cones are readily described in simple cases in two dimensions. It may be worth pausing on Minkowski's use of the square root of $-1$ in his formulation, which again raises issues of physicist versus mathematician and the significance to be attached to mathematical notation. Students may enjoy debating whether the square root of $-1$ in relativity and elsewhere in physics can possibly have true physical significance or whether it is merely a shorthand convenience.

## General Relativity

Here, alas, the mathematics becomes very hard, but the topic is so important that hand-waving is excusable where necessary. An appealing geometric approach is to begin with the idea that by a suitable change of reference system you can "explain" forces. If you are driving a car with the steering wheel on the left and make a sharp turn to the right you feel a force pushing you into the car door. It certainly feels real in your reference frame. But from the point of view of a physicist observing you from a fixed point on the earth this is a fictitious force, explained by your turning; that is, accelerating. Einstein applied a similar approach to the force of gravity, noting in a famous thought experiment that an observer in a box falling freely in a gravitational field would feel no force of gravity at all. In other words, a gravitational field can be removed by transformation to a proper accelerated system.

Again Taylor and Wheeler's book is an excellent source [13, p. 184]. They give a neat two-dimensional analogy of Einstein's four-dimensional spacetime which seems to help student understanding. Consider two people, $A$ and $B$, a certain distance apart on the equator of a sphere. If $A$ and $B$ both travel due north they will find themselves drawing closer and closer together. We can interpret this either as the action of some mysterious force or as merely the effect of the curvature of the sphere. As Taylor and Wheeler emphasize, Einstein's theory of gravity is a local theory. Abandoning Newton's single but unrealizable Euclidean reference frame extending throughout space, Einstein argued that gravitational physics should be done using many local reference frames fitted together, thus making up the global structure of spacetime. Mathematical sophisticates may use this as the jumping off point for delving into the intricacies of differential geometry or differential topology, but almost all students can at least appreciate the basic issues involved. At the least one can discuss the advent of non-Euclidean geometry in the early nineteenth century and its role in Einstein's work.

## Quantum Theory

In pondering the theory of relativity, both special and general, it is not hard to fall into the illusion that one is engaging in a purely mathematical exercise. The aesthetic satisfaction of relativity often appeals greatly to those who enjoy mathematics, to the extent that they may feel it must be true, simply from reading about it. Care should therefore be taken to explain that relativity is grounded in experimental facts, without which the theory would have no standing in physics. Quantum theory presents the mathematical aficionado with a different problem. Here the experimental facts seem much more constraining, and merely reading about the theories that explain the facts does not provide the same degree of confident belief as in relativity. (I speak for myself, but suspect I am not alone.) It is hard to develop an intuition about the quantum nature of the world, and indeed some knowledgeable physicists have stated that it is useless to try. The world simply is this way. Some more attracted to mathematics than to physics can find this disconcerting, and it is one factor contributing to the difficulty of using quantum theory as a vehicle for teaching mathematics or history of mathematics.

Another factor is the abstruse nature of the mathematics involved, to which no real justice can be done in an undergraduate history of science survey. Nevertheless, it is important to attempt to convey a few essential points, especially the distinction between Schrödinger's wave mechanics and Heisenberg's matrix mechanics. One can explain that Schrödinger's approach was in the mainstream tradition of employing partial differential equations in mathematical physics, going back to the eighteenth-century work of Euler and D'Alembert on the vibrating string. Possibly one can even convey something about the meaning of an eigenvalue for a differential equation. In any case, it should be emphasized that Schrödinger's wave equation seemed to promise, vainly as it turned out, that one could retain a visual intuition regarding the behavior of an electron. Heisenberg's approach, on the contrary, explicitly disclaimed pictorial representation and instead relied on encapsulating observable quantities in matrices, a mathematical novelty for most physicists in the 1920s. Here one can take the opportunity to say something about the nineteenth-century roots of matrices in the work of Cayley and others.

There is also an opportunity here to say something about the influence on physics of two great mathematicians of the late nineteenth and early twentieth centuries: Henri Poincaré, a blindingly brilliant lone investigator, and Felix Klein, an institution builder par excellence. In the early parts of their careers they were working on the same research problems in the theory of automorphic functions. This competition was resoundingly won by Poincaré, and helped to turn Klein's energies more towards administration, pedagogy, and history. Poincaré's influence on quantum theory was directly intellectual: his work on perturbation theory in celestial mechanics was important to several of the young quantum theorists, notably Heisenberg. Klein meanwhile had established at Göttingen the preeminent mathematics department of that era. This department trained notable physicists, including Max Born and Arnold Sommerfeld, and also provided the institutional base permitting David Hilbert to flourish. Hilbert's 1924 book with Richard Courant, *Methods of Mathematical Physics*, was a key source for the mathematics of quantum mechanics, especially the analysis of partial differential equations with boundary values. A discussion of Poincaré and Klein can go in many mathematical directions, touching on topics in geometry, algebra, and analysis.

Jammer [8] provides a history of quantum theory up to about 1930, touching on a variety of physical, mathematical, and philosophical issues and referencing all the original papers. Guillemin [6] covers more recent developments, but with much less detail. One result of considerable interest, Bell's theorem of 1964, is given a charming and mathematically elementary exposition by McAdam [11].

## Conclusion

The foregoing examples have been intended to give an idea of how mathematical topics may arise in a history of physical science course covering the 1800s and 1900s. There are two primary ways in which these cases may be useful for an undergraduate instructor teaching a history of mathematics course with a strong technical component, or teaching a mathematics course with a strong historical component. First, they may be a source of problems for practicing mathematical techniques being covered in the class. This will vary markedly with the specific course and the background of the students. Second, they offer a way of explaining where certain mathematical ideas have originated, or have found significant use. This may placate the student who is not content to contemplate mathematics purely for its own sake, and may help to convey the important fact that mathematics is not an entirely independent intellectual discipline. Indeed, these episodes can open up profound debate on the nature of mathematics, and the features that distinguish it from and yet make it applicable to the sciences. A mathematics instructor will likely have limited time to digress from technical details to consider such broad questions, but where such time is available one or more of the topics discussed here can offer avenues for stimulating exploration.

## References

1. Michael J. Crowe, *A History of Vector Analysis*, Dover Publications, Mineola, NY, 1994.
2. Albert Einstein, *Relativity: The Special and General Theory*, Henry Holt, New York, 1920.
3. Ivar Ekeland, *Mathematics and the Unexpected*, University of Chicago Press, Chicago, 1988.
4. Richard Feynman, *The Character of Physical Law*, MIT Press, Cambridge, MA, 1967.
5. K. O. Friedrichs, *From Pythagoras to Einstein*, MAA, Washington, DC, 1965.
6. Victor Guillemin, *The Story of Quantum Mechanics*, Charles Scribner's Sons, New York, 1968.
7. I. N. Herstein, *Topics in Algebra*, Xerox College Publishing, Waltham, MA, 1964.
8. Max Jammer, *The Conceptual Development of Quantum Mechanics*, McGraw-Hill, New York, 1966.
9. Robert Bruce Lindsay, *Basic Concepts of Physics*, Van Nostrand Reinhold, New York, 1971.
10. James Clerk Maxwell, *A Treatise on Electricity and Magnetism*, vol. 1. Dover Publications, New York, 1954.
11. Stephen McAdam, "Bell's Theorem and the Demise of Local Reality," *American Mathematical Monthly* 110 (2003) 800–811.
12. David Ruelle, *Chance and Chaos*, Princeton University Press, Princeton, 1991.
13. Edwin Taylor and John Archibald Wheeler, *Spacetime Physics*, W. H. Freeman, San Francisco, 1966.

# 18

# Building a History of Mathematics Course from a Local Perspective

**Amy Shell-Gellasch**
*Grafenwoehr, Germany*
*Formerly of the United States Military Academy, West Point, New York*

## Introduction

One of the challenges of developing a history of mathematics course is deciding what material to cover. The history of mathematics is far too broad a subject to cover in a year-long course, much less a one-semester course. Some options might be to focus on topics such as ancient mathematics, the history of the calculus, great moments in mathematics, great people in mathematics, and so forth. Any course developed must also take into account the audience and intent of the course. Are the students math majors or education majors? Will this be a general education course or a course for the mathematics major? These questions need to be answered before one can start to choose a focus for the course.

The United States Military Academy (USMA) at West Point, New York, does not have a standing history of mathematics course. So when I proposed a history of mathematics course to be offered only once, I wanted it to be unique. Given that the Academy has a long and important history, especially as the first engineering school in the nation (founded 1802), I wanted my students to get a feel for how mathematics has been an integral component of the Academy for over two hundred years. The course I devised is entitled *The History of Mathematics from the West Point Perspective*. This course uses the history of the USMA, the Department of Mathematical Sciences, and the mathematics curriculum to motivate the mathematical topics covered. I will discuss how the topics were chosen, as well as student assignments that merge concepts in the history of mathematics with the history of the institution and its faculty.

## General Design of a Course

Every school and mathematics department has a rich and interesting history, including important faculty members, both past and present, that students should be aware of. For example, private schools usually were founded by some prominent person or group for a specific purpose. Many state institutions started out as land grant or normal schools that have since evolved or changed focus over the years. Issues such as these can be looked at early in the course. From there, following the history of curricular change at your institution will provide topics for you to cover in a history of mathematics course. The net can be cast even wider to include historical events in your city or state which are of relevance to mathematics and the history of your institution.

For example, an early unit might cover Euclidean geometry since this topic was an integral component of a classical curriculum. This can lead to some interesting discussions about what education is for and how that has changed over time, as well as discussions of the role of philosophy in mathematics. A unit on the calculus might include when and how that topic was brought into the curriculum. Later units may cover computers or electives such as topology, based on when they were introduced into the curriculum and why. Along the way, a look at faculty members and their work lets the students experience mathematics as a vibrant, evolving discipline.

This is just a brief introduction to give a feel for how the course might be organized. This type of history course can be used at any institution. If you choose to conduct a more traditional history course, or feel that you do not have enough material or time to develop a full course of this nature, the history of your department and institution can be incorporated either as a thread running through the course, or as a special unit. In addition, many of the topics and ideas can be incorporated into a traditional history of mathematics course or regular math course as anecdotes or special assignments. The best way to get started is to make friends with your librarian and archivist. They have old books and papers that possibly no one has shown an interest in for years, if ever, and would love to share their knowledge and expertise.

## The Course at West Point

The USMA was one of the first schools in the nation to break away from the traditional classical curriculum of ancient languages and Euclidian geometry aimed at educating gentlemen and the clergy, and to focus on a course of study that would be beneficial to future engineers who, as Army officers, would help build the infrastructure of the United States. To this end, and to this day, every cadet takes a heavy dose of mathematics. By the second decade of the nineteenth century, West Point was teaching calculus and descriptive geometry. Since English-language text books on these topics were rare, many important texts used throughout the United States in the nineteenth century were originally written by USMA staff for use at the Academy. Given this rich mathematical tradition, I developed a course that used the history of the Academy as a point of departure. The history of the Academy, its faculty and curriculum would be the warp and woof onto which topics in the history of mathematics would be woven.

What follows is a detailed description of my course at USMA. Keep in mind that the topics I have chosen are based on the history of the Academy and its mathematical curriculum. I have provided them to give you a framework to build your own course upon.

### Course title: The History of Mathematics from the West Point Perspective

### Course description

This course will explore the history of mathematics and related topics as they have impacted USMA through its curriculum, texts and faculty. The history of the Academy, and in particular the Department of Mathematical Sciences, will be used as a point of departure to study topics such as the history of the calculus, engineering drawing, geometry and descriptive geometry, calculating and mathematical devices, computers and their uses, and topics in modern mathematics. We will also look at original mathematical sources to gain an understanding of the evolution of mathematical thought.

### Texts

*Great Moments in Mathematics Before 1650*, Howard Eves, MAA, 1983

*Great Moments in Mathematics After 1650*, Howard Eves, MAA, 1983

*History of West Point*, Theodore Crackel, University Press of Kansas, 2002

Selected readings

Since this course is open to sophomores[1] and above, I chose texts that are easy to use. I want my students to have good general references that they might use after the course is over. Eves' books provide nice stand-alone chapters. This is beneficial since not all of the standard topics usually found in a history of mathematics course are covered. The students can read the other chapters if they are interested. Eves provides good general introductory material to a topic that can then be supplemented by selected readings, as well as problems at the end of each chapter. William Dunham's *Journey Through Genius* or *The Mathematical Universe* would be good books along these lines. For the history of the Academy, I am lucky enough to have a wonderful book that covers many academic topics as well as the general history of West Point. Many colleges do have book length histories. If not, there may be articles in the alumni magazine or school newspaper of relevance. Again, the archives of your school will also have material you can use.

## Course Topics

To organize the course, I wrote a rough chronology of USMA and the Department of Mathematical Sciences including the mathematics taught at various times. I then included the topics in the history of mathematics that corresponded to these stages. This gave me a rough outline of what topics in the history of the department and mathematics to cover and in what order. For example, at the founding of the Academy, what were other colleges in the United States teaching? This lead to two discussions, the first being traditional classical curriculum being taught at schools like Harvard, which lead into units on Euclidian Geometry and algebraic solutions to equations. Second, in our case, how was the curriculum that was being taught at USMA different from that of other colleges, which included units on descriptive geometry and surveying. When and how the calculus is added to the curriculum will naturally lead to a unit on the history of the calculus. If your institution once had a class in fluxions taught before calling it calculus, all the better.

Near the end of the semester, a unit or several short units can be created around the first electives taught. When did topology first come into the curriculum? statistics? computer science? Since these courses are comparatively recent, records of their inclusion in the curriculum should be easier to find. Also, some of the original teachers of those topics, and maybe even early practioners, may still be around for you to utilize. If your school has a large teacher training program, you may want to look at how the training of mathematics teachers and what is taught in school mathematics has changed over time. Many good social-political discussions may arise here.

At every point, discussions about the books and teaching methods used are relevant. See if your library archives have copies of old texts or information about teaching faculty or methods. Students love to compare "old school" to "new school". Old exams, especially entrance exams are goldmines of problems and discussions, and are usually in your library archives. You will find that the old texts are usually harder and the methods less interactive and more focused on rote memorization.

The hardest part of making the syllabus for this history course was scheduling, which depended on when guest speakers could come and other such restrictions. As you will see below, I was able to include most of the major topics in the history of mathematics, albeit at a fairly cursory level. Since you will be combining two lines of inquiry in one course, the depth that you can go into may be limited. However, if

---

[1]All students at USMA take calculus their first year, so I was able to assume a certain level of mathematical knowledge. Based on the background of your students, you will have to choose a course such as calculus as a prerequisite, or choose topics that they have had experience with.

your school has historically had a strong tradition in a certain area of research or teacher training, your course could be narrower in scope while focusing more deeply on these topics. Since my goal was to let the students see how history affected what was taught at USMA, my course was very general and broad based. This can be seen in my course outline below.

## Topics Outline

I. Founding of the Academy (1801–approx. 1817)

    A. Engineering Corps and the state of engineering c. 1800

        1. Mathematical training c. 1800—the classical approach

            a. Euclid's Elements

            b. Philosophy and the Greeks

            c. Solving equations and the Islamic scholars

            d. Surveying

        2. Early texts at USMA

            a. Hutton and algebra—the English influence

            b. Vauban and engineering drawing

II. Thayer's Academy (1817–approx. 1900)

    A. Ecole Polytechnique and French mathematics
       (Thayer Collection at USMA library—Fred Rickey)

        a. Fluxions and calculus

        b. Mechanics

        c. Geometry (Davies' texts)

        d. Descriptive Geometry (Monge, Olivier, Crozet)

    B. Teaching and learning at the Academy

III. The 20th Century

    A. Calculating and Mathematical Devices

        1. From the slide rule to the TI-89

        2. Computers

            a. History

            b. Introduction at USMA

            c. Uses

        3. Mathematical Modeling

    B. Electives at USMA

        1. Introduction at USMA (majors)

        2. Operations Research (faculty guest lecturer)

        3. Modern mathematics (faculty guest lecturer)

## Guest Lecturers and Presentations

As mentioned above, any faculty members who have been around a while are a wealth of information. Invite them to come in as a guest lecturer to talk about either their research or the evolution of the department. Your history department may have someone familiar with the founding and early history of your institution, so invite them in early on. Also seek out mathematicians, historians and historians of mathematics from outside your school to give special lectures. Give the students reading material about the topic of the lecture and the lecturer ahead of time. In particular, if the guest lecturer has a published paper on the topic they will be speaking about, let the students read that. Getting the students familiar with some of the recent literature in the history of mathematics is an important and often neglected component of a history course. Below is a list of the guest lectures planned for my course. You will notice a mix of faculty members, former faculty members, and outside experts. I feel this mix is beneficial. In particular, former faculty members, especially retired ones, are usually eager to come back.

## Guest Lectures

*The Thayer Collection of the USMA Library,*
   Fred Rickey, USMA
*George Baron: West Point's 0th Professor,*
   Fred Rickey, USMA
*Alden Partridge at West Point,*
   Dick Jardine, Keene State College (formerly of USMA)
*The History of Discrete Dynamical Systems,*
   Dick Jardine, Keene State College
*Charles Davies: Text Book Author at West Point,*
   Amy Ackerberg-Hastings, Anne Arundel Community College
*Thomas Jefferson's Other Academy: the Founding of the University of Virginia,*
   Adrian Rice, Randolph-Macon College
*Mathematics on the Plain* (the mathematical walking tour of West Point),
   Chris Arney, College of St. Rose (former department head, USMA Dept. of Mathematical Sciences)
*The History of Slide Rules and Other Calculating Devices,*
   Peggy Aldrich Kidwell, Smithsonian Institution
*History of Computing at West Point,*
   COL Eugene Ressler, USMA
*West Point's 200 at 200,*
   Chris Arney, College of St. Rose
*A Short History of Operations Research,*
   LTC Darrel Henderson, USMA

I wanted my course to be as interactive and student motivated as possible. In order to do this, at the start of many lessons, I had a student present a short biography (ten minutes at most) of a person relevant to that day's topic. In relation to the above discussion on guest speakers, the biography of the day for the class meeting prior to the guest lecture could be of the speaker. That student could also introduce the speaker at the lecture. Opening any of these guest lectures to the whole college is wonderful for the students and the department.

For the short biographical presentations I had a prepared list of names, both from the history of mathematics and Academy history, which is longer than the number of lessons so that students have some choice in who they research. I make sure ones that are essential to the course are covered. This list may include more recent or even living mathematicians depending on the focus of your course. I recommend that you include a few members of your department who are active researchers in one or more of the areas you will cover. This gives the students a connection to the department and a feel for the activity of mathematics, as well as the realization that mathematics is an evolving subject.

I asked each student to turn in a short electronic copy (we are avid users of PowerPoint slides at the Academy) of their presentation which I then published on the course website for future reference. This list may also provide ideas for any research projects the students are asked to do.

Below is a partial list of persons of relevance to the West Point course.

## List of Subjects for Daily 10 Minute Biography

| | | |
|---|---|---|
| Euclid | Thayer* | Davies* |
| Aristotle | Mahan* | Church* |
| Archimedes | Fermat | Bledsoe* |
| al-Khowarizmi | Descartes | Von Neumann |
| J. Williams* | Newton | Turing |
| W. Barron* | Leibniz | V. Bush |
| G. Barron* | Cauchy | G. Hopper |
| Mansfield* | Monge | Current research faculty members |
| G. Hutton | Olivier | Guest lecturers |
| Vauban | Crozet | |

* denotes people specific to USMA history.

Note that my list includes a mix of big name mathematicians and names specific to USMA history, as well as lesser known mathematicians. I feel it is important to let students see a broad picture of who does mathematics and the diversity of endeavors that are considered part of mathematics. The final assignment of the term was for each student to seek out a mathematician (from West Point or anywhere else), interview them, and present to the class a short biography of that person and their research. Again, this is to involve each student in the current aspects of mathematics and its ongoing history.

## In Class

I think one of the hardest parts of teaching a history of mathematics course for those of us accustomed to teaching straight mathematics courses is what to do in class. We can no longer simply lecture on the new material, go over questions, or have the students work in groups while we look over their shoulders. A successful history course depends on lively and interactive discussions and activities.

There are three aspects to achieving this. First, the instructor must do reading in addition to the assigned readings for the class in order to add depth and diversity to the discussion. The most interesting classes that I have been in, in any field, were the ones in which the instructor brought information from other sources to the discussion. I also felt inspired by their commitment and breadth of knowledge. This is especially interesting in a history course if this additional information is controversial or promotes a different view from the one in the course reading.

Second, and related to the above, the instructor needs to have interesting questions to ask during class. Of course, the instructor should do the readings well ahead of the students. When doing this, compile a list of questions, of all styles, difficulty and abstraction. Some of these can be given to the students before they do the reading, others can be asked in class. The more the questions ask the students to interpret or evaluate the material, the livelier the discussion. Also encourage your students to make their own list of questions. Asking them to lead part of the discussion makes for a more interactive course, as well as taking the pressure off of you and giving them a sense of ownership of the course. On many days I started the discussion by simply asking them what they thought was interesting in the reading. I was often surprised by what they chose. My genuine curiosity about their thoughts lead to wonderful discussions, though they may have differed from my planned discussion topics.

Prior to the start of the semester, I emailed the students enrolled in the course and asked them a few questions. I asked why they enrolled in the course and what they hoped to get out of it. This made me aware of their goals and helped me focus the in-class discussions. I also asked them if they had any particular topics they were interested in. I then took this into consideration when finalizing the syllabus. In the few cases when a topic a student was interested in didn't fit into the syllabus or might not be of interest to the whole class, I suggested they look into that topic for their term-end project.

The third path to success is diversity. Though discussion is important, there are many in-class activities to make the course interesting and interactive. These activities again help give the students a sense of ownership in the course. Below is a list of possible activities. Many of these activities can be combined with or used as graded events such as papers. Again, providing an introductory reading prior to class allows the class to get more out of the planned activity and discussions. It was very clear to me that the days that I had my students actually try to do problems using old methods were the most enjoyable and engrossing for them.

## Class Activities

Debate different sides of an issue
Read/discuss original sources
Work with old texts
Examine old mathematical devices and models
Read/discuss current research or history articles
Solve problems, either old problems or current ones using old techniques
Videos (followed by discussions): Archimedes Palimpsest, Fermat's Last Theorem (Nova), Great Theorems (MAA)
Poster presentations
Research presentations
Guest lectures
Field trips (if you are lucky enough to have a science museum nearby)
Time lines
Old entrance exams (these are always eye opening for students)

## Assignments

As mentioned above, many of these topics are also wonderful assignments. Any history course should include the student doing their own research. I advocate large and integrated projects. This might include a major paper with a presentation. Term-end projects give the students the chance to research a topic in depth and explore many styles of expression. Also, several smaller research projects (such as the biographical presentations mentioned above) allow the students to explore several areas that interest them. This also allows you in some sense to turn the teaching over to the students. Below is a list of possible research ideas. Any of these can be used for papers, presentations or timelines. In any course of this type, I assign research projects of varying length. In this course, I assigned one term paper with presentation, and three shorter assignments. These shorter assignments included an interview, a book review and a biography. I also assigned short homework sets from the book to make sure they were getting their hands dirty with the mathematics. You will choose assignments based on the length of your course, your focus and your audience.

### List of Research Topics

Topics in the history of mathematics
Topics in related fields such as physics and astronomy
Topics in current mathematics
Biographies
Topics related to your institution
Societal/Cultural/Ethnomathematical topics
Philosophy of mathematics
Applied mathematics
Critical book or article reviews
Most influential mathematician
Compare/contrast schools of thought/eras/geographical differences in mathematics
Interview a current researcher or faculty member
Mathematics and art/architecture/music
Evaluation of old texts
Report on attending a mathematical conference
Evaluate a history of mathematics website
Mathematical devices or teaching aides

A history course, in fact any course, runs much more smoothly when the instructor and students are exploring ideas that interest them. So pick a style and focus for your course that you are interested in. That does not mean it has to be one that you are well versed in. A history course is a wonderful venue for you and your students to explore together. Let them tell you what topics they are interested in and incorporate that into your syllabus. And do not be afraid to say "I'm not familiar with that area, let's find out together." There is no reason you could not even present a term paper along with your students!

## The Results

The course was very diverse. The various guest lectures, campus and library tours, and video/discussions were welcome changes to the students' (and my) routine. The "regular" class days were diverse as well. On some days we only discussed West Point history, on others we only discussed the history of mathematics, and some days we drew the two together. The student presentations did give the students ownership in the course. As the course progressed, the students were just as likely to refer to what they had learned from each other as to what I had presented.

Several of the students voiced to me that they had only been marginally interested in taking a history of mathematics course. However, the added component of learning something of the history of the Academy is what really caught their interest.[2] As the semester progressed, the word got out about the course, and

---

[2]In the end of term survey, a few commented that they ended up being more interested in the mathematical parts of the course, as opposed to the Academy history.

other students are now interested in taking the course the next time it is offered. When my students talked to other students about the course, they were truly proud of "their course."

## Conclusion

Many students graduate from college with little or no knowledge of the history of the institution from which they graduate. They also learn very little about the department they are studying in, the people they are learning from and what they do. One way to address these issues in a mathematics department is through a course that combines the history of the department and university with the history of mathematics. The history of your department and school are often untapped treasures. Exploring these will give both you and your students a new view of mathematics, how it is practiced, and the part your school has played in this most human of human endeavors.

# 19

# Protractors in the Classroom: An Historical Perspective

**Amy Ackerberg-Hastings**
*Anne Arundel Community College*

## Introduction

The view of mathematics presented in the school supplies aisles of discount stores revolves around concrete aids: flash cards, rulers, protractors, graph paper, and the like. These inexpensive objects, although snapped up by the general public, are often completely foreign to the daily lives of contemporary mathematicians and mathematics educators. Meanwhile, parents, students, and teachers might assume their classroom tools have some sort of eternal existence outside of historical context. However, beneath surface appearances, there are links between mathematical teaching aids and professional mathematics. The history of mathematics can reveal such commonalties. For example, at the turn of the twentieth century, E. H. Moore required students to prepare graphs on squared paper, easing computation and revealing the ever-present relationship between abstract principles and concrete applications, when he advocated a "laboratory method" of teaching mathematics. [19] [33, pp. 318–319, 327] Yet, the connection between Moore, whose primary reputation was as a research mathematician, and the ubiquitous graph paper of today's elementary mathematics classroom is no longer recognized by many teachers and mathematicians. Similarly, the lowly protractor has a history that transcends abstract mathematics, applied sciences, and mathematical pedagogy and that deserves to be more widely known. Aspects of this story may assist those teachers who are trying to communicate mathematical ideas to students who may not have a background in abstract reasoning but who do have protractors wedged in their backpacks. Therefore, the purpose of this chapter is to provide an overview of the history of the protractor, incorporating suggestions for applying this story to classroom activities at a variety of academic levels.

## The Protractor: A Biographical Sketch

The origins of the protractor are somewhat murky, which is typical for the history of many drawing and measuring instruments. Sighting instruments such as quadrants and astrolabes were commonly marked by degrees well before the sixteenth century. (See Figures 1 and 2 for examples.) Around 1590, English mathematical practitioners began to suggest that such a "divided circle"—with the pinhole sights removed—could be employed along with a plotting board to replace the rhumb and compass method for reading a ship's course at high latitudes. Thomas Blundeville (fl. 1560–1602) described a semicircular divided instrument for this purpose; he called it the "Mariners Flie" in an appendix to his 1589 *Briefe Description of Universal Mappes & Cardes*. This short treatise led David W. Waters, a historian of navigational practice, to

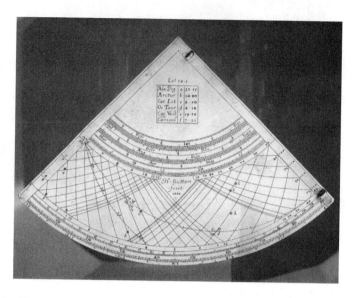

Figure 1. Brass quadrant by Henry Sutton, c. 1660. Negative number p-63375-a. Catalogue number 320152. Courtesy of the Smithsonian's National Museum of American History.

Figure 2. Brass planispheric astrolabe by Georg Hartmann of Nuremberg, 1537. Negative number 79-1769. Catalogue number 336117. Courtesy of the Smithsonian's National Museum of American History.

Figure 3. Theodolite by Fauth and Company, c. 1874–1887. Negative number 75-7235. Catalogue number 334950. Courtesy of the Smithsonian's National Museum of American History.

deem Blundeville the inventor of the protractor, although the concept of dividing an arc for use in measuring angles was apparently on the minds of several mathematical practitioners. [5] [37, p. 212] Indeed, the direct influence of Blundeville's treatise is unclear, for William Barlow (1544–1625), John Davis (1552–1605), and other English writers were also championing the use of the plotting board together with instruments for laying out and measuring angles. By the 1610s, sailors generally had adopted this method for navigating short voyages. Meanwhile, the triangulation technique for surveying land promoted from 1533 by Gemma Frisius (1508–1555), the Flemish astronomer and mathematician who taught Gerhardus Mercator, also depended on accurate measurement of angles. Over the next century, mathematical practitioners engaged in surveying gradually developed the theodolite and transit to accomplish triangulation. (See Figure 3 for the version of this instrument that was typical by the nineteenth century.) During the same period, surveyors began to use cartographic drawing tools, including the protractor, to record their triangulated measurements on portable charts. [2, pp. 7–26] [15, p. 120] [36] [37, pp. 64, 122, 211–212, 218–219, 347]

In general, then, early modern European mathematical practitioners used mathematics—broadly defined —to transform the techniques employed in surveying, navigation, and related arts over the course of the seventeenth century. Mathematicians, engineers, and craftsmen relied upon mathematical instruments for precision; they in turn used these recorded observations to increase the body of theoretical knowledge. As an aid for measurement, the protractor was a key part of these developments, gaining a name as well as a standard shape and the formal enunciation of its functions between 1600 and 1700. For example, in his navigational journal for 1612–1613, the famous explorer, William Baffin, characterized the device's function as "protracting," from the medieval Latin for "extending" or "prolonging." [23] [37, pp. 277–278] This word and the activity of setting off and measuring plane angles had become permanently linked by the time Edward Phillips wrote this definition for his 1658 dictionary, *The New World of English Words*: "*Protractor*, a certain Mathematical instrument made of brasse, consisting of the scale and semicircle, used in the surveying of Land." Indeed, the 180-degree format most familiar to contemporary American students was established as most convenient and popular early on. Yet, even though numerous influential manuals on mathematical instruments appeared in the seventeenth century, the authors of these works almost always ignored even relatively recent drawing instruments such as the protractor to focus on the more glamorous sector and quadrant—Edmund Gunter, who wrote *De sector & radio* in 1623, is one of the most familiar names in this regard. (One exception was Giovanni Pomodoro's 1603 book, *Geometria practica*, which contained the first illustration of a full range of drawing instruments. [31])

Even though young university men across Europe were studying instruments as part of their mathematics curriculum as early as 1626 [36, pp. 193–198, 287], modern textbooks in practical geometry or any other subject did not exist before publishing and education both became more affordable and widespread during the Industrial Revolution. It is not known whether protractors are depicted in any of the surviving student notebooks—where dictated lessons were recorded—scattered throughout American and European archival repositories. By the eighteenth century, students and their instructors could be exposed to protractors through practical geometry manuals that often served as production guidebooks as well. French books were some of the most notable of the period, including Sébastien Le Clerc's 1690 *Traité de géométrie theorique et pratique: a l'usage des artistes*, Nicolas Bion's 1709 *Traité de la construction et des principaux usages des instruments de mathematique*, and Alexis Clairaut's 1741 *Elemens de geometrie*. The makers of drawing instruments, meanwhile, began to experiment with less expensive materials, such as animal horn, and to include small protractors in the sets of instruments they sold to surveyors. They no longer had to mark the divisions on protractors by hand after Jesse Ramsden invented the dividing engine for mechanical marking in 1773. [13] [21]

The books by Le Clerc, Bion, and Clairaut all contained verbal descriptions as well as illustrations of the protractor. For example, after proving the elementary constructions and demonstrating propositions with practical utility such as dividing the line into parts and making similar plane figures, Bion explained how to design and use all of the mathematical instruments that existed during his lifetime. He noted that the protractor could be employed to make angles, to replicate a given angle, to inscribe a regular polygon in a circle, and to draw a regular polygon upon a line. [4, pp. 25–28] (Regular polygons were the fundamental unit of military fortifications; early engineers often worked on military projects.)

Throughout the nineteenth century in Europe and the United States, a clear distinction was drawn between the mathematical practice of workmen and the mathematics taught in college in order to impart mental discipline. Thus, books explaining protractors and other drawing instruments were mainly targeted toward children and mechanics. Artisans transformed some protractors into specialty tools; protractors were a standard part of the engineering drawing toolkit until computer-aided drawing became universal in the second half of the twentieth century. By the end of the nineteenth century, protractor measurement was taught in American junior high schools as part of students' early exposure to geometry. The remaining sections of the chapter will deal with these aspects of protractor history in additional detail.

**Research and Reflection Project.**    Search for protractors in the catalog entries on "Epact: Scientific Instruments of Medieval and Renaissance Europe" or in the collections database of the Museum of the History of Science, Oxford. [14] [26] Ask students to discuss how the protractors in these collections are similar to and/or different from the familiar, semicircular, plastic form. How does the craftsmanship involved in making instruments reflect on mathematics as a human activity? (See the Smithsonian's "Educator's Toolkit" for additional guidelines on teaching with historical artifacts. [34]) If middle school students can agree that these tools have visible beauty, try challenging them to start looking for the abstract beauty in mathematics.

## The Protractor and Practical Geometry

Nineteenth-century college and high school courses in basic plane and solid geometry were highly structured and formal in the United States. Students memorized proofs from textbooks based on Euclid's *Elements of Geometry* and repeated them to the instructor during classroom recitations. If the students were permitted to use aids to prepare the figure accompanying each proof, they were limited to ruler and compass. In the second half of the nineteenth century, we find occasional explanations of protractors included in the first wave of "reform" textbooks aimed at secondary school students. For example, Thomas Hill, president of Harvard University, discarded the formal structure of Euclid in his 1863 *A Second Book of Geometry: Reasoning Upon Facts*, written for thirteen- to eighteen-year-olds. Hill's treatise as a whole was a hodgepodge of abstract geometrical principles, classical logic, and instrumental constructions. Although he used mental reasoning to develop the Pythagorean Theorem, he spent the second half of this textbook arguing that students should use concrete aids when they needed geometry in their daily lives. Therefore, he explained how readers could construct their own protractors by dividing a 60-degree arc into fourths and then into fifths, probably drawing on earlier manuals similar to the one by Bion. [18, pp. 61–68] In general, though, protractors were commonly employed in the teaching only of practical geometry, which often took place in settings more informal than college recitations, such as common school classrooms, mechanics' institutes, or homes. Practical geometry textbooks usually resembled Charles Davies's 1846 *Elements of Drawing and Mensuration Applied to the Mechanic Arts*. This textbook, prepared "for the instruction and use of practical men," asked readers to use protractors to bisect lines, erect perpendiculars, and construct triangles. [11, pp. 38–52]

Protractors could also be found in the surveying textbooks written by Davies and others for the American engineering schools. These institutions were modeled on European military academies which, like the colleges, catered to a select group mainly composed of young, upper-class men. These students generally were first taught the abstract, proof-based form of Euclidean geometry and then introduced to the applied disciplines. In engineering courses, the students learned to lay off angles with semicircular protractors while preparing technical drawings, and they used circular protractors with navigational charts. [12, pp. 21–34, 165–168] [35, p. 156] Mass production, a wider range of inexpensive materials, and universal public education finally paved the way for the protractor to enter the school classroom at large in the twentieth century. Some twentieth-century high school textbooks even included a protractor inside their covers. [16] [17] [22] [29] By the 1920s, schoolteachers also had models they could use at the blackboard, such as a fiberboard version (fifteen inches in diameter) sold by the Eugene Dietzgen Company. Nevertheless, many twentieth-century high school geometry teachers isolated protractor measurement from their treatment of formal proof if they included these devices at all. (The author, for instance, was admonished that true mathematics students did not resort to protractors at the outset of her own high school geometry course in 1985.)

**Activity.** Challenge students to divide a straight angle into 180 degrees. After they try on their own, distribute instructions for constructing a protractor from Hill's *Second Book of Geometry*. (See Appendix 1.) Those involved in "writing across the curriculum" programs may wish to deconstruct the directions as a sample of expository writing: Are the instructions clear? For what audience were they written? Are the students able to understand and follow the steps? From a mathematical standpoint, do the students think it is "cheating" to use a mechanical aid to trisect the angle? Discuss with prospective secondary teachers whether dividing the angle is an appropriate topic for a high school geometry class.

# Marketing the Protractor

In the nineteenth and early twentieth centuries, there were hundreds of instrument makers and dealers in Europe and North America. They sold their wares to engineers, architects, surveyors, machinists, colleges, and schools. In order to stand out from their competitors, most firms published catalogs. These booklets contained advertisements, pictures and descriptions of instruments, prices, and sometimes even information on the history or creator of the instruments. [6] [7] [20] [25] In these catalogs of instruments, protractors could be found in cases of drawing instruments, packaged with dividers, pens, pencils, and rulers. Protractors were also sold separately. They were made from sterling silver, animal horn, plastic, or paper. Like mathematical instruments in general, the items of best quality came from England, Germany, and Switzerland. In addition to semicircular and circular protractors, several English makers produced rectangular ivory or wooden protractors that could fit in a mechanic's pocket next to a ruler but still contained all 180 degrees marked around three edges. (See Figure 4.)

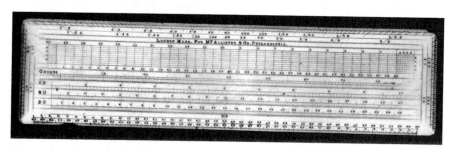

Figure 4. Ivory rectangular protractor sold by William Y. McAllister, c. 1836–1853. Negative number 72-9870. Catalogue number 310743. Courtesy of the Smithsonian's National Museum of American History.

Although many of their finer mathematical instruments would continue to be imported from Europe into the twentieth century, nineteenth-century Americans did turn some of their creative attention to the protractor—which could be manufactured in the United States after William J. Young developed the first American dividing engine for mechanically marking degrees around 1830. Several inventors filed for patents on their designs and improvements: seven between 1790 and 1873, and forty-nine from 1875 to 1900. For example, the draftsman's protractor was patented on February 21, 1887, by Alton J. Shaw, an apprentice of Samuel Darling, who manufactured machinists' measuring tools in association with the Brown & Sharpe Manufacturing Company of Providence, Rhode Island. (See Figure 5.) The draftsman's protractor consisted of two pieces of sheet steel fitted together on a grooved edge; the square frame around this reversible protractor could be placed against a T-square. Darling added a vernier, which read to an accuracy of five minutes, through a patent granted on December 2, 1890. The eight-and-one-half-inch protractor was sold from 1887, when it cost $6.50 plus $1.25 for the morocco leather and velvet case, to at least 1952, with its price at $17.50 in the late 1930s. [1] [9] Thus, the draftsman's protractor was relatively expensive for an individual drawing instrument. The drafter who purchased it, however, would have felt he was making a lifetime investment in his toolkit.

Most of the protractors sold in the twentieth century, however, were inexpensive semicircular instruments that were not expected to last forever. Some of these were also invented by Americans, such as the "contact goniometer" patented by Samuel Louis Penfield (1856–1906) in 1900. (See Figure 6.) Penfield was the head of the mineralogy department at Yale's Sheffield Scientific School. [30] [38] He developed several pieces of apparatus for solving crystallographic problems, including a six-inch protractor printed on a rectangular piece of paper along with a ruler, three foot-to-the-inch scales, and a diagonal scale. Penfield affixed a celluloid arm to an eyelet at the center of his goniometer. The Central Instruments Corporation of Chicago sold this protractor from, at latest, 1909 (when it was priced at 80 cents) to at least 1950 (when

Figure 5. The Brown & Sharpe Draftsman's Protractor, with case and instruction sheet, 1887–1890. Negative number 2000-11239-7a. Catalogue number 336072. Courtesy of the Smithsonian's National Museum of American History.

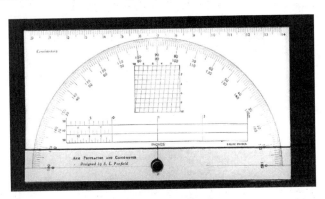

Figure 6. Goniometer patented by Samuel L. Penfield and sold by Central Instruments Company, c. 1909-1950. Negative number 83-5935. Catalogue number 1982.0147.02. Courtesy of the Smithsonian's National Museum of American History.

it cost 65 cents). William C. Marshall, formerly of the Sheffield Scientific School, required the instrument for students using his 1912 *Elementary Machine Drawing and Design* textbook. [24, p. 8]

In its 1909 catalog, the Keuffel and Esser company sold xylonite (an early plastic made of cellulose nitrate) semicircular protractors in a range from 45 cents for a five-inch semicircular protractor to $5.50 for a model with a ten-inch diameter and beveled edge. (See Figure 7.) In addition, merchants and individuals could purchase a brass semicircular protractor for as little as 9 cents, one made from horn for 14 cents, or a protractor printed on cardboard or tracing paper for 20 cents. [7, pp. 174–175, 214–215] Like other plastic products from the early twentieth century, any surviving "xylonite" protractors are highly unstable. They

**Figure 7.** Plastic semicircular protractor, Keuffel and Esser, c. 1909–1936. Negative number 83-13954. Catalogue number 1982.0386.05. Courtesy of the Smithsonian's National Museum of American History.

may be warped, yellowed, and brittle. More recent plastics are chemically inert and just as popular—by the 1990s, the Safe-T Company was reporting that it alone sold millions of protractors every year.

**Team Teaching Opportunity.** The development of the mathematical instrument trade in the United States mirrors the broad outlines of American history. Throughout the late nineteenth and early twentieth centuries, American industrialists, artists, and political and social leaders increasingly asserted American economic and technological power on the world stage. This period was characterized by emphasis on individual improvement, system, and efficiency. Prospective middle school mathematics teachers could work with social studies, history, and English instructors to develop lesson plans bringing together the mathematics and science, technology and industry, literature, and politics of the Second Industrial Revolution and Progressive Eras.

**Research Project.** Although much of the history of the protractor remains to be written, substantial information is readily available on a number of other mathematical, engineering, and commercial instruments. For instance, secondary students could report on the origins and development of the slide rule, transit, astrolabe, electronic calculator, cash register, or computer.

## The Protractor and Educational Standards

If American students of formal geometry did not need all of those inexpensive school protractors, then who did buy them? The number of college students trained in surveying or civil engineering remained relatively small through the end of the nineteenth century, but the system of public elementary and secondary schooling was established during this time period and served an ever-larger percentage of the population. Educators and mathematicians moved fairly quickly to attempt to standardize classroom practice and curriculum. Although leaders disagreed on how and when to teach arithmetic, algebra, and basic plane and solid geometry, they reached an early and lasting consensus that practical geometry instruction belonged in junior high schools and that protractors were clearly appropriate tools for these classes.

This consensus is evidenced by the American educational reports and standards that have appeared since the late nineteenth century. For example, the Mathematics Conference of the Committee of Ten, convened by the National Educational Association in 1893, determined that systematic geometry instruction should begin at age ten, noting, "This course should include among other things the careful construction of plane figures, both by the unaided eye and by the aid of ruler, compasses and protractor; the indirect measurement of heights and distances by the aid of figures carefully drawn to scale; and elementary mensuration, plane and solid." [28, p. 110] Ten years later, J. C. Packard listed the following tools required for each pupil in a class beginning the study of geometry: ruler, compasses, scissors, drawing board, protractor, parallel ruler, squared paper, T-square, and a triangle of 60 and 30 degrees. [10, p. 409] A committee organized by the Mathematical Association of America in 1923 stated that students in grades 7 through 9 should learn intuitive geometry, including the "direct measurement of distances and angles by means of a linear scale and protractor." The students would leave protractors behind when they learned formal proof-based plane and solid geometry in grades 10 through 12. [27, pp. 22, 32–42] Even as late as 2000, the National Council of Teachers of Mathematics Measurement Standard for grades 6 through 8 recommended that teachers show students how to determine "benchmark" angles by sight and then teach them how to use protractors to measure angles directly to a precision of one-half degree. [32, pp. 242–243]

**Debate.**   Ask undergraduate students to take sides on whether secondary school geometry students should be allowed to use protractors. It may be helpful to draw the constructions necessary in the proof of a theorem with a protractor and then repeat the process by developing an unmeasured diagram so the students can begin to weigh the strengths and weaknesses of each technique. After they research and prepare their positions, hold a formal debate in class. As a related issue, discuss computer software, graphing calculators, and other instructional aids. Do these tools help or hinder understanding?

## Conclusion

The story of the protractor is generally a narrative of applied mathematics. The instrument originated in navigational and surveying practice around 1600. It proved as well to be a useful tool for drawing geometrical objects that represented concrete situations, such as the components of machines. By the nineteenth century, protractors were discussed in most manuals that covered practical geometry, surveying, and civil engineering. Instrument makers and dealers sold various types of protractors to professionals whose work relied upon mathematical instruments and to students who were first learning about geometry and its applications. In the twentieth century, protractors were a standard component of the introduction to geometry provided to children of middle school age. Although historians still must address the numerous gaps that remain in the history of the protractor, such as how the object was used in vocational classes and for instruction in the visual arts, many of the primary sources described in this chapter are available through university and rare book libraries for faculty or student research and can also be used to enrich understandings of mathematics. Indeed, simply through familiarizing themselves with the history of the protractor presented here, mathematicians, education professors, and teachers can develop an appreciation for the material culture of mathematics and determine how to exploit these objects as pedagogical aids. Who knew those ubiquitous back-to-school sales could inspire the use of history of recent mathematical activities to teach mathematics?

**Discussion.**   This chapter thus raises the age-old question of where—or whether—to draw the line between pure and applied mathematics. After all, major "pure" mathematicians such as Carl Friedrich Gauss were very interested in dividing the circle because they knew angular measure was essential to the "applied" disciplines of astronomy, navigation, and surveying. Thus, the construction and use of mathematical instruments, including the protractor, represent an intersection between practical and theoretical concerns. Mathematics majors and teaching candidates alike can benefit from considering this aspect of the history of modern mathematics as a step toward working out their own definitions of what mathematics is and who mathematicians are.

## Acknowledgements

The author is indebted to Peggy Kidwell, the Smithsonian's National Museum of American History, and the staff of the Dibner Library, Smithsonian Institution Libraries, for assistance with the preparation of this paper. She also thanks the editors, an anonymous reviewer, Wendy Ackerberg, and Krista Olson Brady for commenting on a preliminary draft. Other work related to this research was presented to the Canadian Society for History and Philosophy of Mathematics (2002) and the Joint Meetings of the American Mathematical Society and Mathematical Association of America (2003).

## Appendix

**Thomas Hill, *A Second Book of Geometry*, Boston, 1863, pp. 63–64.**

184. The formation of a protractor. —Take a piece of hard, smooth card, draw a fine, straight line, as AB (see Fig[ure 8]), and with a convenient radius, say three inches, draw the arc BC. Measure carefully the arc B of 60° by having the compasses, while yet unaltered from the radius with which you drew the arc, step from B to C. Divide the arc as accurately as possible into four equal arcs of 15° each, and set off two such arcs beyond C, so as to make the whole arc 90°. Divide each arc of 15° carefully into three equal parts, which will each be 5°. Divide each also into five parts, each of which will be 3°. By stepping over the whole arc with the compasses open for three degrees, first stepping over it lightly to make sure that twenty steps will exactly make 60°, and then with a heavier step, so as to leave footprints; repeating this heavier stepping from each point of division of the 5° arcs, you can divide the prolonged arc into 90 equal degrees. The first divisions, starting from B, will give 3, 6, 9, 12, 15, 18, 21, &c. The second, starting from the 5° point, will give 2, 5, 8, 11, 14, 17, 20, &c. The third, starting from the 10° point, will give 1, 4, 7, 10, 13, 16, 19, 22, &c.; and these three series evidently embrace all numbers. Mark each fifth point with a longer mark, and number them from B towards C.

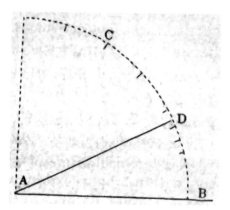

**Figure 8**. Diagram of divided arc from Thomas Hill, *A Second Book in Geometry* (Boston, 1863), p. 63.

185. The graduated arc and its centre . . . is called a protractor, and may be found for sale, engraved on wood, ivory, or brass. It is used for measuring angles, and also for drawing angles of a given size.

## References

1. Amy Ackerberg-Hastings, "The Brown & Sharpe Draftsmen's Protractor," *Rittenhouse*, 15 (2001) 31–38.

2. James A. Bennett, *The Divided Circle: A History of Instruments for Astronomy, Navigation and Surveying*, Phaidon and Christie's Limited, Oxford, 1987.

3. James K. Bidwell and Robert G. Clason, eds., *Readings in the History of Mathematics Education*, National Council of Teachers of Mathematics, Washington, DC, 1970.

4. Nicolas Bion, *Traité de la construction et des principaux usages des instruments de mathematique*, Paris, 1709.

5. Thomas Blundeville, *A Briefe Description of Universal Mappes & Cardes, and of Their Use*, London, 1589; reprint, Theatrum Orbis Tevarvum Ltd., Amsterdam, 1972.

6. *Catalogue & Price List of Eugene Dietzgen Co.*, Chicago, 1904.

7. *Catalogue of Keuffel & Esser Co.*, New York, 1909.

8. Allan Chapman, *Dividing the Circle: The Development of Critical Angular Measurement in Astronomy*, 2nd ed., Praxis Publishing, 1995.

9. Kenneth L. Cope, intro., *A Brown & Sharpe Catalogue Collection, 1868 to 1899*, The Astragal Press, Mendham, NJ, 1997.

10. Byron Cosby, "Efficiency in Geometry Teaching. A Study—An Experiment—A Result," *School Science and Mathematics*, 12 (1912) 406–415.

11. Charles Davies, *Elements of Drawing and Mensuration Applied to the Mechanic Arts*, A. S. Barnes & Co., New York, 1846.

12. ——, *Elements of Surveying and Navigation; With a Description of the Instruments and the Necessary Tables*, 7th ed., A. S. Barnes & Co., Philadelphia, 1843.

13. H. W. Dickinson, "A Brief History of Draughtsmen's Instruments," *Transactions of the Newcomen Society*, 27 (1949–1951) 73–84.

14. Epact: Scientific Instruments of Medieval and Renaissance Europe, http://www.mhs.ox.ac.uk/epact.

15. Maya Hambly, *Drawing Instruments 1580–1980*, Sotheby's Publications, London, 1988.

16. Walter Wilson Hart, Veryl Schult, and Henry Swain, *New Plane Geometry*, D. C. Heath and Company, Boston, 1964.

17. Herbert Edwin Hawkes, William Arthur Luby, and Frank Charles Touton, *New Plane Geometry*, Ginn, Boston, 1929.

18. Thomas Hill, *A Second Book in Geometry*, Brewer and Tileston, Boston, 1863.

19. Peggy Aldrich Kidwell, curator, Slates, Slide Rules, and Software: Teaching Math in America, Smithsonian's National Museum of American History, http://americanhistory.si.edu/teachingmath.

20. *A Manual of the Principal Instruments Used in American Engineering and Surveying, Manufactured by W. & L. E. Gurley*, 37th ed., Troy, NY, 1903.

21. Anthony J. Leiserowitz, The Dividing Engine in History, Virtual Museum of Surveying, http://www.surveyhistory.org/the-dividing-engine.htm.

22. Edith Long and William Charles Brenke, *Plane Geometry*, The Century Co., New York, 1916.

23. Clements R. Markham, ed., *The Voyages of William Baffin, 1612–1622*, Hakluyt Society, London, 1881; reprint, B. Franklin, New York, [196-].

24. William C. Marshall, *Elementary Machine Drawing and Design*, McGraw-Hill, New York, 1912.

25. William Y. McAllister, *A Priced and Illustrated Catalogue of Mathematical Instruments*, Philadelphia, 1867.

26. Museum of the History of Science, Oxford, http://www.mhs.ox.ac.uk.

27. National Committee on Mathematical Requirements, *The Reorganization of Mathematics in Secondary Education*, Mathematical Association of America, 1923.

28. National Educational Association of the United States, *Report of the Committee of Ten on Secondary School Studies*, American Book Company, New York, 1894.

29. Claude Irwin Palmer and Daniel Pomeroy Taylor, *Solid Geometry*, ed. George William Myers, Scott, Foresman and Company, Chicago and New York, [1918].

30. L. V. Pirsson, Samuel Lewis Penfield, *American Journal of Science*, 172 (1906) 345–367.

31. Giovanni Pomodoro, *Geometria practica, dichiaratá da Giovanni Scala sopra le tauole . . . Opera per generali da guerra, capitani, architetti, bombardieri e ingenieri cosmografinon che per ordinarij professori di misure*, Rome, 1603.

32. *Principles and Standards for School Mathematics*, National Council of Teachers of Mathematics, 2000. Electronic version: Measurement Standard for Grades 6–8,  http://standards.nctm.org/document/chapter6/meas.htm.

33. David Lindsay Roberts, Mathematics and Pedagogy: Professional Mathematicians and American Educational Reform, 1893–1923, Ph.D. diss., The Johns Hopkins University, 1997.

34. Smithsonian Center for Education and Museum Studies, Educator's Toolkit, http://educate.si.edu/ut/educat_fs.html.

35. Edward W. Stevens, Jr., *The Grammar of the Machine: Technical Literacy and Early Industrial Expansion in the United States*, Yale University Press, New Haven and London, 1995.

36. Anthony Turner, *Early Scientific Instruments: Europe 1400–1800*, Sotheby's Publications, London, 1987.

37. David W. Waters, *The Art of Navigation in England in Elizabethan and Early Stuart Times*, Hollis and Carter, London, 1958.

38. H. L. Wells, Samuel Lewis Penfield, *National Academy of Sciences Biographical Memoirs*, 6 (1909) 119–146.

# 20

# The Metric System Enters the
# American Classroom: 1790–1890

**Peggy Aldrich Kidwell**
*National Museum of American History, Smithsonian Institution*

## Introduction

The nineteenth century saw an enormous expansion in American mathematics education. Publicly funded elementary or common schools were established, first in the northern states and then throughout the country. By the second half of the century, public high schools also were becoming usual. The extension of engineering education, much of it modeled after the École Polytechnique in Paris, also encouraged mathematics instruction. At the same time, improvements in printing, cheaper paper, and the national markets created by railroads made it possible to supply students in the new schools with relatively inexpensive, uniform textbooks.

Several teachers and former teachers sought to supply the new schoolrooms with texts. Publishers such as A. S. Barnes & Company of Hartford and Appleton's of New York developed and marketed entire series of books [1, 18, 24]. Authors were mindful both of information included by their competitors and of requirements of school committees and newly established state boards of education. To remain competitive, many arithmetic textbooks came to include a discussion of the relatively recently formulated metric system of weights and measures. Discussions of the metric system in these books well illustrate the difficulty of making new mathematical structures part of the educational canon.

## The Origins of the Metric System

In the wake of the revolution of 1789, French citizens called for uniform weights and measures throughout their country. With the approval of the National Assembly and subsequent national governments, a commission of the Paris Academy of Sciences and its successor, the Institute of France, developed totally new units for measuring distance, volume, weight, angles and even time. These units were interconnected. Units of one quantity, such as length, increased by powers of ten (millimeters, centimeters, decimeters, meters). The liter was the volume of a cube 10 centimeters on a side. One kilogram was the weight of one liter of water at a standard temperature. No such simple relations exist in English weights and measures among units of length (inches, feet, yard, miles) or among units of length, volume and weight. In short, the French introduced not only national standards, but a system of standards. This system survives today, in modified form, as the metric system.

The United States Constitution explicitly grants the federal government the power to regulate weights and measures, without specifying what units are to be used. In a 1790 report, ignored by Congress, Secretary of State Thomas Jefferson proposed that the country either adopt weights and measures based on English use or introduce new units which were decimal multiples and submultiples of one another. Some information about the metric system was soon available from almanacs and encyclopedia articles, but almost all commerce was carried out in British units. Great Britain established non-metric national standards of measurement in 1824 and the U.S. followed suit in 1834. The U.S. Coast Survey soon supplied both custom houses and the states with standard pounds, yards and bushels [14].

## Introducing the Metric System in American Textbooks

From colonial times, many arithmetic textbooks published in the United States included tables of weights and measures. Some school districts required that students memorize such tables. Thus Ohio teacher Joseph Ray (1807–1855) [20] entitled the final chapter of his popular textbook *Ray's Arithmetic, First Book* simply "Tables." Here Ray explained: "In the schools of some of the larger cities where this book is extensively used, the pupils studying it are required to learn the tables. For the use of such schools, the following pages have been inserted." [29, p. 81]. Ray's tables suggest the complexity of the weights, measures and coinage then in use. They included English money, Troy, apothecaries' and avoirdupois weights; volume measures for crops, for liquids, and for solids; measures of length and surface area; units of time; and angular measure.

A few textbook authors included mention of metric units in their tables. One of the first to do so was the Boston school teacher and school superintendent Frederick Emerson (1788–1857). In the third and most advanced part of his *North American Arithmetic* (1834), Emerson discussed not only the weights and measures of the United States, but those of several foreign nations. He devoted considerable attention to French units of measure, these "being more nicely adjusted than those of any other country" [13, p. 250]. Emerson described the metre, litre, are (an area measure), stere (a volume measure) and gramme and their multiples and submultiples as interrelated units. Emerson's example may well have inspired rival textbook author Charles Davies to mention the metric system as well. Davies (1798–1876) taught at West Point from 1816 to 1837, and prepared a full series of texts that ranged from arithmetic to calculus and descriptive geometry [17]. In the version of his *University Arithmetic,* that was copyrighted in 1859, Davies discussed several measured quantities including length, surface area, volume, weight, and time. For each quantity, Davies described units used to measure it. Thus, for example, in talking about distance, Davies mentioned the inch, foot, yard, and French "metre." Talking about weight, he mentioned the avoirdupois pound, the apothecaries' pound and the "kilogramme." He did not, however, try to talk about systematic relationships between units of the metric system [11, p. 383–406]. Similarly, Dana Colburn, a teacher in Connecticut and Rhode Island, included in his *Arithmetic and Its Applications* (1862) all the units in *Ray's Arithmetic, First Book*. He then added a note on the metre and its multiples, the gramme and its multiples, and French money, commenting that the metric units were "often referred to in this country, especially in scientific works" [7, p. 40–41]

## Metric Units Made Legal

By the 1860s, several American reformers believed that the metric system should be used in the United States. For example, John A. Kasson (1822–1910), a Vermont-born Iowa lawyer and Republican activist, thought that metric units would ease international affairs. Appointed First Assistant Postmaster by President Lincoln, Kasson learned of the complex system of fees used to compute charges for international mail. Mindful that the U.S. was losing money in these transactions, Kasson called for an international conference of postal officials that would standardize rates and procedures. Meeting in Paris in May, 1863, the delegates recommended that international mail be weighed in metric units, a practice soon widely adopted. Kasson attended the conference as the American representative. Later, as a member of the House of Representatives, Kasson advocated legislation to make the metric system legal in the U.S. During the same period, a committee of the American National Academy of Sciences recommended that the country adopt a decimal system of weights and measures, and Parliament voted to legalize metric units in Britain. Kasson drew this evidence together, and in 1866 his legislation legalizing metric weights and measures passed both houses of Congress without discussion and was signed into law by President Johnson. Soon the Office of Weights and Measures was preparing metric standards for distribution to the states. Metric units were now legal, though not mandatory, in the U.S. [19].

The Committee on Coinage, Weights and Measures of the U.S. House of Representatives also sought to encourage instruction about the metric system in the schools. In a resolution of February 2, 1867, the committee "observed with gratification the efforts made by the editors and publishers of several mathematical works, designed for the use of Common Schools and institutions of learning, to introduce the METRIC SYSTEM of Weights and Measures, as authorized by Congress, into the system of instruction of the youth of the United States, in its various departments." Hoping "to extend further the knowledge of its advantages, alike in public education and in general use by the people," it specifically asked Charles Davies "to confer with Superintendents of Public Instruction, and teachers of Schools, and others interested in a reform of the present incongruous system, and by lectures and addresses to promote its general introduction and use" [10, no page number [inside cover]]. Never one to ignore congressional authorization, Davies promptly prepared a short pamphlet outlining the metric system, which was published that very year. He also extended the discussion of the metric system in his *University Arithmetic* and *Practical Arithmetic*.

Other textbooks incorporated the metric system in several different ways. The well-known Cincinnati author Horatio N. Robinson died in 1867, and his publisher arranged to have an account of the metric system by another author appended to his books [22, 23]. Two college professors, Edward Olney of the University of Michigan and Shelton P. Sanford of Mercer University in Georgia, added their own appendices [26, 32]. Such appendices were not confined to books on arithmetic. For example, in his *Plane Trigonometry and Functional Analysis*, Alfred H. Welsh included appendices on mensuration, surveying, angle measurement, and the metric system. Welsh thought study of the metric system would prepare his high school students for their physics course and, more generally, that "It needs no seer to assure us that the metric system is that of the future…" [33, p. 9]. Other authors incorporated the metric system into the text of their books, either in special chapters on the subject [28, p. 89–95] or in chapters on weights and measures [12, p. 198–233; 27, p. 191–193]. Metric units also were incorporated into books for those needing to master business arithmetic. Thus David White Goodrich, a former "lightning calculator" for the Erie Railway, included metric measures in the chapter on standard measurement at the end of his book, *The Art of Computation, Designed to Teach Practical Methods of Reckoning with Accuracy and Rapidity* (1873). Goodrich's brief account of metric units began with the prescient comment, "As Congress has authorized the use of this system in the United States, it is possible that it may gradually creep into general use. It is so simple and comprehensive, and so extensively used in scientific measurements, that we are glad to give its general features" [15, p. 81].

A few authors even went so far as to present metric measures before English common measures. For example, in 1879 William F. Bradbury (1829–1914), an Amherst College graduate, a former mathematics teacher and the headmaster of Cambridge Latin School in Cambridge, Massachusetts, published a new edition of an arithmetic by James S. Eaton. Bradbury had been an advocate of the metric system for some time and *Bradbury's Eaton's Practical Arithmetic* reflects this. Metric measures were treated as part of the section on decimal arithmetic, and the book includes a full set of problems on measuring areas and volumes using metric units. In the following chapter, Bradbury discussed "customary" units of measure, though he omitted Troy and apothecaries' weights from the usual tables. He also included a section on reducing customary measures to the metric system [3].

The Unitarian minister Thomas Hill and high school mathematics teacher George A. Wentworth took a similar approach in their book, *A Practical Arithmetic* (1882). In the preface, they confidently predicted that "The Metric System in a few years will be in common use, and will supersede other systems, as dollars and cents have superseded pounds, shillings, and pence" [16, p. iv]. Hill and Wentworth incorporated metric units in much of their book, even before discussing the metric system as a whole in the ninth chapter. Tables and problems using older units of measure were confined to the final chapter.

## Organizations and Objects—The American Metrological Society and the Metric Bureau

In 1873, a small group of metric enthusiasts, most of them from New York and New England, joined together to form the American Metrological Society and to advocate general adoption of the metric system. Early members of the society included Columbia University President Frederick A. P. Barnard, William F. Bradbury, Amherst College librarian Melvil Dewey, and the secretary to the state of Connecticut's board of education, Birdsey G. Northrop. These men firmly believed that students should learn about the metric system from objects as well as textbooks. As Northrop put it "The metric tables are given in nearly every series of Arithmetics. But when only the tables are used, the impression is vague and the figures are soon forgotten. The object itself always makes a clearer and more lasting impression than any verbal description of it" [25, p. 45]. In 1864, even before the legalization of the metric system in the U.S., the General Assembly of the state of Connecticut had recommended that metric weights and measures be taught along with established units in the public schools of the state. Northrop, who held his position in Connecticut from 1867 to 1883, arranged by the mid-1870s for the manufacture and distribution at cost to Connecticut teachers of rules of several lengths that were divided metrically on one side and in inches on the other. With his annual report for 1877 he distributed a chart of the metric system copyrighted by A. & T. W. Stanley of New Britain early the previous year. It showed a meter divided to centimeters and millimeters next to a yard divided to inches and eighths of an inch. Below these was a general account of the metric system, above them scenes from the Philadelphia Centennial Exposition of 1876, described in metric units. Northrop also arranged to have metric scales made by the Fairbanks Scale Company of St. Johnsbury, Vermont, exhibited in the Connecticut educational exhibit at the Centennial.

Melvil Dewey and William Bradbury believed that there was a more general market for metric teaching devices. In March of 1876, Bradbury applied for a patent for an "improvement in devices for teaching the metric system." Bradbury's invention was a wooden cube, 10 centimeters on a side, that was designed to weigh exactly one kilogram. It fit snugly into a metal box that held exactly one liter of water. Bradbury received patent 176,735 for this invention on May 2, 1876. A few weeks earlier, Dewey left his position at Amherst to set up business in Boston. Dewey was a great believer in increasing efficiency, be it through decimal classification of books, spelling reform, or the use of metric weights and measures. He established the Library Bureau to distribute card catalogs and other library supplies, the American Metric Bureau to sell apparatus and charts for teaching the metric system, and the Spelling Reform Association to promote simplified spelling. The American Metric Bureau sold not only meter sticks and other rules, charts of the metric system, metric scales, and Bradbury's demonstration apparatus, but metric volume measures made by the Shakers of Sabbathday, Maine, metric weights, centigrade thermometers, and a wide range of metric stationery—a total of some 323 items by 1877. Publications of the American Metric Bureau included not only introductions to the metric system, but a journal entitled *The Metric Bulletin* (later *The Metric Advocate*). These publications adopted some of Dewey's simplified spelling—the French gramme, metre and litre became the gram, meter and liter. Some products of the Metric Bureau were offered by better-known instrument makers such as J. W. Queen & Company of Philadelphia [19, 34].

Dewey asked the American Metrological Society to cooperate with his efforts to introduce metric teaching apparatus into the schools. A committee of the society duly prepared a circular addressed "To the Teachers of the United States" calling for instruction in the metric system that was illustrated by suitable apparatus. F. A. P. Barnard was pleased to report at the May, 1878, meeting of the society that New Jersey had appropriated $2000 to supply objects for teaching the metric system to every public school in the state [2, p. 66, 71, 99]. However, the attention of the metrological society seems to have shifted away from teaching devices to matters such as standard time, and a survey of other volumes of its *Proceedings* reveals no similar later announcements.

In addition to influencing textbooks, teachers and teaching apparatus, metric advocates encouraged nineteenth-century colleges to make knowledge of the metric system a requirement for admission. In April, 1876, the American Metrological Society sent a circular to 350 colleges urging that the metric system be used in the classroom and that the schools report whether it was being used. By mid-May, 16 colleges had replied. Eight reported that knowledge of the metric system was required for admission, and 5 others expected to require such knowledge soon [2, p. 58–63]. In 1890, Florian Cajori found that some knowledge of the metric system was required for admission to Yale, Princeton, Cornell, and the University of Texas, but mentions no such requirement for West Point, Harvard, or several other colleges. Metric units were taught at 41 of 45 normal schools Cajori surveyed and at nearly all of 91 academies, institutes and high schools which responded to his queries [6].

Cajori also commented that several abstruse topics had generally been dropped from elementary arithmetic textbooks, and noted approvingly that "at least one new subject has been quite generally and, we think, appropriately introduced into our books—the metric system" [6, p. 109]. To be sure, in 1887 the Boston School Board had voted to exclude the metric system—along with methods for taking cube roots, techniques for finding volumes and surface areas of various solids, and certain commercial topics—from the required course in arithmetic. However, Cajori looked forward to the time when "this country will wheel in line with the leading European nations and adopt this [the metric] system to the exclusion of the wretched systems now in use among us" [6, p. 359].

## Second Thoughts

The change Cajori envisioned did not come to pass. Alterations in textbooks, meter sticks, liter blocks, college requirements and the propaganda of the American Metrological Society proved insufficient to bring about general use of metric units. The United States had a large population, a stable government, and relatively homogeneous weights and measures throughout the country. France had no such uniformity when metric measures were introduced there except in the case of timekeeping, and the metric calendar did not long endure. Other European nations that adopted the metric system also had diverse measures and/or small populations. Moreover, neither Great Britain nor the United States made metric measurements mandatory.

At the same time, some authors campaigned actively against metric units. One surprising source of opposition came from the University Convocation that met annually at the State University of New York in Albany. In the summer of 1866, the Convocation appointed textbook author Charles Davies, soon to be authorized by Congressional committee to teach the metric system, along with two regents of University, to prepare a report on "what measures, if any, the Convocation should adopt in regard to a Uniform System of Weights and Measures" [9, p. 5] Davies was to collect materials for the report; the committee expected to find unanimously in favor of introducing the metric system into general use. On further examination, however, the committee, especially Davies, developed doubts about the wisdom of this idea. On reporting this to the University Convocation in 1869, the committee was discharged and replaced by a new committee made up of Davies, Hale and another textbook author, James B. Thomson. Thomson took no part in the proceedings of the committee, but by the summer of 1870 Davies had prepared a lengthy report outlining several objections to the introduction of metric weights and measures. Appended to the report were comments of John Adams and John Herschel on appropriate units of measurement.

Davies raised thirteen specific objections to the metric system. Most of these might be summarized as objections to change—a new system would supersede familiar measures; require people to abandon the tried and true, learn a new language and become used to new prices; require restatement of surveys and property records; and obliterate existing senses of distance, area and volume. At the same time, Davies, like Herschel, believed that common measures reflected familiar concepts such as the length of the foot. He also objected to the small size of the gram and the length of the meter that were the base units of the system, and preferred to give distinct units distinct names (inch, foot, yard, mile) rather than distinguishing them by prefixes. Finally, Davies cited the advantages of remaining connected to Britain, both because of its industrial importance and because English was the language most widely spoken in the United States. It would be a shame, he argued, to abandon "short, sharp Saxon words, for their equivalents expressed in a foreign language" [9, p. 49]. In order to suggest the foreign origins of metric units, throughout this book he used the French spellings gramme, metre and litre—in his 1867 pamphlet on the metric system he had spelled these words gram, meter and liter.

Clearly, at least for Davies, introducing the metric system into the American classroom was about far more than simplifying calculations. Further obstacles to the introduction of metric units came from three quite different sources, none of which can be discussed fully here. First of all, from 1864 the Scottish astronomer Charles Piazzi Smyth argued on the basis of extensive measurements of the Great Pyramid of Giza that the ancient Egyptians had designed that building using a unit of length very similar to the British inch [5]. Smyth found a small but vocal number of disciples in the United States, who believed strongly in the ancient and indeed Biblical if not God-given character of British measures, in contrast to those proposed by republican France. This group, organized as the International Institute for Preserving and Perfecting Weights and Measures, waged an active campaign against the metric system in the 1880s and 1890s [8]. Second, some engineering societies, mindful of the value of the non-metric standards they had established, objected to the use of metric units in manufacturing and building [21]. Finally, mathematicians worked to rationalize the curriculum taught in American schools, changing arithmetics that had become small compendia for the aspiring merchant into textbooks for masses of young children. Among the topics easily discarded were discussions of little-used units of measure—including metric units. To give one small example, in 1895 William F. Bradbury, that metric advocate and inventor of metric demonstration apparatus in the 1870s, copyrighted an arithmetic textbook entitled *Sight Arithmetic*. Some problems in this book have problems using weights and measures, but there are no tables of units of measure and none of the units used in problems are metric [4]. It is ironic that these reforms, which were designed in part by mathematicians who were fully aware of the international status of their discipline, and which accompanied the establishment of a community of American research mathematicians of international renown, led in part to the disappearance from arithmetic textbooks of what have come to be international standards of measurement [30, 31].

In the 1970s, after Great Britain decided to adopt metric units, several American reformers once again turned their attention to metric reform. Many of the arguments from that decade show curious echoes of the discussion I have just described.

## Implications for the Classroom

Most of the papers in this volume describe ways in which discoveries and events in the history of mathematics have been used to illuminate mathematics teaching. What are the implications of the nineteenth-century discussion of the metric system for twenty first-century teaching? As one who is not a teacher by profession, I cannot answer this question with authority. I would suggest two themes that emerge from this story that teachers might wish to emphasize in discussing many aspects of the history of mathematics teaching. First of all, the content of mathematics textbooks is and has long been influenced by factors external to mathematics. Social and economic needs, the laws of the United States, the policies of various government bodies and reform groups, and the business considerations of textbook authors and publishers all helped determine what, if anything, was available for American students to learn about the metric system. What some believed was most rational was not necessarily what was taught. Second, evidence of the incorporation of new ideas into the classroom can come not only from printed and manuscript sources but from objects. Particularly in the second half of the nineteenth century, educational theories that encouraged object learning, improvements in transportation and manufacturing, and the growing number of schools in the U.S. combined to provide a market for devices like those used in teaching metric units. Even when metric units were not adopted, that market remained. Both historians and mathematicians are usually better trained in the study of printed sources than in the examination of objects. However, particularly in an era when visual culture has come to play such a large role in our lives, paying attention to objects and their context is both useful and instructive.

## Acknowledgments

This paper has benefited from ideas of Amy Ackerberg-Hastings and Deborah Jean Warner. I also thank Joan Nichols as well as librarians and archivists at American University and at the Smithsonian Institution for their assistance, and an anonymous reviewer for helpful comments.

## References

1. Amy Ackerberg-Hastings, "Mathematics is a Gentleman's Art: Analysis and Synthesis in American College Geometry Teaching, 1790–1840," PhD Dissertation, Iowa State University, 2000.

2. American Metrological Society, *Proceedings*, 1 (1876–1878).

3. William F. Bradbury, *Bradbury's Eaton's Practical Arithmetic*, Thompson, Brown and Company, Boston, 1879.

4. William F. Bradbury, *Sight Arithmetic*, Thompson, Brown, and Company, Boston, 1895.

5. H. A. Brück and M. T. Brück, *The Peripatetic Astronomer : The Life of Charles Piazzi Smyth*, A. Hilger, Bristol [Avon] and Philadelphia, 1988.

6. Florian Cajori, *The Teaching and History of Mathematics in the United States*, Government Printing Office, Washington, D.C., 1890.

7. Dana Colburn, *Arithmetic and Its Applications*, H. Cowperwait & Co., Philadelphia, 1862.

8. Edward F. Cox, "The International Institute: First Organized Opposition to the Metric System," *Ohio Historical Quarterly*, 68 #1 (1959) 54–83.

9. Charles Davies, *The Metric System Considered with Reference to Its Introduction into the United States*, A. S. Barnes & Co., New York and Chicago, 1871.

10. ——, *The Metric System Explained and Adapted to the Systems of Instruction in the United States*, A. S. Barnes & Co., New York, 1867.

11. ——, *New University Arithmetic*, A. S. Barnes & Co., New York, 1860 (1856 copyright).

12. Philotus Dean, edited J. P. Cameron, *The High School Arithmetic*, A. H. English & Co., Pittsburgh, 1874

13. Frederick Emerson, *The North American Arithmetic, Part Third*, Russell, Odione & Metcalf, Boston, 1834.

14. A. H. Frazier, *United States Standards of Weights and Measures: Their Creation and Creators*, Smithsonian Institution Press, Washington, D.C., 1978.

15. David White Goodrich, *Goodrich's Lightning Calculator The Art of Computation, Designed to Teach Practical Methods of Reckoning with Accuracy and Rapidity*, D. W. Goodrich & Co., New York, 1873.

16. Thomas Hill and George Albert Wentworth, *A Practical Arithmetic*, Ginn, Heath, Boston, 1882.

17. Keith Hoskin, "Textbooks and the Mathematisation of American Reality: The Role of Charles Davies and the U.S. Military Academy at West Point," *Paradigm*, 13 (May, 1994) 11–41

18. Louis C. Karpinski, *A Bibliography of Mathematical Works Printed in America through 1850*, Arno Press, New York, 1980 (this is a reprint of the 1940 edition).

19. Peggy A. Kidwell, "Publicizing the Metric System in America from F. R. Hassler to the American Metric Bureau," *Rittenhouse*, 5 (1991) 111–117.

20. David E. Kullman, "Joseph Ray The McGuffey of Mathematics," *Ohio Journal of School Mathematics*, 38 (1998) 5–10.

21. *Manufacturer and Builder*, 13 (June, 1881) 126.

22. M. McVicar, "The Metric System of Weights and Measures," in Horatio Nelson Robinson, *The Progressive Higher Arithmetic*, Ivison, Phinney, Blakeman & Co., New York, 1868.

23. ——, "The Metric System of Weights and Measures," in Horatio Nelson Robinson, *Robinson's Progressive Arithmetic*, Ivison, Blakeman Taylor & Co., New York, 1871.

24. John A. Nietz, *Old Textbooks: Spelling, Grammar, Reading, Arithmetic, Geography, American History, Civil Government, Physiology, Penmanship, Art, Music, as Taught in the Common Schools from Colonial Days to 1900*, University of Pittsburgh Press, Pittsburgh, 1961.

25. Birdsey G. Northrop, *Report of the Secretary of the Connecticut Board of Education for 1877*, The State, Hartford, 1877.

26. Edward Olney, *The Elements of Arithmetic...*, Sheldon & Company, New York, 1875.

27. George Payn Quackenbos, *A Higher Arithmetic*, D. Appleton and Company, New York, 1874.

28. ——, *A Mental Arithmetic*, D. Appleton and Company, New York, 1871.

29. Joseph Ray, *Ray's Arithmetic, First Book*, Van Antwerp, Bragg & Co., Cincinnati and New York, 1857 (first introduced about 1837).

30. David L. Roberts, "E. H. Moore's Early Twentieth-Century Program for Reform in Mathematics Education," *Mathematical Monthly*, 108 (2001) 689–696.

31. ——, "Mathematics and Pedagogy: Professional Mathematicians and American Educational Reform, 1893–1923," PhD. Dissertation, Johns Hopkins University, 1997.

32. Shelton P. Sanford, *Common School Arithmetic on the Analytic System...*, University Publishing Company, New York and New Orleans, 1872.

33. Alfred H. Welsh, *Plane Trigonometry and Functional Analysis*, G. J. Brand & Co., Columbus, 1878.

34. Wayne A. Wiegand, *Irrepressible Reformer: A Biography of Melvil Dewey*, American Library Association, Chicago and London, 1996.

# 21

# Some Wrinkles for a History of Mathematics Course

**Peter Ross**
*Santa Clara University*

## Introduction

In teaching an upper-division history of mathematics course using a traditional text—recently I used Howard Eves'
*An Introduction to the History of Mathematics, Sixth Edition*, 1990—I found it helpful to incorporate several wrinkles
in the course. Aside from enlivening the course they permit the inclusion of some recent history, as my examples
below illustrate. This inclusion is especially useful in teaching with a traditional text such as Eves, where one is
hard-pressed to get to even 18th century mathematics in a single term. I will discuss three such wrinkles, each of
which involves some recent history of mathematics:

- Using a handout "Some Questions to Whet your Appetite" in class on the first day;

- Starting class with an historical fact for that date;

- Showing a biographical video for an outstanding modern mathematician.

## Some Questions to Whet your Appetite

To start the course with gusto, I passed out in the first class meeting a list of sixteen questions, distributed as
follows: Mathematical People (8), Mathematics (5), Mathematical History (3). Students worked in class on the
questions individually or with their neighbors for twenty minutes or so, then we discussed possible answers at the
end of class. (Students did not hand in their answers.) The complete list of (slightly-modified) questions is in the
appendix. I discuss below each of the ten questions that involve some *recent* history, and I include some students'
responses and comments, along with several observations of my own.

*Which three mathematicians are generally considered (e.g., by G. H. Hardy) to be the greatest ever?*

Many students suggested Euclid while a couple said Fibonacci. One person said Newton but no one named Gauss or
Archimedes. (One student later recalled that she was "stunned" that the last three were considered to be the greatest.)
Later in the course I used the following related quotation from Felix Klein, "The most eminent mathematicians, as
Archimedes, Newton, Gauss, have always uniformly included both theory and applications." [1]

*Which 20th century mathematician (he died in Germany in 1944) is considered the most influential of that
century?*

Einstein was given as an answer, but not Hilbert. In fact, only a few students had even heard of David Hilbert.
Later in the course I had occasion to refer back to some of the questions on the first-day list. This occurred for the
above question when I discussed Hilbert's famous address to the International Congress of Mathematicians in 1900,
in which he presented twenty-three open problems for the next century.

*Many mathematicians were also famous scientists. Which were important philosophers?*

Students liked this question as it was open-ended, and they had some good answers such as Newton, Descartes, and Leibniz. Plato and Einstein were also given as answers. One student who had had three philosophy courses asked, "Weren't most mathematicians philosophers anyway?"

> *Name an important woman mathematician in history.*

The class found this question interesting but the only name given was Sophie Germain, by a student who had heard of her in a mathematics education course. No student had heard of Emmy Noether when I mentioned her. A woman student later told me privately that she thought it was "pathetic" that most of the class, including herself, couldn't name a single woman mathematician. In suggesting term projects in a history of mathematics course it is tempting not to single out the work of women mathematicians, since too little is known about early women mathematicians like Hypatia while the work of more recent women mathematicians appears to be too sophisticated and challenging. But there are good biographies of both Emmy Noether and Sofia Kovalevskaya, for example, and some of their mathematics is accessible to better students.

> *In the last decade the most famous open problem in mathematics was solved.*
>   a. *What was the conjecture (now a theorem)?*
>   b. *Who first formulated it, and about when?*
>   c. *Who proved the conjecture?*

Several students correctly answered parts a and b, but no one could recall Andrew Wiles' name. Near the end of the course we saw in class about half, due to time constraints, of the PBS NOVA special "The Proof" (of Fermat's Last Theorem) [2]. Students found it quite engrossing, and said they would have liked to have seen the whole hour-long video. They seemed quite impressed with Wiles' dedication and passion for mathematics, which was most apparent in the touching moment in the video when he was on the verge of tears and couldn't speak.

> *You met the Fundamental Theorem of Calculus in freshmen calculus.*
>   a. *What does the Fundamental Theorem of Arithmetic say?*
>      (No, you didn't see it in grade school!)*
>   b. *What does the Fundamental Theorem of Algebra say?*
>      (No, you probably didn't see it in high school!)*

Several students knew the answer to part a from our transition course that involves abstract algebra and number theory. No one knew the answer to part b, but students felt that it was important to know; the adjective "Fundamental" may have had some influence here. I found it a bit difficult to explain the Fundamental Theorem of Algebra briefly on the first day of the course. However, it was helpful to have planted the seed, so that much later I could mention the Fundamental Theorem of Algebra when we encountered complex solutions to cubic and quartic equations and, near the end of the course, Gauss' work.

> *Which theorem of high school geometry is the most significant, in the sense of having the widest applications and generalizations (even to infinite-dimensional spaces)?*

Students later said that this was a "tough" question, but someone did get Pythagoras' Theorem after I gave a hint. Initial answers given included SAS, AAS, and even "completing the square"! One student chose proofs and generalizations of Pythagoras' Theorem for her term project, which turned out to be an excellent choice as the topic is both content-rich and accessible.

> *Name one of the three famous geometrical problems from antiquity that involved using Euclidean tools. (All three were shown much later to be impossible to solve.)*

One student answered squaring the circle, but trisecting a general angle and duplicating the cube were not mentioned. The significance for the history of mathematics of the three classical problems didn't become apparent to students until much later in the course, but including the question on the first day was helpful as an "advance organizer."

> *Give the approximate century in which the following subjects were first developed:*
>   a. *calculus*          b. *linear algebra*          c. *group theory*
>   d. *analytic geometry*    e. *Euclidean geometry*    f. *non-Euclidean geometry*

This was the most difficult question, with only part e getting plausible answers. Yet every student had had courses in at least four of the six subjects, thus indicating how small a role history plays in our traditional mathematics courses. One student later told me that she had "totally guessed" answers to this question, but it was OK to include it so that students could see how little of mathematical history they knew. It is somewhat surprising that mathematics majors know exact dates like 1066, 1492, and 1861 from their history courses, yet not know even the millennia in which geometry and calculus were developed!

*Why is there no Nobel prize in mathematics?*

Most students did not even know that there was **not** a Nobel Prize in mathematics. This question provoked the most discussion after class ended. As students were walking out the door I gave them a copy of my one-page article "Why Isn't There a Nobel Prize in Mathematics?" [3] This question is a nice springboard to the Fields Medals, which in turn can lead to discussions of *very* recent history of mathematics.

Including the "questions to whet your appetite" in the first class meeting might not have had a dramatic effect on the rest of the course. But, they did stimulate student interest and let students know that there was much worth learning in the history of mathematics. And, many of the questions and their answers were useful for referring to later in the course when we encountered the related topics.

# Beginning Class with an Historical Item for that Date

About once a week I started class with an interesting or significant historical fact involving that very date. Here are several examples I have used when I taught the course in the spring:

- April 9, 1810   Laplace announced his central limit theorem.
- April 30, 1633   Galileo forced by the Inquisition to recant his support of the Copernican Theory.
- May 1, 1631   Fermat got his law degree.
- May 3, 1831   Cayley called to the Bar.
- May 29, 1832   Almost certain he would die in a duel the next day, Galois wrote all night on the solvability of polynomial equations and groups of substitutions.
- May 30, 1832   Galois was mortally wounded in the duel and died of peritonitis the next day.
- June 4, 1919   Emmy Noether at age 37 finally received the right to teach at Göttingen.
- June 7, 1954   Alan Turing committed suicide because he was persecuted by the British government for his homosexuality.

I usually amplified the above bits of information, for example, by explaining what the central limit theorem was or by giving some background information on modern mathematicians like Noether and Turing. In this manner I could insert more recent history into the course as well as give students the feeling that mathematics is a living subject.

I should acknowledge here that the above examples were taken from the unpublished 133-page "A Calendar of Mathematical Dates" that V. Fred Rickey kindly sent me. This calendar includes a number of birth dates and death dates, some of which are quite informative. Here are two such that could be used in class:

- April 21, 1951   Michael H. Freedman was born in Los Angeles. In 1986 he received a Fields Medal for his proof of the four-dimensional Poincaré conjecture.
- April 26, 1920   Srinavasa Aaiyangar Ramanujan died at age 32. This self-educated mathematician, who was discovered by G. H. Hardy of Cambridge, is remembered for his notebooks crammed with complicated identities.

Fred Rickey's History of Mathematics Page on his home page at the United States Military Academy [4] has a wealth of other information that is useful for teaching a history of math course.

As a wrinkle on a wrinkle I'd like to mention that beginning a class with a **mathematical** instead of an historical item can also be valuable. Here is an amusing example that I used after hearing it at a number theory talk given by Carl Pomerance.

*Proof that there are infinitely many composite numbers.*   Suppose that there are finitely many. Multiply them together, but **do not** add 1!

## Showing Biographical Videos

There are several excellent PBS and MAA videotapes or DVDs that chronicle the life and work of outstanding modern mathematicians. I mentioned above "The Proof" involving Andrew Wiles and his proof of Fermat's Last Theorem. Two other PBS videos that I would recommend for a history of math course are "A Brilliant Madness," an American Experience documentary on John Nash, and "The Man Who Loved Numbers," a NOVA special on Ramanujan. Information about all three videos, including how to order them, can be found at the PBS website. The MAA Bookstore website [5] also has some video documentaries worth showing, such as "N is a Number, A Portrait of Paul Erdős."

Instructors who do show one or more of these tapes might want to first get up to snuff on the person profiled. For example, for Nash one can read Sylvia Nasar's illuminating biography [6] or, if pressed for time, just read my four-page review of this book [7].

## Conclusion

Of course there are many other ways than the three wrinkles above for incorporating recent history of mathematics into a traditional course. One way is to encourage students towards topics involving it in their choice of term projects. Student projects that did involve 19th or 20th century mathematics in my course were Determinants and Matrices, How Probability Can Help You, The Gauss-Bonnet Formula, How Big Is Infinity?, Ramanujan's Math & His Genius, and The origins of probability: Why winners avoid Las Vegas.

Incorporating recent history into a general history of mathematics course **is** difficult. But the payoff is well worth it, even if it is at the expense of some traditional topics. Students can easily learn older history on their own later by consulting references such as their textbook or the NCTM's yearbook on Historical Topics for the Mathematics Classroom [8], or by visiting the St. Andrews website [9]. But learning recent history is difficult since the mathematics itself is demanding, and since much recent history is not yet well-researched or summarized in readable form. Thus, including some of it in a history of math course may be more valuable than just covering additional traditional topics. Finally, including some recent history helps to dispel the popular misconception that mathematics is a dead subject.

## Appendix

### Some Questions to Whet your Appetite

#### Mathematical People

1. Which three mathematicians are generally considered (e.g., by G. H. Hardy) to be the greatest ever?

2. Name one of the two main inventors/discoverers of calculus.

3. Who was the most prolific mathematician ever? Hint: He did not name the number $e$ after himself.

4. Name the mathematician after whom our $(x, y)$ coordinate system is named.

5. Which 20th century mathematician (he died in Germany in 1944) is considered the most influential of that century?

6. Name the two founders of probability theory.

7. Many mathematicians were also famous scientists. Which were important philosophers?

8. Name an important woman mathematician in history.

#### Mathematics

9. In the last decade the most famous open problem in mathematics was solved.

    a. What was the conjecture (now a theorem)?

    b. Who first formulated it, and about when?

    c. Who proved the conjecture?

10. You met the Fundamental Theorem of *Calculus* in freshmen calculus.

    a. What does the Fundamental Theorem of *Arithmetic* say? (No, you didn't see it in grade school!)

    b. What does the Fundamental Theorem of *Algebra* say? (No, you probably didn't see it in high school!)

11. What key axiom of Euclidean geometry fails in non-Euclidean geometry?

12. What theorem of high school geometry is the most significant, in the sense of having the widest applications and generalizations (even to infinite-dimensional spaces)?

13. Name one of the three famous geometrical problems from antiquity that involved using Euclidean tools. (All three were shown much later to be impossible to solve.).

## Mathematical History

14. What is the name of our numeral system, and why did it get this name?

15. Give the approximate century in which the following subjects were first developed:

    a. calculus        b. linear algebra        c. group theory

    d. analytic geometry    e. Euclidean geometry    f. non-Euclidean geometry

16. Why is there no Nobel prize in mathematics?

## References

1. Klein, Felix, *Elementary Mathematics from an Advanced Standpoint*, volume 2 on Geometry, translated by E.R. Hedrick and C.A. Noble, Dover, 1939, p.190.

2. http://www.pbs.org/wgbh/nova/proof/

3. Ross, Peter, Why Isn't There a Nobel Prize in Mathematics?, *Math Horizons* (November 1995) 9. This article is also available at http://mathforum.org/social/articles/ross.html

4. http://www.dean.usma.edu/math/people/rickey/hm/default.htm

5. http://www.maa.org

6. Nasar, Sylvia, *A Beautiful Mind*, Simon and Schuster, New York, 1998

7. Ross, Peter, "Book review of *A Beautiful Mind*," *College Mathematics Journal* 31:3 (May 2000) 240–244.

8. National Council of Teachers of Mathematics, *Historical Topics for the Mathematics Classroom*, Thirty-first Yearbook, NCTM, Washington, DC, 1969.

9. The MacTutor History of Mathematics archive, http://www-groups.dcs.st-and.ac.uk/~history/

# 22

# Teaching History of Mathematics Through Problems

**John R. Prather**
*Ohio University–Eastern*

## Introduction

As teachers, we are always looking for ways to actively engage our students in the learning process. One approach in a history of mathematics course is to have students work on a historically motivated set of problems which are independent of the other requirements of the course. These problems are described, and the effects on the class are discussed.

In devising my history of mathematics course, I had in mind three goals. First, in addition to having a sense of how mathematics *was* developed, it is important that students see how mathematics *is* developed. A history of mathematics class seemed an excellent place for students to learn the process involved in creating mathematics. Second, the class should broaden the students' views of mathematics. In particular, they should see a significant amount of mathematics that is not motivated by geometry, algebra, or calculus. Third, they should understand that mathematics is a dynamic, live subject that has not all been done already. I still remember taking Calculus III, and thinking to myself, "How much more math could there be?" Hopefully, the students would not come out of this class thinking the same thing. To address these goals, I developed an extensive list of problems, and required students to spend at least three hours each week working on the problems between class meetings.

In general, the problems selected satisfy two criteria: They are easily stated, and they lead to mathematical ideas of historical importance. While I bend these rules slightly for particularly interesting problems, almost all of the problems on the list meet these two criteria. The list itself currently contains over two hundred problems. It includes problems that are relatively easy, problems that are relatively hard, problems that are impossible, and problems that are still open. Students are not told which are which. The problems are from a broad array of topics, including a number of problems from recent history. In addition to seeing problems of a historical nature, the students get a sense of the kind of problems that present-day mathematicians consider.

## The Class

I hand out the list of problems on the first day, and students are expected to begin working on them from that point on. In addition to the problems, they are given reading assignments and more standard "do it like they did it" homework exercises. Each class meeting they are required to turn in these more standard homework exercises, but I do not grade them except to see that the students made an effort. In addition, they must attach a cover page. On it they must state that they did the required readings (from [1]) and that they spent the required time on the problem set (one or two hours depending on the day that the class meets). I trust that they are honest, and give them credit for doing this work if they say that they did it. If they did not do it, I deduct an appropriate number of points from their homework grade. On their cover page, they must also write approximately half a page about the problems on which they worked, and on the progress they made or did not make. This half page is not graded, but I do read it, and make comments or suggestions about how they might attack individual problems.

Usually we spend about a quarter to a half of each class period discussing the problems, but I carefully avoid giving out solutions. Instead, I talk more generally about the problems, and give them an opportunity to share the progress they have made with the rest of the class. The rest of each class session is spent discussing either the history or the mathematics that the students encountered in their reading for the session. For the most part, I try to let student interest drive the discussions. If we will encounter the solutions to problems on which students are working in the readings, I try to warn the students, and allow them to do those readings at a later date.

## The Problems

The problems themselves are organized by topic. Within topics, they are organized around a major problem. Preceding each of these major problems is a set of similar questions which are designed to facilitate the students' understanding of the major problem. They may also give some insight into how to attack the major problem. Listing all of the topics covered here would be too cumbersome, but suffice it to say that there are problems relating to arithmetic and different bases, lots in number theory, some in geometry, some in the development of algebra, a few relating to calculus and linear algebra, a few in probability and statistics, and several in discrete mathematics. I am always looking for more.[1]

To give an example of how the list is constructed, here is a set of problems leading to the impossible problem of trisecting an angle with a compass and straightedge. You should notice that there is no indication of when the difficulty increases.

1. Using only a straightedge and a compass, bisect a given line segment.
2. Using only a straightedge and a compass, bisect a given angle.
3. Using only a straightedge and a compass, trisect a given line segment.
4. Using only a straightedge and a compass, trisect a given angle.

Perhaps the problems that work the best are the ones in number theory, and the ones in discrete mathematics. In these two areas, there are lots of problems where students can generate examples, make conjectures, and possibly even prove some results. Even difficult problems can be attacked with very few tools. Many of these problems, like the ones mentioned below, lead to ideas that are still being developed to this day. As a result, students not only get a sense of the process involved in solving a mathematical problem, they see mathematics that is still being developed.

Having taught this class now on four separate occasions, one particular problem has emerged as a clear favorite. Currently it is number 66 on the list, and it states:

A number is said to be perfect if the sum of its proper divisors is equal to the number. For example, the proper divisors of 6 are 1, 2 and 3, and $1 + 2 + 3 = 6$. So 6 is perfect. On the other hand, the proper divisors of 12 are 1, 2, 3, 4 and 6, and $1 + 2 + 3 + 4 + 6 = 16 \neq 12$. So 12 is not perfect. Can you find another perfect number?

---

[1]If you would like to obtain a current copy of the complete set, please contact me at prather@ohiou.edu. I would be happy to provide them.

A related problem follows this, and asks simply, "How many perfect numbers are there?" By trial and error, students quickly find that the next such number is 28. At this point, they get hooked. They attempt to find the third perfect number. When trial and error fails miserably, they typically test multiples of 6 and 28. Unfortunately, the next perfect number is not a multiple of the earlier ones. Now at this point, they start to feel a bit frustrated, but they have already spent so much time on the problem that few will give up. In class, I will usually tell them that with perseverance, they will find the next one, but never give them any hint as to how they might do it. They will then either go back to pure trial and error (which will work in about two weeks), or they will start to guess at patterns. Sometimes they noticed that both 6 and 28 are triangular, which cuts down significantly on the time. Many will restrict their attention to only numbers that end in a 6 or 8, and almost none will be trying odd numbers. Two students found the key characterization (i.e., that all even perfect numbers have the form $2^{n-1}(2^n - 1)$) based only on the first two perfect numbers. One way or another, however, they will eventually determine that the next perfect number is 496.

By now they are ready to start looking for patterns. In fact, if they ask, I tell them not to attempt to find the fourth perfect number by trial and error. Of course, they are in no mood to do that anyway. After looking at the three numbers for a while, most students will find the characterization above. Usually, however, they write it in a different form, often $2^{n-1}(1 + 2 + 2^2 + \cdots + 2^{n-1})$. Some will have difficulty writing down any general formula even though they see the pattern. Using this formula, they will find the next perfect number 8128. Still, they will be bothered when they find out that the formula does not always yield a perfect number. For example, if $n = 4$, the formula yields 120. Fairly quickly, they realize that $n$ must be prime in the formula. They are almost there, but then they find out that when $n = 11$, the formula yields $2,096,128$ which is not perfect. With a little more thought, they realize that the second factor above must be prime for the formula to work, and they have successfully characterized the even perfect numbers.

At this point most students tire of this problem, and move on. Many, however, start to ask more questions which lead to some very recent mathematical work. They will ask if there are any odd perfect numbers — a problem which is still open. I think it surprises them that no one knows the answer to this question. They will also ask about how one can tell if numbers of the form $2^n - 1$ are prime. This leads to a discussion of the difficulty of determining whether or not a number is prime. One can easily discuss cryptography at this point. In addition, I mention Mersenne primes and the role that computers are playing in the search for large prime numbers. On rare occasions, students will attempt to prove the result, but to date none have succeeded.

A short aside here may be in order. Since I do not force students to work on any particular problem, I do not ask the students to prove all of their results. Typically, the students that I have are not quite ready to take that step anyway, and they would probably become frustrated if this issue were pushed too hard. After all, most have only had Calculus I or II when they take this class. Rather than forcing this issue, I let the students decide how far they will push a particular problem. I do, however, discuss issues related to proof, and prove many results myself (especially when students ask about problems that are not on my list). Hopefully, the simple fact that I always ask "why?" will motivate them to do the same when they approach problems outside of school. On the other hand, if the course had more sophisticated students, having them prove their results might be desirable.

Another problem that often gets students interested is the Königsberg Bridge Problem [2, pp. 573–580]. It is stated as follows:

A town in Prussia in the 18th century called Königsberg consisted of four areas divided by a river and connected by seven bridges. If we label the areas A, B, C and D, two of them (say A and B) were in the middle of the river. North of A and B was C and south was D. A and C were connected by two bridges; A and D were connected by two bridges; A and B were connected by one bridge; B and C were connected by one bridge; and, finally, B and D were connected by one bridge. Is it possible for a person to start at one point, walk across each bridge once, and return to the starting point without walking across any bridge more than once?

This is another problem that students enjoy because it allows them to get a handle on it without really knowing much mathematics. After trying for a while, most students will conclude that it is not possible to create such a path. Many will even identify the reason it is not possible. In the process, however, most will "abbreviate" their work, and will start to represent the land masses as points. In addition, they will start to consider other maps, and ask themselves why it is possible to find what is now referred to as an Euler circuit in some, but not in others. This creates an opportunity to discuss more recent developments in combinatorics and optimization. One would rarely have an opportunity to discuss such topics if the course were taught in a more chronological manner. As I develop the problem list further, I would like to add more problems of this type.

I would like to briefly mention two other problems that give the students a clearer understanding of what mathematicians might be working on today. One is the set of problems which asks the students to consider Fermat's Last Theorem, and some related results. I like this problem because of the recent proof by Andrew Wiles. It sends students a clear message that the fact that no one has ever been able to do something, does not mean it is impossible. The other problem is the four color problem [3]. Again, this is a good problem where students can generate a lot of examples. Some will even look at the problem in terms of graphs in order to isolate key features. Now while this problem is beyond them, it does eventually give me an opportunity to discuss how computers might be used in a proof. Once again, the fact that the result was proved recently gives students a perspective on mathematics as a dynamic, changing field.

## Effects of the Problems

Using these problems has many benefits. First, and foremost, the students genuinely enjoy working on these problems. While they are expected to put a significant amount of time and energy into the problems, there is no pressure on them to get a correct answer. After a short period the students become genuinely engaged. Even students who would otherwise just "put in their time" see other students asking interesting questions in class, and become involved themselves. In fact, the class as a whole becomes invigorated, and the problems drive the class meetings. At the beginning of each class, the students are asked about their progress, and this discussion sets the tone for the rest of the class. The energy generated by the problems carries over into later class discussions about history or other mathematics. Moreover, students become accustomed to asking the question "Why?" in all of the work that they do for the class.

Through the problems, the students also have the opportunity to work "like mathematicians." In the course of working on these problems, the students create examples, test conjectures, and revise them. More mathematically mature students could be encouraged to try to prove their results. After going through this process, the students end up with a strong sense of what mathematicians do, even if they are doing it at a fairly elementary level. One other nice result that comes from this process is that students encounter actual problems that cannot be solved quickly. Unlike most algebra or calculus classes, there are no examples to follow. The students are left totally on their own, and they see problems that take them days to solve. As a result, the students gain a clear understanding that just because they cannot solve a problem in fifteen minutes does not mean that they cannot solve it at all. On top of that, they are pleasantly surprised when they find that they can actually solve difficult problems without much assistance.

Another benefit is that students see a huge variety of problems. Not only do they see the problems that they work on themselves, they see most of the problems that other students in the class are working on.

They also see a lot of problems that are still unsolved. The idea that mathematics is still developing, and will continue to develop is a powerful concept. In fact, it is that understanding that will motivate many of them to pursue studies and careers in mathematics. Most of the students in this class are pre-service teachers. Hopefully, when they teach their own classes, they will have the tools necessary to convey to their students the idea that mathematics is a broad and vibrant field.

## Improvements

While I am quite happy with the problems as they stand, there can always be improvements. The most obvious improvement would be the addition of more problems. In particular, it would be nice to have even more modern problems. One major issue with the inclusion of current problems is that they often are difficult to state. For example, the Riemann hypothesis is an important problem that would be nice for them to see, but just stating the problem requires a significant amount of complex analysis. Similarly, it is difficult to effectively state problems exploring the idea of an uncountable set. Many other interesting current problems share this difficulty. A second issue is that while there are many nice problems in some areas of mathematics like graph theory, too many similar problems can keep students from getting an adequate breadth of experience. Students will sometimes gravitate to particular kinds of problems, and it is good to make sure that they eventually "run out." That way, without officially stopping them, the students will be forced to look at other types of problems.

## Conclusion

I have been thrilled with the results of this course. The students all work hard, and learn a lot. Even more interesting, they all learn different things. They start to understand the process of creating mathematics, get an idea of what mathematicians are working on today, and encounter many different problems that they would not otherwise see. Many students have told me that this class is the best, or one of the best classes that they have had as an undergraduate. Moreover, the students seem to retain much of what they learn in this class. Often in higher level classes, students will say, "Didn't we see this in History of Math?" It is at these times in particular that I realize that the effort that both they and I have gone through has been worthwhile.

### References

1. Burton, William, *The History of Mathematics: An Introduction*, WBC/McGraw Hill, Boston, 1999.
2. Newman, James R., *The World of Mathematics*, Simon and Schuster, New York, 1956.
3. Wilson, Robin, *Four Colors Suffice: How the Map Problem Was Solved*, Princeton University Press, Princeton, 2003.

## Appendix: Sample Student Work

After students have spent the required time outside of class, they must write a paragraph describing the work they have done. Two examples of student work are given below. I have chosen these two because of how well they describe the process that most of the students go through. Certainly, not all students give this much detail.

The first student was working on perfect numbers.

I think I may have found the pattern for perfect numbers. After looking through past results and reading the Fundamental Theorem of Arithmetic, I came up with some new ideas. I attached the summary of my work. Also, when reviewing for the test, I almost ended up reading the entire book over, so that definitely helped.

First: I already knew this:

$6 = 1 + 2 + 3$

$28 = 1 + 2 + 4 + 7 + 14$

$496 = 1 + 2 + 4 + 8 + 16 + 31 + 62 + 124 + 248$

Second: I noticed there was one prime factor of each number and 1 such as

$6 = 1, 3$

$28 = 1, 7$

$496 = 1, 31$

Third: If you add these two factors together for each number, you get

$6 = (1 + 3) = 4 = 2^2$

$28 = (1 + 7) = 8 = 2^3$

$496 = (1 + 31) = 32 = 2^5$

You must be able to multiply a power of 2 by something to reach one of these numbers. That's when the F T of A gave me a clue. Since these numbers so far are not primes, they must be expressed as a product of primes.

Fourth: If you divide the prime factor into the number, you get this

$6 \div 3 = 2$

$28 \div 7 = 4$

$496 \div 31 = 16$

so 6 must be 2·something etc. 28 must be $2^2$·something etc.

BUT I NEED A PRIME FACTOR OF SOME SORT!

So trial and error led me to

$2(n - 1) = 2n - 2 = 6$

$2n = 8$

$n = 4$

this is not right!

$2^2(n - 1) = 4n - 4 = 28$

$4n = 32$

$n = 8$

this is not right!

but some more trial and error (with a lot of things I would be embarrassed to show you) led me to using 2 powers of 2, 1 even and 1 odd.

Result:

$2(2^2 - 1) = 2 \cdot 3 = 6$

$2^2(2^3 - 1) = 4 \cdot 7 = 28$

$2^4(2^5 - 1) = 16 \cdot 31 = 496$

so if this is right the next number would be $2^6(2^7 - 1) = 64 \cdot 127 = 8128$

So if this is right, GREAT. I can work on some different things. Although, through my experiments, I noticed some neat things I would like to explore.

The second student worked on the Königsberg Bridge Problem.

For my first hour, I began working on the Königsberg bridge problem. I began by drawing a map of what the area might look like. This in itself is somewhat difficult because the directions are not detailed. Not everyone's interpretation of the map will be the same. After labeling the regions and bridges, I began trial and error. I was quick to realize that I should document each combination I try. As I was trying the combinations, Bridge 5 seems to be a trouble area. Once you cross that bridge, it seems you need to cross it a second time to get back to where you started from.

For my next two hour period, I decided to work again on the bridge problem. After some time, I started to draw lines around the path because I could no longer visualize [it] in my head. I found several ways to complete the problem using one bridge twice. I recreated my map on the computer to better visualize. I noticed a problem with region B. It only has three bridges and all connect to a different region. In many of my trials, if I start in B, I can get out, in, and out again. I need one more bridge to get in again. However, if I start elsewhere, I get in, out, and in region B, but I need to finish in another region. At this point, I am beginning to think there is no way to solve this problem. My notes are a disaster. Interestingly enough, my father was curious about what I was doing. So I explained it, and he tried his hand at it. After half of an hour, or so he gave up saying he had to go back to work.

In both of these cases, the students are going through similar processes. Both did a lot of experimentation at first, but after getting a handle on the problem realized that they needed more general ideas. After a number of failures, both also had some success. This is exactly the process that mathematicians go through when confronting unsolved problems.

# About the Authors

**Francine F. Abeles** is a professor of mathematics and computer science, and is in charge of the department's graduate degree programs at Kean University (Union, NJ). She has remained on its faculty except for a year when she was a visiting member of the Courant Institute of Mathematical Sciences at New York University, and for another year at Stevens Institute of Technology (NJ) to obtain a postdoctoral degree in computer science. Dr. Abeles is the reviews editor of the *Review of Modern Logic*, and a regular contributor to the abstracts section of *Historia Mathematica*. Her main interests in mathematics and its history are in linear algebra, geometry, logic, new developments in the mathematical aspects of computing, and the British mathematicians Charles L. Dodgson and Henry J.S. Smith.

**Amy Ackerberg-Hastings** is currently at Anne Arundel Community College. Amy conducts research into the history of American geometry education. She earned a PhD in the history of technology and science from Iowa State University in 2000 for a dissertation titled, "Mathematics is a Gentleman's Art: Analysis and Synthesis in American College Geometry Teaching, 1790–1840." Her article, "Analysis and Synthesis in John Playfair's *Elements of Geometry*," appeared in the *British Journal for the History of Science* in 2002.

**Eisso J. Atzema** received a PhD in History of Mathematics from Utrecht University (NL) in 1993. Since 1996 he has worked in the United States, for the last 6 years at the University of Maine. His research interests include the history of geometry, particularly in the 19th century, and the history of high school and college-level mathematics teaching in North America.

**William Calhoun** obtained his PhD in mathematical logic from the University of California Berkeley in 1990. His research is in computability theory and combinatorics. He is an associate professor in the Department of Mathematics, Computer Science and Statistics at Bloomsburg University of Pennsylvania, where he teaches mathematics and theoretical computer science courses. He is an avid reader of the history of mathematics, particularly the history of mathematical logic and computing.

**Lawrence D'Antonio** currently teaches mathematics and computer science at Ramapo College of New Jersey. His thesis, from Syracuse University, was in the field of summability methods for Fourier series. In recent years his research interests have turned to the history of mathematics. His current projects are investigating the history of the Fundamental Theorem of Calculus and the history of elliptic integrals starting with Euler.

**Sarah J. Greenwald** received her PhD in 1998 from the University of Pennsylvania and received her BS from Union College in Schenectady NY in 1991. She is currently an assistant professor of mathematics at Appalachian State University in Boone, NC. Her research area is in the Riemannian geometry of orbifolds and she is also interested in the geometry of the universe. She enjoys creating classroom activity sheets in order to engage students with mathematics related to "The Simpsons" and has taught a class on women and minorities in mathematics with portions dedicated to both mathematical content and equity issues.

**David Henderson**    was born in Walla Walla, WA, and graduated from Ames (Iowa) High School, Swarthmore College (mathematics, physics, philosophy), and the University of Wisconsin (PhD in geometric topology under R.H. Bing). After a two-year stint at the Institute for Advanced Study in Princeton, he joined the mathematics faculty at Cornell University in 1966 and has been there ever since.

David's great love in mathematics is geometry of all sorts. His interests have widened into issues of what he calls educational mathematics. He has been directing PhD theses in both mathematics and mathematics education.

These interests led him to be invited to the ICMI Study Conference on the Teaching of Geometry in Sicily in 1995 where he met Daina. David had written four text books on geometry (three with Daina) and they have collaborated in many other activities. David has had visiting academic positions in India, Moscow, Warsaw, West Bank (Palestine), South Africa, USA, and Latvia. Currently, he is Professor of Mathematics at Cornell University.

**Holly Hirst**   received her PhD in mathematics from The Pennsylvania State University in 1987. She has been a member of the faculty at Appalachian State University since 1990 where she has been nominated several times for the College of Arts and Sciences Academy of Outstanding Teachers and was the recipient of the Alumni Association's Outstanding Teaching Award in 1995. She teaches a wide variety of courses which include mathematical modeling, including the Annenburg/CPB INPUT award winning liberal arts course developed by Dr. Hirst and a team of colleagues from the mathematical sciences department. As Co-Principal Investigator on the NSF Funded National Computational Science Institute, she is actively involved in developing and coordinating workshops for faculty interested in incorporating computational science and modeling into undergraduate mathematics and science courses. The work on using population models in the classroom detailed in this monograph received the Undergraduate Computational Science Award from the Department of Energy in 1995.

**Patti Wilger Hunter**   is Assistant Professor of Mathematics at Westmont College in Santa Barbara, California. She has published papers on the history of the American statistics community, including "An Unofficial Community: American Mathematical Statisticians before 1935," which appeared in *Annals of Science* in 1999. She is currently investigating the life and work of Gertrude M. Cox.

**Dick Jardine**   dreamed of becoming a mathematics teacher while sitting in a high school algebra class in his home state of New Hampshire. In a roundabout way, he has finally come to live his dream. He left New Hampshire to attend the Military Academy at West Point, where he earned his BS degree. Subsequent Army assignments around the world included an opportunity to earn a PhD in mathematics from Rensselaer Polytechnic Institute and to conclude his military service teaching mathematics at West Point. Since leaving the military, Dick has been teaching and is chair of the Mathematics Department at Keene State College, back home in New Hampshire. His interests are in applied mathematics and the effective use of technology and the history of mathematics in teaching and learning mathematics.

**Jeff Johannes**   earned a BA from Cornell University. He went on to complete a PhD in knot theory at Indiana University. He is currently an Assistant Professor at SUNY Geneseo, where he teaches a wide variety of courses ranging from calculus and geometry to "What is Reality?" and mathematics and music. In addition to mathematics, he enjoys music and winter.

**Peggy Aldrich Kidwell**  is Curator of Mathematics at the Smithsonian's National Museum of American History. She obtained her PhD in history of science from Yale University and has published on a variety of topics relating to the histories of mathematics, computing devices, and astrophysics.

**Jerry Lodder**  received his AB from Wabash College, Crawfordsville, Indiana, and PhD from Stanford University, Stanford, California. He has held visiting positions at the Institut des Hautes Études Scientifiques near Paris, the Centre National de la Recherche Scientifique in Strasbourg, and the Mathematical Sciences Research Institute in Berkeley. His teaching with original source material as well as his research in geometry and topology have been supported through grants from the National Science Foundation. Presently he is Professor of Mathematics at New Mexico State University.

**Matt D. Lunsford**  is Associate Professor of Mathematics at Union University in Jackson, TN. He was born in Ruston, LA on December 13, 1964. He received a bachelor's degree in 1987 from Louisiana Tech University, a master's degree in 1989 from the University of Nebraska, and a doctorate in mathematics from Tulane University in May 1993. He has been a faculty member at Union University since August 1993. His research interests include classical Galois theory and history of mathematics. He and his wife Deanna have three children: Cara, Thomas, and Emma.

**Linda McGuire**  received her undergraduate degree in mathematics from Seton Hall University. She then earned an MS and PhD in mathematics from Stevens Institute of Technology. Linda joined the Mathematical Sciences Department faculty of Muhlenberg College in the fall of 1999 after having spent two years as a visiting professor at Gettysburg College. Her teaching interests center on implementing cooperative learning strategies in the classroom, emphasizing the development of strong writing skills, and integrating technology into all levels of mathematics education. Linda's research interests are in the fields of combinatorics, network analysis and graph theory. She is always interested in involving her students in her continuing studies and enjoys working with them on research projects in this area. Linda's primary avocations are hiking, practicing yoga, reading, watching movies (all varieties), cooking, traveling, and embarking on spontaneous adventures with friends.

**David Pengelley**  is professor of mathematics at New Mexico State University. In addition to ongoing research in algebraic topology, he is one of the creators of a teaching program utilizing student projects in calculus classes, yielding the book "Student Research Projects in Calculus" (MAA, 1992). Dr. Pengelley has also collaborated on developing the use of original historical sources in teaching mathematics, leading to two recent companion books of annotated original sources (http://www.math.nmsu.edu/~history). He has developed a graduate course on the role of history in teaching mathematics, and worked with school teachers to incorporate original historical sources into student class work. This focus on teaching with original sources has led to research in history of mathematics, such as the article "Gauss, Eisenstein, and the 'Third' Proof of the Quadratic Reciprocity Theorem: Ein Kleines Schauspiel", and work in progress on Sophie Germain's manuscripts on Fermat's Last Theorem. Dr. Pengelley received the 1993 Award for Distinguished Teaching from the Southwestern Section of the Mathematical Association of America, loves backpacking, is active on environmental issues, and has become a fanatical badminton player. Academic Training: University of California, Santa Cruz, B.A., Mathematics, 1973; Oxford University, Fulbright-Hays Fellowship, 1978–79; University of Washington, PhD, Mathematics, 1980.

**John Prather**  received his PhD in mathematics from the University of Kentucky in 1997 in complex analysis. He is an Assistant Professor of Mathematics at Ohio University–Eastern. His current interests include the history of mathematics, number theory, and teacher preparation.

**David Lindsay Roberts** specializes in studying the history of American mathematics education. He earned an AB in mathematics from Kenyon College in 1973, followed by an MA in mathematics and an MS in industrial engineering, both from the University of Wisconsin-Madison. After eleven years as an operations research analyst, he entered Johns Hopkins University, completing a PhD in the history of science in 1998. Since then he has been an adjunct professor of history at the University of Maryland, and a postdoctoral fellow at the Smithsonian and the National Academy of Education. He is currently serving as chair of the Oral History Task Force of the National Council of Teachers of Mathematics.

**Robert Rogers** received his BS in mathematics with certification in secondary education from SUNY College at Buffalo, his MS in mathematics from Syracuse University, and his PhD in mathematics from SUNY Buffalo. He is currently Associate Professor of Mathematics at SUNY Fredonia. His current interests are analysis, the history of mathematics (as it pertains to teaching pre-service teachers), the history of calculus and analysis, and the application of mathematics to political science. He is a former recipient of the SUNY Fredonia President's Award for Excellence in Teaching and the MAA Seaway Section's Distinguished Teaching Award.

**Peter Ross** became interested in the history of mathematics when he audited a course on it in the 1960s that was taught by Kenneth O. May. He completed both his MA and PhD at the University of California at Berkeley, with interludes for teaching in India as a Peace Corps Volunteer (1963–1965) and working on new "new math" projects for the government (1969–1974). He has taught in the Department of Mathematics and Computer Science at Santa Clara University since 1982. Since 1985 he has been writing Media Highlights for the College Mathematics Journal, often highlighting articles on the history of mathematics. He also has enjoyed writing book reviews for the Journal of biographies of Steve Smale, John Nash, and Jaime Escalante.

**Amy Shell-Gellasch** was formerly an Assistant Professor in the Department of Mathematical Sciences at the United States Military Academy, West Point, New York, where she conducted research with V. Fred Rickey on the history of the Department of Mathematical Sciences at USMA. Currently she lives in Germany with her husband and new son and is working as a freelance historian of mathematics. She received her BSEd from the University of Michigan in 1989, her MA from Oakland University in 1995, and her DA from the University of Illinois at Chicago in 2000. Her dissertation was a historical piece on mathematician Mina Rees.

**Shai Simonson** received his BA in mathematics from Columbia University and his MS and PhD in computer science from Northwestern University. He has taught mathematics and computer science at the middle school, high school, college, and graduate levels for over 20 years, at locations all over the world. His research interests include theoretical computer science, computer science education, mathematics education, and history of mathematics. Recently, he was director of ArsDigita University, an experimental post-baccalaureate program in computer science (http://aduni.org). He is currently professor of computer science at Stonehill College in Easton, MA, and director of the advanced mathematics program at the South Area Solomon Schechter Day School in Stoughton, MA. He spends his free time raising three boys, playing Go and bridge, and cycling to work.

**Daina Taimina** was born and received all her formal education in Riga, Latvia. In 1977 she started to teach at the University of Latvia and continued for more than 20 years. Her PhD thesis was in theoretical computer science (under Rusins Freivalds) but later she got more involved with geometry, history of mathematics, and mathematics education. These interests led to her being invited to the ICMI Study

Conference on the Teaching of Geometry in Sicily in 1995 where she met David. She has written a book on the history of mathematics (in Latvian) and (with David) three recent geometry textbooks. Daina was a Visiting Associate Professor in Cornell 1997-2003. Currently, Daina is a Senior Research Associate at Cornell University.

**Homer White**   studied philosophy as an undergraduate before receiving his PhD in ergodic theory in 1991 from the University of North Carolina at Chapel Hill. He is currently Associate Professor of Mathematics at Georgetown College in Kentucky. His interests in the history of mathematics include the geometry of Leonhard Euler, and Indian mathematics during the classical period.